Innovative Methods for Numerical Solutions of Partial Differential Equations

Innovative Methods for Numerical Solutions of Partial Differential Equations

edited by

M. M. Hafez
J.-J. Chattot

Universtity of California, Davis

World Scientific
New Jersey • London • Singapore • Hong Kong

Published by

World Scientific Publishing Co. Pte. Ltd.

P O Box 128, Farrer Road, Singapore 912805

USA office: Suite 1B, 1060 Main Street, River Edge, NJ 07661

UK office: 57 Shelton Street, Covent Garden, London WC2H 9HE

British Library Cataloguing-in-Publication Data
A catalogue record for this book is available from the British Library.

INNOVATIVE METHODS FOR NUMERICAL SOLUTION OF PARTIAL DIFFERENTIAL EQUATIONS

ISBN 981-02-4810-5

Printed in Singapore by Mainland Press

Contents

Dedication

This volume consists of papers presented at a symposium honoring Phil Roe on the occasion of his 60th birthday and in recognition of his original contributions to Computational Fluid Dynamics (CFD) over the past twenty years.

The symposium entitled "Progress in Numerical Solutions of Partial Differential Equations" was held in Arcachon, France, on July 11-13, 1998. The authors from U.S., U.K., France, Italy, India and Japan, are internationally known researchers in this field.

The book covers many topics including theory and applications, algorithm developments and modern computational techniques for industry.

Phil Roe was born on May 4, 1938 in Derby, U.K. He received his B.A. in 1961 and Diploma in Aeronautics in 1962 from Cambridge University, Department of Engineering. He worked at the Royal Aircraft Establishment, Bedford, U.K. from 1962 to 1984. He joined the Cranfield Institute of Technology as a professor of aeronautics from 1984 to 1990 and he has been a professor in the Department of Aerospace Engineering, University of Michigan since 1990.

Prof. Roe became internationally known in the CFD community immediately after he published a paper on his flux differencing and averaging technique entitled "Approximate Riemann Solvers, Parameter Vectors, and Difference Schemes" in the *Journal of Computational Physics* in 1981. Since then this paper has been cited over 500 times and it has been selected for reprinting in the 25th Anniversary issue of *Journal of Computational Physics.* Prior to this publication, he had written several Royal Aircraft Establishment reports on high speed aerodynamics. He also wrote three reports on shock capturing and numerical algorithms for the linear wave equation.

Prof. Roe made many important contributions to CFD during the last two decades, covering many aspects of this field, including grids, schemes and solvers. In particular, one should mention his work on acceleration of Runge-Kutta integration algorithms, his optimal smoothing multistage schemes and soft walls and remote boundary condition for unsteady flows, his characteristic-based schemes and multidimensional upwinding, his limiters and high resolution schemes for structured as well as unstructured grids together with preconditioning techniques and his recent work on vorticity preserving schemes.

He worked with many researchers in U.S. and abroad and in many areas of aeronautics and beyond. His research interests include robust algorithms with applications to stiff flow problems, two phase flows and magnetohydrodynamics, where he has recently made fundamental contributions.

He has supervised, so far, 11 M.Sc. and 25 Ph.D. students and has been an external examiner for Ph.D. candidates in over 20 British, French and Swiss universities.

Prof. Roe has received many awards, including NASA Group Achievement Award "for work... which has formed the foundation of modern computational fluid dynamics" in 1993, and the University of Michigan College of Engineering Research Excellence Award in 2000–2001. He was awarded, jointly with B. van Leer and K. Powell, $750,000 from W. M. Keck Foundation to establish the Laboratory of Computational Fluid Dynamics in 1994 and recently he was part of a team selected by NASA Goddard for a $1,500,000 contract to develop a computational model of solar wind. He was elected AIAA Fellow in 1996. A complete list of his publications and professional activities are included in the next article.

Prof. Phil Roe has influenced many people besides his students, his colleagues and his friends. He is a remarkable intellectual and a scholar of highest calibre and his pleasant personality and deep insight are simply outstanding. We wish Prof. Roe an active and productive career for many good years to come.

M. M. Hafez
J.-J. Chattot

Contributions of Philip Roe

Prof. Roe was born on May 4, 1938 in Derby, U.K. He received his B.A. in 1961 and Diploma in Aeronautics in 1962 from Cambridge University, Department of Engineering. He worked at the Royal Aerospace Establishment, Bedford, U.K. from 1962 to 1984. He joined the Cranfield Institute of Technology as a professor of aeronautics from 1984 to 1990 and he has been a professor in the Department of Aerospace Engineering, University of Michigan since 1990. In the following, his professional activities, lists of graduate students he supervised as well as his publications are included.

Professional Activities

- Organising Committee, International Conference on Computational Fluid Dynamics, Kyoto 2000 and Sydney 2002.
- Joint organiser, American Mathematical Society Symposium on Simulation of Transport in Transition regimes, May 2000.
- Visiting Research Fellow, University of Reading, 1998-1999
- Advisory Editor-Journal of Computational Physics,
- Editor-in-Chief-Journal of Computational Physics, 1992-1994.
- Consultant, ICASE, NASA Langley.
- Reviewer for numerous journals and funding agencies.
- Organiser, short course on Computational Fluid Dynamics, Cranfield, 1984-1989.
- External examiner for Ph.D. candidates in over twenty British, French and Swiss universities.
- Visiting Scientist, NASA Ames, 1989.
- Visiting Professor, University of Bari, 1988.
- Consultant, European Space Agency.

Honours and Awards

- NASA Group Achievement Award, 'for work.. which has formed the foundation of modern computational fluid dynamics', 1993.

- Departmental Research Award, Aerospace Engineering, University of Michigan, 1994

- Award of $750,000 from W.M. Keck Foundation to establish Laboratory in Computational Fluid Dynamics, 1994 (jointly with B. van Leer and K. G. Powell).

- Elected AIAA Fellow, 1996,

- 1981 paper 'Approximate Riemann solvers, parameter vectors and difference schemes' (cited over 500 times) selected for reprinting in 25th Anniversary issue of *Journal of Computational Physics.*

- Part of team selected by NASA Goddard for $1,500,000 contract to develop a computational model of solar wind.

- Honored by 60th Birthday Symposium "Innovative Numerical Methods for Partial Differential Equations", Arcachon, France, June, 1998.

- University of Michigan College of Engineering Research Excellence Award 2000-2001 (shared with John P. Boyd)

Current Research Interests

High-Resolution Methods Exploitation of developed computational methods (based on Riemann solvers, limiters, finite volumes) in new areas such as magnetohydrodynamics, rarefied flows, sound generation, elastodynamics, micromanufacturing.

High-Order Methods Development of techniques offering improved accuracy for long-range propagation of linear or low-amplitude waves. Candidate methods include Discontinuous Galerkin and Upwind Leapfrog methods. Also high-order (Hermite) cell-vertex schemes.

Multidimensional Algorithms Development of algorithms directly modelling genuinely multidimensional aspects of the governing equations, including the division into elliptic, parabolic and hyperbolic modes, and methods especially adapted to vortical flows.

Magnetohydrodynamics Algebraic structure of the MHD equations, nonlinearities, degeneracies and their computational implications.

Robust Algorithms Development of codes guaranteed never to violate physical criteria such as positivity of mass or energy. I am looking to merge ideas from Godonov-type schemes and Bolzmann-type schemes.

Adaptive Grid Generation Cell-vertex methods implemented on grids which minimise local truncation error. These methods may form a natural link with techniques of automatic design and shape optimisation.

Stiff Flow Problems Efficient computation of flows in which the timescale of reaction, relaxation, etc differs greatly from the residence time.

Two-Phase Flow Mathematical modelling of two-phase flows such as bubbly liquids, with special attention to possible ill-posedness and the implications for computation.

Radiation Transport Application of new advection schemes to radiative flows. News versions of, and alternatives to, discrete-ordinate methods.

Mathematical modelling of debris dispersal (with K. G. Powell) Probabilistic description of the dispersal of debris from airborne explosions, leading to partial differential equations for probability of encounter.

Graduate Student Supervision
Masters Projects with date of Completion

Smadar Karni Numerical solutions of the Euler equations in a non-conservative formulation. University of Tel-Aviv, 1985.

George Vrizalas Redesign of a leading-edge slat to avoid compressibility effects associated with high suction peaks. College of Aeronautics, Cranfield, 1986.

Hong-Chia Lin Comparison of two computational methods for the Euler equations. College of Aeronautics, Cranfield, 1987.

Nikola Gagovic Computation of flow fields with forward blowing. College of Aeronautics, Cranfield, 1987.

Robert Townshend Design of submarine control surfaces. (jointly supervised with A. Boyd) College of Aeronautics, Cranfield, 1988.

Mark Bannister Computing the effect of wingtip devices. College of Aeronautics, Cranfield, 1988.

Rolf Reinelt The accuracy of Euler codes for transonic flow. College of Aeronautics, Cranfield, 1988.

Detlef Schultz Experiments with far-field boundary conditions. College of Aeronautics, Cranfield, 1988.

Steven Rham Development of a edge-centered scheme for the Euler equations. College of Aeronautics, Cranfield, 1989.

Martin Clark A first-order 3D Euler code for hypersonic waverider design using an upwind space marching technique. College of Aeronautics, Cranfield, 1989.

Christophe Corre Experiments on cell-centre and cell-vertex schemes in the case of the Ringleb flow, University of Michigan, 1991.

Doctoral Theses with Dates of Completion and Current Employment

Doctoral Students Advised

Peter K. Sweby Flux-difference splitting methods for the Euler equations. (jointly supervised with M.J. Baines) University of Reading, 1982.
(Senior Lecturer, University of Reading)

Smadar Karni Far-field boundary conditions in aerodynamics. College of Aeronautics, Cranfield, 1989.
(Associate Professor, Mathematics, University of Michigan)

Hong-Chia Lin Topics in the computation of hypersonic viscous flow. College of Aeronautics, Cranfield, 1990.
(Lecturer, Nan-Rong Institute of Technology, Taiwan)

David W. Levy Use of a rotated Riemann solver for the two-dimensional Euler equations. (jointly supervised with K.G. Powell, B. van Leer) University of Michigan, 1990.
(Design Engineer, Cessna Aircraft)

Christopher L. Rumsey Development of a grid-independent Riemann solver. (jointly supervised with B. van Leer, K.G. Powell) University of Michigan, 1990
(Research Scientist, NASA Langley)

James J. Quirk Adaptive mesh refinement for steady and unsteady shock hydrodynamics. College of Aeronautics, Cranfield, 1991.
(Research Scientist, Los Alamos National Laboratory)

Karim Mazaheri Numerical wave propagation and steady-state solutions. University of Michigan, 1992.
(Lecturer, Sharif University, Teheran)

George T. Tomaich A genuinely multi-dimensional upwinding algorithm for the Navier-Stokes equations on unstructured grids using a compact, highly-parallelizable spatial discretization, University of

Michigan, 1995.
(Exa Corporation, Boston)

Jens-Dominik Müller On triangles and flow, (jointly with H. Deconinck, von Karman Institute, Brussels) University of Michigan, 1995.
(Research Fellow, University of Oxford)

Lisa-Marie Mesaros Multi-dimensional fluctuation-splitting schemes for the Euler equations, University of Michigan, 1995.
(Team Leader, FLUENT, Ann Arbor)

Shawn L. Brown Approximate Riemann solvers for moment models of dilute gases, University of Michigan, 1995.
(Lecturer, Wright State Unuversity)

Creigh McNeil Efficient upwind algorithms for solution of the Euler and Navier-Stokes equations, (jointly with N. Qin) Cranfield University, 1995.
(Researcher, Centre for Turbulence Research, Stanford University)

Robert B. Lowrie Compact higher-order numerical methods for hyperbolic conservation laws. (jointly with B. van Leer), University of Michigan, 1996 (Research Scientist, Los Alamos)

Mohit Arora Explicit Characteristic-based high-resolution algorithms for hyperbolic conservation laws with stiff source terms, University of Michigan, 1996 (Morgan Stanley Bankers, Houston)

Rho-Shin Myong Theoretical and computational investigations of nonlinear waves in magnethydrodynamics. University of Michigan, 1996
(Assistant Professor, Gyeongsang National University, Korea)

Brian T. Nguyen Three-level time-reversible schemes for acoustic and electromagnetic waves. University of Michigan, 1996
(Research Scientist, Lawrence Livermore National Laboratory)

Jeffrey P. Thomas Investigation of upwind leapfrog schemes for acoustics and aeroacoustics, University of Michigan, 1996.

(Assistant Professor, Duke University)

Cheolwan Kim High-order upwind leapfrog schemes for advection,acoustics and aeroacoustics, University of Michigan, 1997.
(General Motors Research Laboratory, Detroit)

Timur Linde A three-dimensional adaptive multifluid model of the heliosphere, (jointly supervised with T. I. Gombosi, awarded University of Michigan distinguished dissertation prize) University of Michigan, 1998
(Research Fellow, University of Chicago)

Dawn D. Kinsey Toward the Direct Design of Waveriders, University of Michigan, 1998.
(Team Leader, MathSoft, Seattle)

Jeffrey A. F. Hittinger Foundations for the extension of the Godunov method to hyperbolic systems with stiff relaxation, (jointly supervised with A. Messiter) University of Michigan, 2000
(Postdoctoral Fellow, Lawrence Livermore National Laboratory)

Current Supervision of Research Students

Suichi Nakazawa Dissipation-free algorithm for elastic wave propagation. (jointly supervised with P. D. Washabaugh)

Hiroaki Nishikawa Simultaneous flow solver and mesh optimisation.

Mani Rad Genuinely multidimensional flow solver

Edward Wierzbicki Partial differential equations modelling the probability of debris encounters. (jointly with K. G. Powell)

List of Publications

1 Books

P.L.Roe(ed), *Numerical Methods in Aeronautical Fluid Dynamics*, Academic Press, 1982.

2 Review Articles

P.L.Roe, Characteristic-based schemes for the Euler equations, in **Annual Review of Fluid Mechanics, 1986**, eds M.van Dyke, J.V.Wehausen, J.L.Lumley, Annual Reviews,Inc., 1986.

P.L.Roe, A survey of upwind differencing techniques, 11th International Conference on Numerical Methods in Fluid Dynamics, Williamsburg, 1989, in **Lecture Notes in Physics, vol 323**, eds D.L.Dwoyer, M.Y.Hussaini, R.G.Voigt, Springer, 1989.

P.L.Roe, Modern numerical methods applicable to stellar pulsation, NATO Advanced Study Institute, Les Arcs, France, 1989, in **The Numerical Modelling of Nonlinear Stellar Pulsations-Problems and Prospects**, ed J.R.Buchler, Kluwer, 1990.

P.L.Roe, Beyond the Riemann problem, in *Algorithmic Trends in Computational Fluid Dynamics*, eds M.Y.Hussaini, A. Kumar, M.D. Salas, Springer 1993.

K.G.Powell, **P.L.Roe**, J.J.Quirk, Adaptive-Mesh Algorithms for Computational Fluid Dynamics, in *Algorithmic Trends in Computational Fluid Dynamics*, eds M.Y.Hussaini, A. Kumar, M.D. Salas, Springer 1993.

P.L.Roe, A brief introduction to high resolution schemes, Technical introduction to *Upwind and High resolution Schemes* eds M.Y.Hussaini, B. van Leer, J van Rosendale, Springer, 1997.

P.L.Roe, Est-ce-qu'il-y-a une fonction de flux ideale pour les lois de conservation hyperboliques? CANUM 98.

P.L.Roe, Shock Capturing (90 page chapter) in *Handbook of Shockwaves*, G. Ben-Dor *et.al, eds,* Academic 2000.

3 Refereed Articles in Journals

J.G.Jones, K.C.Moore, J.Pike, **P.L.Roe**, A method for designing lifting configurations for high supersonic speeds, using axisymmetric flow fields. *Ingenieur Archiv*, **37** no.1, 1968.

P.L.Roe, Approximate Riemann solvers, parameter vectors, and difference schemes, *J. Comput. Phys*, **43**, no.2, 1981.

J.Pike, **P.L.Roe**, Accelerated convergence of Jameson's finite volume Euler scheme using van der Houven integrators, *Computers and Fluids*, **13**, 1985.

P.L.Roe, Discrete models for the numerical analysis of time-dependent multi-dimensional gas dynamics, *J. Comput. Phys*, **63** no.2, 1986.

P.L.Roe, Remote boundary conditions for unsteady multidimensional aerodynamic calculations, *Computers and Fluids*, **17**, 1989.

B.Einfeldt, C.D.Munz, **P.L. Roe**, B. Sjögreen, On Godonov-type methods near low densities, *J. Comput. Phys.*, **92** no.2 1991.

P.L.Roe, Discontinuous solutions to hyperbolic problems under operator splitting, *Numerical Methods for Partial Differential Equations*, **7** pp 277-297, 1991.

P.L.Roe, Sonic flux formulae, *SIAM J. Sci. Stat. Comput.*, **13**, no. 2, 1992.

P.L.Roe, D.Sidilkover, Optimum positive linear schemes for advection in two and three dimensions, *SIAM J.Num.Anal*, **29** No 6, 1992.

B.van Leer, W-T Lee, **P.L.Roe**, K.G. Powell, C-H. Tai, Design of Optimally-Smoothing Multi-Stage Schemes for the Euler Equations,

Comm Appl. Num. Math, **8**, p 761, 1992.

P.L.Roe, M. Arora, Characteristic-based schemes for dispersive waves I. The method of characteristics for smooth solutions, *Num Meth for PDEs*, **9**, p 459, 1993

H.Deconinck, **P.L.Roe**, R.Struijs, A multidimensional generalisation of Roe's flux difference splitter for the Euler equations, *Computers and Fluids*, **22**, p215, 1993.

C.L. Rumsey, B. van Leer, **P.L.Roe**, A multidimensional flux function with applications to the Euler and Navier-Stokes equations, *J.Comput.Phys*, **105**, p 306, 1993.

J.F. Clarke, S. Karni, J.J. Quirk, **P.L.Roe**, L.G. Simmonds, E.F. Toro, Numerical computation of two-dimensional unsteady detonation waves in high-energy solids, *J.Comput.Phys*, **106**, p 215, 1993.

R.B.Lowrie, **P.L.Roe**, On the numerical solution of conservation laws by minimizing residuals, *J.Comput.Phys*, **113**, p 304, 1994.

P.L.Roe, Reduction of certain wave operators to locally one-dimensional form, *Applied Math Letters*, **8**, 3, 1995.

J-C Carette, H. Deconinck, **P.L.Roe**, Multidimensional Upwinding-Its Relation to Finite Elements, *Int. J. Num. Meth. in Fluids*, **20**, 8/9, p 935, 1995.

P.L.Roe, D.S. Balsara, Notes on the eigensystem of magnetohydrodynamics, *SIAM J. App. Math.*, **56**, p 57, 1996.

M. Arora, **P.L.Roe**, A well-behaved limiter for high-resolution calculations of unsteady flow. *J.Comput.Phys*, **128**, p 1, 1997.

M. Arora, **P.L.Roe**, On post-shock oscillations due to shock-capturing schemes in unsteady flow. *J.Comput.Phys*, **130**, p 25, 1997.

K. Mazaheri, **P.L.Roe**, Numerical Wave Propagation and Steady-State Solutions- Soft Wall and Outer Boundary Conditions, *AIAA*

Jnl. **35**, no 8, p 965, 1997.

R.S. Myong, **P.L. Roe**, Shock waves and rarefaction waves in magnetohydrodynamics, I. The model system, *J. Plasma Physics*, **58-3**, pp 485-519, 1997.

R.S. Myong, **P.L. Roe**, Shock waves and rarefaction waves in magnetohydrodynamics, II. The MHD system, *J. Plasma Physics*, **58-3**, pp 521-552, 1997.

T. J. Linde, T. I. Gombosi, **P.L.Roe**, K. G. Powell, D. L. DeZeeuw, Heliosphere in the magnetized local interstellar medium: results of a three-dimensional MHD simulation, *J. Geophys. Res.* **103, A2**, pp 1889-1904, 1998.

P.L.Roe, Linear bicharacteristic schemes without dissipation, *SIAM J. Sci.Comp.* **19**,5, p 1405, 1998

T. Linde, **P.L. Roe**, On a mistaken notion of "proper upwinding", *J.Comput.Phys*, **142**, pp 611-614, 1998.

R.S. Myong, **P.L. Roe**, Godunov-type schemes for magnetohydrodynamics, I. a model system, *J.Comput.Phys*, **147**, pp. 545-567 1998.

K. Mazaheri, **P.L.Roe**, Numerical Wave Propagation and Steady-State Solutions- Artificial Bulk Viscosity, *AIAA Jnl.*, submitted.

K. G. Powell,*P. L. Roe*, T. J. Linde, T. I. Gombosi, D. L. De Zeeuw , Solution-Adaptive Upwind Scheme for Ideal Magnetohydrodynamics, *J.Comput.Phys*, **154**, pp. 284-309, 1999.

M.Hubbard, **P.L. Roe**, An algorithm for high-resolution advection on unstructured grids *Int. J. Num. Meth, in Fluids*, **33** p 711-736, 2000.

K. W. Morton, **P.L. Roe**, Vorticity-preserving Lax-Wendroff schemes for the system wave equation, *SIAM J. Scientific Computing*, to appear

G. Toth, **P.L. Roe**, Divergence- and curl-preserving prolongation and restriction operators, *J.Comput.Phys*, submitted.

4 Invited Conference Papers

P.L.Roe, Numerical modelling of shockwaves and other discontinuities, Institute of Mathematics and its Applications conference, Reading, U.K. March,1981, in **Numerical Methods in Aeronautical Fluid Dynamics**, ed. P.L.Roe, Academic Press, 1982.

P.L.Roe, Fluctuations and signals - a framework for numerical evolution problems, Institute of Mathematics and its Applications conference, Reading, U.K., March 1982, in **Numerical Methods for Fluid Dynamics**, eds. K.W.Morton, M.J.Baines, Academic Press, 1983.

P.L.Roe, Upwind schemes using various formulations of the Euler equations, INRIA Workshop, Rocquencourt, 1983, in **Numerical Methods for the Euler Equations of Fluid Dynamics**, eds F.Angrand, A.Dervieux, J.A.Desideri, R.A.Glowinski, SIAM, 1985.

P.L.Roe, Some contributions to the modelling of discontinuous flows, Am. Math.Soc Symposium, San Diego, 1983, in **Large-scale Computations in Fluid Mechanics**, eds B.E.Engquist, S.Osher, R.C.J. Somerville, Lectures in Applied Mathematics, vol 22, Am.Math.Soc.,1985.

P.L.Roe, A basis for upwind differencing of the two-dimensional unsteady Euler equations, Institute of Mathematics and its Applications conference, Oxford, 1986, in **Numerical Methods for Fluid Dynamics II**, eds K.W.Morton, M.J.Baines, Oxford University Press, 1986.

P.L.Roe, Finite-volume methods for the compressible Navier-Stokes equations, Montreal,1987, in **Numerical Methods for Laminar and Turbulent Flow**, eds C.Taylor, W.G.Habashi, M.M.Hafez, Pineridge Press, 1987.

P.L.Roe, Mathematical problems associated with computing flow of real gases, GAMNI/SMAI/IMA conference on Computational Aeronautical Dynamics, Antibes, 1989, Academic Press 1993.

P.L.Roe, Modern Shock-Capturing Methods, 18th International Symposium on Shockwaves, Sendai, Japan, 1991.

P.L.Roe, A New Class of Shock-Capturing Scheme, Workshop on Computational Fluid Dynamics for Unsteady Flow, Sendai, Japan, 1991.

P.L.Roe, Waves in Discrete Fluids, Institute of Mathematics and its Applications Conference, Reading, 1992, in **Numerical Methods for Fluid Dynamics III**, eds K.W.Morton, M.J.Baines, Oxford University Press, 1992.

P.L.Roe, Technical Prospects for Computational Aeracoustics, AIAA/DGLR Meeting on Aeroacoustics, Aachen, 1992.

P.L.Roe, Long-range numerical propagation of high-frequency waves, 5th International Symposium on Computational Fluid Dynamics, Sendai, Japan, 1993.

P.L.Roe, Mathematics and Numerics in Hyperbolic Conservation Laws. Conference on Mathematics and Computers in Simulation, Missisipi State University, 1993.

C.T.P. Groth, **P.L. Roe**, T.I. Gombosi, S.L. Brown, On the nonstationary wave structure of a 35-moment closure for rarefied gas dynamics, AIAA paper 95-2312,AIAA Fluid Dynamics Meeting, San Diego, June, 1995.

P.L.Roe, Multidimensional upwinding, Workshop on the Physics and Numerics of High-speed Flow, Bordeaux, May, 1996.

P.L.Roe, Cell-vertex methods, past, present and future, Meeting to honour the retirement of Professor K. W. Morton, Oxford, April, 1997.

P.L.Roe, Physical reasoning in computational fluid dynamics, *Godunov's Method for Gas Dynamics: Current Applications and Future Developments*, University of Michigan, May 1997.

P.L.Roe, Compounded of many simples, reflections on the role of model problems in CFD, Workshop on Barriers and Challenges in Computational Fluid Dynamics, NASA Langley, August, 1996, eds Venkatakrishnan, Salas and Chakravarthy, Kluwer, 1998.

P.L.Roe, New applications of upwind differencing, First Ami Harten Memorial Lecture, Manchester, UK, May, 1995, in 'Numerical Methods for Wave Propagation Problems' ed E. F. Toro, Kluwer, 1998.

P.L.Roe, Est-ce-qu'il-y-a une function de flux ideal pour les lois de conservation hyperboliques? CANUM 98, Arles, 1998.

P.L.Roe, Three lectures in 'Modelisation numérique des plasmas magnetisés', Summer School, Carqueiranne, September 1998.

P.L.Roe, Identifying the unstable modes of some two-phase flow problems, CEA Saclay, Paris, January 1999.

P. L. Roe, J.A. F. Hittinger, Toward Godunov-type methods for hyperbolic systems with stiff relaxation, *An international Conference to honour Professor S K Godunov, in the year of his 70th birthday*, October 1999. ed E. F. Toro, Numeritek, to appear.

P. L. Roe, K. W. Morton. Preserving vorticity in finite-volume schemes, *Finite Volumes for Complex Applications*, eds Vilsmeier, Benkhaldoun, Hanel, Duisberg, 1999, Hermes.

P.L. Roe, Computing mixed conservation laws by elliptic/hyperbolic decomposition,*Eighth International Conference on Hyperbolic Problems Theory, Numerics, Applications*, Otto-von-Guericke-Universität Magdeburg Feb-Mar, 2000.

P. L. Roe, Computational Aspects of Rarefied Flows, IMA Workshop on Rarefied Flows, Minneapolis, May, 2000.

P. L. Roe, Al-Khwarizmi's contribution to fluid dynamics, First International Iranian Aerospace Conference, Teheran, December, 2000.

P. L. Roe, Title to be decided, International Workshop on Hyperbolic and Kinetic Problems, Catania, Sicily, February 2001.

P. L. Roe, Title to be decided, International Conference on Numerical Methods in Fluid Dynamics, Oxford, March, 2001.

5 Published Conference Papers Subject to Selection

P.L.Roe, The use of the Riemann problem in finite-difference schemes, Seventh Int. Conf. Num. Meth. in Fluid Dyn., Stanford, 1980, in **Lecture Notes in Physics, vol 141**, eds W.C.Reynolds, R.W.MacCormack, Springer, 1981.

P.L.Roe, M.J.Baines, Algorithms for advection and shock problems, Fourth GAMM Conference on Numerical Methods in Fluid Mechanics, Paris, 1981, in **Notes on Numerical Fluid Mechanics, vol 5**, ed H.Viviand, Vieweg, 1982.

P.L.Roe, M.J.Baines, Asyptotic behaviour of some non-linear schemes for linear advection problems, Fifth GAMM Conference on Numerical Methods in Fluid Mechanics, Rome, 1983, in **Notes on Numerical Fluid Mechanics, vol 7**, eds M.Pandolfi, R.Piva , Vieweg, 1984.

P.L.Roe, J.Pike, Efficient construction and use of approximate Riemann solvers, Sixth Int. Symp. Comp. Meth. in Appl. Sci. and Eng., Versailles, 1983, in **Computing Methods in Applied Science and Engineering, VI**, eds R.Glowinski, J-L.Lions, North-Holland,1984

P.L.Roe Upwind differencing schemes for hyperbolic conservation laws with source terms, in **Lecture Notes in Mathematics, vol 1270, Nonlinear Hyperbolic Problems**, eds C.Carasso, P-A.Raviart, D.Serre, Springer-Verlag, 1986.

B.van Leer, J.L.Thomas, **P.L.Roe**, R.W.Newsome, A comparison of numerical flux formulas for the Euler and Navier-Stokes equations. AIAA 8th CFD conference, Honolulu,1987. AIAA paper 87-1104 CP.

E.F.Toro, **P.L.Roe**, A new numerical method for hyperbolic conservation laws, Meeting on Combustion and Detonation Phenomena, Warsaw,1987.

P.L.Roe, B.van Leer, Nonexistence, non-uniqueness and slow convergence of discrete hyperbolic conservation laws, Institute of Mathematics and its Applications Conference, Oxford, 1988, in **Numerical Methods for Fluid Dynamics III**, eds K.W.Morton, M.J.Baines, Oxford University Press, 1988.

E.F.Toro, **P.L.Roe**, A hybrid scheme for the Euler equations using random choice and Roe's methods, Institute of Mathematics and its Applications Conference, Oxford, 1988, in **Numerical Methods for Fluid Dynamics III**, eds K.W.Morton, M.J.Baines, Oxford University Press, 1988.

P.L.Roe, Momentum analysis of waverider flow fields, in *1st International Hypersonic Waverider Symposium*, University of Maryland, 1990.

P.L.Roe, H.Deconinck, R.Struijs, Recent progress in multidimensional upwinding, Twelfth Int. Conf. Num. Meth. in Fluid Dyn., Oxford,1990, in **Lecture Notes in Physics, 371** , Springer.1991.

C.L.Rumsey, B.van Leer, **P.L.Roe**, A grid-independent approximate Riemann solver with applications to the Euler and Navier-Stokes equations, AIAA Conference, Reno,1991.

P.L.Roe, M.Arora, Design of algorithms for a stiff dispersive hyperbolic problem, AIAA paper 91-1535, AIAA 10th Computational Fluid Dynamics Conference, Hawaii, 1991.

K.Mazaheri, **P.L. Roe**, New light on numerical boundary conditions, AIAA paper 91-1600, AIAA 10th Computational Fluid Dynamics Conference, Hawaii, 1991.

H. Deconinck, K.G.Powell, **P.L. Roe**, R. Struijs, Multidimensional schemes for scalar advection, AIAA paper 91-1532, AIAA 10th Computational Fluid Dynamics Conference, Hawaii, 1991.

C.L.Rumsey, B.van Leer, **P.L.Roe**, Effect of a multidimensional flux function on the monotonicity of Euler and Navier-Stokes computations,AIAA paper 91-1530, AIAA 10th Computational Fluid Dynamics

Conference, Hawaii, 1991.

B.van Leer, W-T.Lee, **P.L. Roe**, Characteristic time-stepping, or, local preconditioning of the Euler equations, AIAA paper 91-1552, AIAA 10th Computational Fluid Dynamics Conference, Hawaii, 1991.

R.Struijs, H.Deconinck, P. de Palma, **P.L. Roe**, K.G.Powell, Progress on multidimensional upwind Euler solvers for unstructured grids, A-IAA paper 91-1550, AIAA 10th Computational Fluid Dynamics Conference, Hawaii, 1991.

B.van Leer, W-T.Lee, **P.L.Roe**, K.G.Powell, C-H.Tai, Design of optimally smoothing multi-stage schemes for the Euler equations, Conference on Multigrid Methods, Boulder,CO, April, 1991.

H.Deconinck, **P.L.Roe**, R.Struijs, A multidimensional generalization of Roe's flux difference splitter for the Euler equations, 4th International Symposium on Computational Fluid Dynamics, Davis, California, 1991.

G.T.Tomaich, **P.L.Roe**, Compact schemes for convection-diffusion equations on unstructured meshes. Forum on Novel Computational Methodology for Transport Equations, Pittsburg, 1992.

J-D Múller, **P.L.Roe**, Experiments on the accuracy of some advection schemes on unstructured and partly structured meshes. Forum on Novel Computational Methodology for Transport Equations, Pittsburg, 1992.

P.L.Roe, L.Beard, A new wave model for the Euler equations, Thirteenth Int. Conf. Num. Meth. in Fluid Dyn, Rome, 1992.

J-D Müller, **P.L.Roe**, H.Deconinck, Delaunay-based triangulations for the Navier-Stokes equations with mimimum user input, Thirteenth Int. Conf. Num. Meth. in Fluid Dyn, Rome, 1992

G.Bourgois, H.Deconinck, **P.L.Roe**, R.Struijs, Multidimensional upwind schemes for scalar advection on tetrahedral meshes., 1st European Conf on CFD, Brussels, 1992.

P.L.Roe, Long-range numerical propagation of high-frequency waves, 5th International Symposium on Computational Fluid Dynamics, Sendai, Japan, 1993.

K.Mazaheri, **P.L.Roe**, Numerical Wave Propagation and Steady-State Solutions- Bulk Viscosity Damping, AIAA paper 93-3331, Orlando, 1993.

J.P.Thomas, **P.L.Roe**, Development of Non-Dissipative Algorithms for Computational Aeroacoustics, AIAA paper 93-3382, Orlando, 1993.

H.Paillière, H.Deconinck, R.Struijs, **P.L.Roe**, L.M.Mesaros, J-D Müller, Computations of inviscid compressible flows using fluctuation-splitting on triangular grids, AIAA paper, Orlando, 1993.

J-C Carette, H.Deconinck, H.Paillière, **P.L.Roe**, Multidimensional upwinding; its relation to finite elements, "Finite Elements Methods in Fluids", Barcelona, September, 1993.

B. Nguyen, **P.L. Roe**, Application of an upwind leapfrog method for electromagnetics, in *10th Annual Review of Progress in Applied Computational Electromagnetics*, vol 1, pp 446-458, Conference of the Applied Computational Electromagnetics Society, Monterey, March 1994.

M. Arora, **P.L. Roe**, On oscillations produced by moving shockwaves, Fourteenth Int. Conf. Num. Meth. in Fluid Dyn, Bangalore, 1994.

T. Linde, D. de Zeeuw, T. Gombosi, K.G. Powell, **P.L. Roe** A 3d model of the heliosphere, American Geophysical Union, San Fransisco,December 1994.

L.M.Mesaros, **P.L. Roe**, Multidimensional fluctuation splitting schemes based on decomposition methods, AIAA CFD Meeting, San Diego, June, 1995.

R.B. Lowrie, **P.L. Roe**, B. van Leer, A space-time discontinuous Galerkin method for the time-accurate numerical solution of hyperbolic conservation laws, AIAA CFD Meeting, San Diego, June, 1995.

J.P. Thomas, C. Kim, **P.L. Roe**, Progress toward a new computational scheme for aeroacoustics, AIAA CFD Meeting, San Diego, June, 1995.

J-D. Müeller, **P.L. Roe**, H. Deconinck, Multigrid implentation of fluctuation-splitting schemes, AIAA CFD Meeting, San Diego, June, 1995.

H. Deconinck, H. Paillèrre, **P.L. Roe**, Conservative upwind residual-distribution schemes based on the steady characteristics of the Euler equations. AIAA CFD Meeting, San Diego, June, 1995.

K.G. Powell, **P.L. Roe**, Rhoshin Myong, T. Gombosi, D. de Zeeuw, An upwind scheme for magnetohydrodynamics, AIAA CFD Meeting, San Diego, June, 1995.

T. Linde, T.I. Gombosi, **P.L. Roe**, K.G. Powell, D. de Zeeuw, A 3d MHD model of the heliosphere: the effects of polar coronal holes on the solar wind, American Geophysical Union, 1995.

C.P.T. Groth, T.I. Gombosi, **P.L. Roe**, S.L. Brown, A new model of the polar wind. American Geophysical Union, Baltimore, Maryland, May 1995. 1995.

M. Arora, **P.L. Roe**, A fresh look at viscous conservation laws via equivalent relaxation systems, SIAM Annual Meeting, 1995.

P.L. Roe, L. M. Mesaros, Solving steady mixed conservation laws by elliptic/hyperbolic splitting, plenary presentation, 13th International Conference on Numerical Methods in Fluid Dynamics, Monterey, July, 1996.

M. Arora, **P.L. Roe**, Characteristic-based numerical algorithms for stiff hyperbolic systems, 13th International Conference on Numerical Methods in Fluid Dynamics, Monterey, July, 1996.

K. G. Powell, **P.L. Roe**, D. DeZeeuw, M. Vinokur, A computational approach for modelling solar-wind physics. 13th International Conference on Numerical Methods in Fluid Dynamics, Monterey, July, 1996.

P.L. Roe, E.Turkel, The quest for diagonalization of differential systems, Workshop on Barriers and Challenges in Computational Fluid Dynamics, NASA Langley, August, 1996, Kluwer, 1998. by Springer, 1997.

M. Arora, **P.L. Roe**, Characteristic-based algorithms for stiff conservation laws. Workshop on Barriers and Challenges in Computational Fluid Dynamics, NASA Langley, August, 1996, Kluwer, 1998.

T. Linde, **P.L. Roe**, On positively-conservative high-resolution schemes, Workshop on Barriers and Challenges in Computational Fluid Dynamics, NASA Langley, August, 1996, Kluwer, 1998.

R.B. Lowrie, **P.L. Roe**, B. van Leer, Space-time methods for hyperbolic conservation laws, Workshop on Barriers and Challenges in Computational Fluid Dynamics, NASA Langley, August, 1996, Kluwer, 1998.

C. Kim, **P.L. Roe**, Solution of aeroacoustic test problems by a fourth-order upwind leapfrog method. 2nd International Workshop on Computation Aeroacoustics, Florida, November 1996

P.L. Roe, Fluctuation splitting on optimal grids, AIAA CFD Meeting, Snowmass, Colorado, June 1997.

C. Kim, **P.L. Roe** A fourth-order upwind leapfrog method for acoustic waves. AIAA CFD Meeting, Snowmass, Colorado, June 1997.

T. Linde, **P.L. Roe**, Robust Euler codes, AIAA CFD Meeting, Snowmass, Colorado, June 1997.

M. Hubbard, **P.L. Roe**, Multidimensional upwind fluctuation distribution schemes for scalar time dependent problems, International Conference on Numerical Methods for Fluid Dynamics, Oxford, 1998.

M. Rad, **P.L. Roe**, A new formulation of potential flow, AIAA CFD Meeting, June 1999,

M. Rad, **P. L. Roe**. An Euler code that can preserve potential flow,

Finite Volumes for Complex Applications, eds Vilsmeier, Benkhaldoun, Hanel, Duisberg, 1999, Hermes.

J. A. F. Hittinger, **P. L. Roe**. On uniformly accurate upwinding for hyperbolic systems with relaxation., *Finite Volumes for Complex Applications*, eds Vilsmeier, Benkhaldoun, Hanel, Duisberg, 1999, Hermes.

H. Nishikawa, Mani Rad **P.L. Roe** Grids and Solutions from Residual Minimisation,*1st International Conference on Computational Fluid Dynamics*, Kyoto, 2000, Springer Verlag.

M. Rad, H. Nishikawa and **P.L. Roe**, Some Properties of Residual Distribution Schemes for Euler Equations, *1st International Conference on Computational Fluid Dynamics*, Kyoto, 2000, Springer Verlag.

6 Royal Aerospace Establishment Publications

P.L.Roe, Some exact calculations of the lift and drag produced by a wedge in supersonic flow, either directly or by interference, RAE TN 2981, 1964. (also Aeronautical Research Council R&M 3478,1967)

D.R.Andrews, **P.L.Roe**, W.G.Sawyer, Preliminary measurements above and below a delta wing and body compination at M=4.0, RAE TR 65032, 1965.

P.L.Roe, An experimental investigation of the flow through inclined circular tubes at a Mach number of 4.0, RAE TR 65110, 1965. (also Aeronautical Research Council CP 884)

P.L.Roe, Exploratory flow measurements in the wing-body junction of a possible Mach four vehicle, RAE TR 65257.

P.L.Roe, Guided Weaons Aerodynamic Study; force and moment measurements on some monoplane and cruciform slender wing-body combinations at M=4.0, RAE TR 66257. (also Aeronautical Research Council CP 972, 1968)

P.L.Roe, A momentum analysis of lifting surfaces in inviscid supersonic flow. RAE TR 67124, 1967. (also Aeronautical Research Council R&M 3576,1969)

L.C.Squire, **P.L.Roe**, Off-design conditions for waveriders. RAE TM 1168, 1969.

P.L.Roe, Proposals for comparing prediction methods for high-speed lifting shapes, RAE TM 1249, 1970.

P.L.Roe, A simple treatment of the attached shock layer on a plane delta wing, RAE TR 70246, 1970.

P.L.Roe, L.Davies, L.C.Squire, Report on papers presented at EUROMECH 20 on the aerodynamics of bodies at high supersonic speeds, RAE TR 71054, 1971.

P.L.Roe, Estimating the slope of an experimental graph, RAE TM 1379, 1971.

L.Davies, **P.L.Roe**, J.L.Stollery, L.H.Townend, Configuration design for high-lift reentry, RAE TM 1379, 1971.

P.L.Roe, A result concerning the supersonic flow beneath a plane delta wing, RAE TR 72077, 1972. (also Aeronautical Research Council CP 1228, 1972)

P.L.Roe, Some aspects of shock-capturing algorithms, RAE TM 1708, 1977.

P.L.Roe, An improved version of MacCormack's shock-capturing algorithm. RAE TR 79041, 1979.

P.L.Roe, Numerical algorithms for the linear wave equation, RAE TR 81047, 1981.

7 Miscellaneous Publications

P.L.Roe, *Aerodynamics at Moderate Hypersonic Mach Numbers*, AGARDograph 42, 1967.

P.L.Roe, *An Introduction to Numerical Methods Suitable for the Euler Equations*, von Karman Institute Lecture Series 1983-02.

P.L.Roe, Generalised formulation of TVD Lax-Wendroff schemes, ICASE Report 84-53 1984.

P.L.Roe, Error estimates for cell-vertex solutions of the Euler equations, ICASE Report 87-6 1986.

P.L.Roe, The Influence of Mesh Quality on Solution Accuracy, Commonwealth Advisory Aeronautical Research Council Specialists Meeting, Bangalore, CC.AE. 1002, 1988.

P.L.Roe, Upwind Differencing, Commonwealth Advisory Aeronautical Research Council Specialists Meeting, Bangalore, CC.AE. 1002, 1988.

P.L.Roe, The best shape for a tin can *Mathematical Spectrum*, 1990.

R.Struijs, H.Deconinck, **P.L.Roe**, *Fluctuation Splitting schemes for multidimension convection problems, an alternative to finite-volume and finite-element schemes*, von Karman Institute Lecture Series 1990-03.

R.Struijs, H.Deconinck, **P.L.Roe**, *Fluctuation Splitting schemes for the 2D Euler equations*, von Karman Institute Lecture Series 1991-01.

P.L.Roe, 'Optimum' upwind advection on a triangular mesh, ICASE Report 90-75, 1990.

H.Deconinck, R.Struijs, H.Bourgeois, H.Paillère, **P.L.Roe**, Multidimensional Upwind Methods for Unstructured Grids, AGARD R787 (Proceedings of AGARD/NASA Special Course), 1992.

J-D Müller, **P.L.Roe**, H.Deconinck, A Frontal Approach for Node Generation in Unstructured Grids, AGARD R787 (Proceedings of A-GARD/NASA Special Course), 1992.

H.Deconinck, R.Struijs, H.Bourgeois, **P.L.Roe**, *Compact Advection Schemes on Unstructured Grids*, von Karman Institute Lecture Series 1993-04.

P.L.Roe, Multidimensional upwinding, motivation and concepts, von Karman Institute Lecture Series 1994-04.

P.L.Roe, Upwinding without dissipation, von Karman Institute Lecture Series 1994-04.

P.L.Roe, New aplications of upwind differencing, von Karman Institute Lecture Series 1994-04.

"A One-Sided View:" the real story

Bram van Leer
University of Michigan

with a post-script by

Ken Powell
University of Michigan

Abstract

The circumstances under which the paper "A One-Sided View" by Roe, LeVeque and Van Leer (1983), consisting entirely of limericks, was produced, and its failure to get published, are scrutinized. The paper then follows, after all these years.

1 Historic backdrop

It is 1983, a great year for CFD. The concepts of approximate Riemann solvers and limiters have empowered numerical analysts, and research in these subjects is burgeoning. TVD conditions[1] have just been introduced, the Harten-Lax-Van Leer review[2] on upwind differencing and Godunov-type schemes is appearing in SIAM Review, and the Woodward-Colella review[3] on computing flows with strong shocks, submitted to JCP, is circulating as a preprint. In the footsteps of an active "older" generation - Van Leer, Woodward, Harten, Colella, Roe, Osher, Engquist - a new generation of bright numerical analysts is emerging, dedicating their careers to CFD: LeVeque, Sweby, Tadmor, Berger, Mulder. And at NASA's research centers, engineers actually are *listening* to all these numerical types and their fancy ideas. This

[1] A. Harten, "High-resolution schemes for hyperbolic conservation laws," *J. Comput. Phys.* **49** (1983), pp. 357-393.

[2] A. Harten, P. D. Lax and B. van Leer, "On upstream differencing and Godunov-type schemes for hyperbolic conservation laws," *SIAM Review* **25** (1993), pp. 35-61.

[3] P. R. Woodward and P. Colella, "The numerical simulation of two-dimensional fluid flow with strong shocks," *J. Comput. Phys.* **54** (1994), pp. 115-173.

year at NASA Langley, for instance, the basis of the CFL3D code is laid by Jim Thomas and Kyle Anderson[4].

These are ideal conditions for a grand inspirational gathering of all the new talent and ideas. The opportunity for such a meeting arrives with the *15th AMS-SIAM Summer Seminar on Large-Scale Computations in Fluid Mechanics*, to be held in La Jolla, June 27 - July 8, 1983. The organizers are Björn Engquist and Stan Osher of UCLA, and Richard Somerville of the Scripps Institution of Oceanography, La Jolla. Engquist and Osher invite all their friends[5], including all members of the upwind-differencing clan, and almost all appear. As a counterweight some innocent computational meteorologists[6] are added, creating an odd mix that leads to some interesting moments[7] during the Seminar.

2 A new passion: limericks

It is at this meeting that a new passtime emerges: *composing CFD limericks.* The exact date of birth of this activity has not been recorded, but the whole thing started with Phil Roe reciting at luncheon the one and only CFD limerick [8] he had ever made (and not a flawless one). This created a challenge among the intelligent, witty and enthousiastic Seminar participants, and soon new limericks on all possible subjects of CFD and numerical analysis in general were being drafted on paper napkins. I volunteered to collect these, copy them neatly and compile them. We soon outgrew the improvisational napkin-stage and I brought a note pad to breakfast and luncheon.

Yes, this became serious business: we started with limericks at the crack of dawn. The La Jolla campus cafetaria offered a splendid Californian breakfast with lots of fresh fruit and other wholesome things, motivating the most active participants to appear at its doorstep at opening time, 7.00 am, and staying in the cafetaria inventing limericks until the lectures would start, two hours later.

Only once was a limerick session held elsewhere, namely, on Black's Beach;

[4]W. K. Anderson, J. L. Thomas and B. van Leer, "A comparison of finite-volume flux-vector splittings for the Euler equations," AIAA Paper AIAA 95-0122.

[5]"A One-Sided View," 1.i

[6]"A One-Sided View," 2.2.i

[7]"A One-Sided View," 2.2.ii

[8]"A One-Sided View," 2.1.i

my 1983 Calender shows this happened on Saturday, July 2. Black's Beach had the reputation that people would bathe there in the nude. We didn't see anything of the sort, but, admittedly, the weather wasn't great that day: the sun was defecting[9] and now and then there was a slight drizzle. Still, I felt like a nerd, bringing a note pad to the beach. This session stands very clearly in my mind, in particular because David Gottlieb was with us, that is, Phil Roe, Randy LeVeque, Pete Sweby and I. David inspired two great limericks: the one on the spectral technique[10] and the one in which the main rhyme is "La Jolla.[11]" That rhyme was David's challenge to Phil when we were leaving the beach. While walking up the sloping path, after some thinking, Phil produced the full limerick without hesitating once. Ah, a great moment in the history of CFD, and I was there.

I also vividly recall the luncheon session where Randy presented his perfect limerick[12] about the scalar conservation law $u_t + f_x = 0$, which in turn inspired me to start one[13] about the system case. This is the most ingenious limerick we made; the cook's nutritional advice: "You've had burgers enough, / try more variable stuff: / may I offer you eggs and Roe tea?" has a double meaning, with the punchline verbalizing the expression $u_x + \rho_t$. This limerick was not perfected until weeks after the La Jolla meeting, at ICASE, where I was spending the rest of the summer.

3 "A One-Sided View"

At ICASE I scrutinized all limericks we had produced, arranged them in the form of a paper, and had it typed. We had been very systematic in our coverage of CFD, the Seminar and its participants, and already in La Jolla we had produced some of the extras that characterize a real paper: one reference, an acknowledgement and a funding blurb. An abstract[14] was graciously mailed to me later by Phil. The title became: "A One-Sided View;" authors were Roe, LeVeque and Van Leer, with an acknowledgement[15] of substantial assistance by Sweby.

---

[9] "A One-Sided View," 1.ii
[10] "A One-Sided View," 2.3.iv
[11] "A One-Sided View," 5.i
[12] "A One-Sided View," 3.i
[13] "A One-Sided View," 3.ii
[14] "A One-Sided View," Abstract
[15] "A One-Sided View," Acknowledgement

The paper appeared in preprint format as an *ICASE Special Internal Report*, number 2 in the so-called *Pink Grundlehrer Series*, established by ICASE Director Milt Rose to absorb the more frivolous creations by ICASE staff. These reports were for private distribution only; on the cover the reader is warned: "Reports in the ICASE Grundlehrer Series have no intrinsic value, scientific or otherwise."

4 Getting it (not) published

I submitted "A One-Sided View" to AMS for inclusion in the Seminar proceedings, along with my regular Seminar contribution. The manuscript proceeded smoothly through the editorial system; I received an edited version for approval of changes made by the text-editor. For instance, the first sentence of the funding acknowledgement[16], "Research was supported in part / by agencies with a kind heart," was altered into "Research was supported in part / by agencies with kind hearts." A grammatical zealot, the editor had not noticed there were a rhyme and a meter to be preserved.

Eventually the paper landed on the desk of Stan Osher, co-editor of the proceedings, who immediately blocked its publication. In the belated rejection letter I received from the Manager of Editorial Services she writes: "[..] the editors [..] believe that it is better suited for some other journal - perhaps, National Lampoon or Punch." Stan's comment per telephone was that the paper was not serious enough for inclusion in the proceedings of a seminar funded by NSF, NASA and, particularly, ARO. He obviously did not want to jeopardize his relations with funding agencies.

It was not until twelve years later that Stan finally admitted to me the paper should have been published. The only real objection he had had was the language in the limerick[17] about himself: "He claims Engquist-Osher / is totally kosher / and runs like a son-of-a-bitch." In La Jolla we thought this was a great pastiche of Stan's manner of speaking; Stan's own suggestion of reworking the limerick such that its last line would become: "and his lifestyle gets posher and posher," was firmly rejected.

In retrospect the non-publication of "A One-Sided View" appears to be the regrettable result of a lack of communication, more precisely, a lack of

[16] "A One-Sided View," title page, footnote
[17] "A One-Sided View," 2.1.vi

experience in negotiating on both sides. May we all learn from tragedies like this.

5 Epilogue

Thus, "A One-Sided View" was never officially published, not in 1985 when the Seminar proceedings[18] appeared (two volumes that are still superb references on many topics), and not in SIAM Review, 1992, as the paper's only reference[19] boasted. May the current volume, dedicated to Phil Roe on his sixtieth birthday, finally provide a haven for this elegant piece of CFD trivia, and at the same time pay homage to Phil's unique spirituality.

One last, apologetic word to the reader. Some of the limericks appear to be self-congratulating, although they were not intended as such. This is the result of our ardor to cover important topics in CFD, combined with the multiple authorship. For instance, a limerick on Roe's linearization[20] absolutely needed to be included; it was composed by me as a tribute to Phil. Likewise, the limerick on MUSCL[21] was Randy's way of complimenting me.

6 Post-script

The paper "A One-Sided View"
was rejected as frivolous when new.
But by happenstance
it was published in France
after only a decade or two.

KEN POWELL

[18] "Large-Scale computations in Fluid Mechanics," B. Engquist, S. Osher and R. 'C. J. Somerville (Eds.), Lectures in Applied Mathematics, Vol. 22, Part 1 and Part 2, American Mathematical Society, Providence, RI, 1985.

[19] "A One-Sided View," Reference 1

[20] "A One-Sided View," 2.1.ii

[21] "A One-Sided View," 2.1.iii

The AMS/SIAM summer Seminar on
Large-Scale Computations in Fluid Dynamics
La Jolla, CA, June 27 - July 8, 1983

A one-sided view

P. L. Roe, Royal Aircraft Establishment Bedford, England
R. LeVeque, University of California, Los Angeles
B. van Leer, Delft University of Technology Delft, The Netherlands

Abstract: A critical study is made of the tricks of the upwinding trade. Five lines, it would seem, can describe any scheme of the class that the authors surveyed.

1 Introduction

I said to my darling; "I may go
to meet at U.C. San Diego
the full upwind clan
invited by Stan,
for a boost of the mutual ego.

By often escaping detection
the sun caused no marked defection.
To swim the Pacific
ain't all that terrific
if you must get dry by convection.

2 Basic numerical techniques

2.1 Conservative difference schemes

Conventional difference equations
give shocks that induce oscillations.
By adding some logic
we get monotonic
numerical representations.

Research was supported in part by agencies with. a kind heart. No proposals were rated, no funds allocated; in fact, no one knew this would start.

A characteristic equation
when differenced defies conservation,
which so badly we need.
But at last we were freed
by the grace of Roe's linearization.

To the podium many will hustle
to enter their claim in the tussle.
For the issue is fame
and it seems such a shame
we can't all take credit for MUSCL.

It's really not easy outsmartin'
the TVD schemes of A. Harten.
The name of the game
is they're all the same
so you'd better give up before startin'.

The sight of the slides of Colella
turns all his competitors yella.
Where others may fail
he's got the detail,
cause the grids are paid for by Ed Teller.

In spite of the entropy glitch
those contracts are making Stan rich.
He claims Engquist-Osher
is totally kosher
and runs like a son-of-a-bitch.

2.2 The state of the art in related areas

The conference could not have been better
except for the following matter:
that out of those listed
some speakers insisted
that they'd give a talk on the weather.

"We all know the problem of Riemann,
the basis of all of our schemin'."
This assertion will get
uninitiates upset
and the meteorologists steamin'.

So, medium-term weather prediction
turns out to be merely a fiction.
It's just anyone's guess,
if you ask how to dress
it will offer no useful restriction.

2.3 Auxiliary techniques

To sort out a boundary procedure
just talk to this elegant Swede here.
Your results will look nice
in the sense of Heinz Kreiss
and your program may even be speedier.

When exhausted by over-refinin'
don't throw up your hands and start whinin'.
No sense in postponin',
just go for rezonin'
(for details please talk to Mac Hyman).

Approximate factorizations
applied to the Euler equations,
are not all that fast,
in fact, they're surpassed
by classical point relaxations.

Now listen and please do not mock:
the spectral technique I'll unlock.
A hundred harmonics
make quite good transonics,
though fifty must die for the shock.

3 Theoretical results

Full proofs are exceedingly rare
except in the simple case where
the f is convex
in $u_t + f_x$
such as $f = \frac{1}{2}u^2$

At dinner on day number three
the cook said to my friend and me:
"You've had burgers enough,
try more variable stuff:
may I offer you eggs and Roe tea?"

4 An observation of Sweby

Now look who we have over here:
it's Roe and LeVeque and Van Leer.
They put all their time
into making things rhyme.
Will their paper [1] get written this year?

5 Conclusions

On returning from sunny La Jolla
I was summoned to see my employer.
"Once out of my reach
your went straight to Black's Beach.
Don't deny it 'cause everyone saw ya."

Acknowledgement

We offer our thanks to Pete Sweby who handed us many a freebee. With a line and a rhyme he was there all the time. You may ask: without him, where would we be?

References

[1] Roe, R. LeVeque, B. van Leer, "Limericks for the bored engineer", in: SIAM Review (1992), submitted, perhaps to appear.

Collocated Upwind Schemes for Ideal MHD

Kenneth G. Powell

W. M. Keck Foundation CFD Laboratory
Department of Aerospace Engineering
University of Michigan
Ann Arbor, MI 48109

Key Words: Magnetohydrodynamics, MHD, Upwind, Parallel, Adaptive

Abstract. This paper presents a computational scheme for compressible magnetohydrodynamics (MHD). The scheme is based on the same elements that make up many modern compressible gas dynamics codes: a high-resolution upwinding based on an approximate Riemann solver for MHD and limited reconstruction; an optimally smoothing multi-stage time-stepping scheme; and solution-adaptive refinement and coarsening. The pieces of the scheme are described, and the scheme is validated and its accuracy assessed by comparison with exact solutions. A domain-decomposition-based parallelization of the code has been carried out; parallel performance on a number of architectures is presented.

1 Introduction

Solving the magnetohydrodynamic (MHD) equations computationally entails grappling with a host of issues. The ideal MHD equations — the limit in which viscous and resistive effects are ignored — have a wave-like structure analagous to, though substantially more complicated than, that of the Euler equations of gas dynamics. The ideal MHD equations exhibit degeneracies of a type that do not arise in gas dynamics and also, as they are normally written, have an added constraint of zero divergence of the magnetic field.

Beginning with the work of Brio and Wu [1] and Zachary and Colella [2], the development of solution techniques for the ideal MHD equations based on approximate Riemann solvers has been studied. In both of those references, a Roe-type scheme for one-dimensional ideal MHD was developed and studied. Roe and Balsara [3] proposed a refinement to the eigenvector normalizations developed in the previous work, and Dai and Woodward [4] developed a nonlinear approximate Riemann solver for MHD. Other approximate Riemann solvers were also developed by Croisille et al [5] (a kinetic scheme) and by Linde [6] (an HLLE-type scheme). In addition Tóth and Odstrcil [7] have compared various schemes for MHD.

One of the issues that remains to be resolved for this class of schemes for ideal MHD is the method by which the $\nabla \cdot \mathbf{B}$ constraint is enforced [8]. One approach is that of a Hodge projection, in which the magnetic field is split into the sum of the gradient of a scalar and the curl of a vector, resulting in a Poisson equation for the scalar, such that the constraint is enforced. Another approach is to employ a staggered grid, such as that used in the constrained transport technique [9]. In this work, an alternative is presented. The ideal MHD equations are solved in their symmetrizable form. This form, first derived by Godunov [10], allows the derivation of an approximate Riemann solver with eight waves [11, 12]. The resulting Riemann solver, described in detail in this paper, maintains zero divergence of the magnetic field (a necessary initial condition) to truncation-error levels, even for long integration times.

In the following sections, the governing equations are given in the form used here, and an eight-wave Roe-type approximate Riemann solver is derived from them. A solution-adaptive scheme with the approximate Riemann solver as its basic building block is described, and validated for several cases.

2 Governing Equations

The governing equations for ideal MHD in three dimensions are statements of

- conservation of mass (1 equation)
- conservation of momentum (3 equations)
- Faraday's law (3 equations) and
- conservation of energy (1 equation)

for an ideal, inviscid, perfectly conducting fluid moving at non-relativistic speeds. These eight equations are expressed in terms of eight dependent variables:

- density (ρ),
- $x-$, $y-$ and $z-$components of momentum (ρu, ρv and ρw),
- $x-$, $y-$ and $z-$components of magnetic field (B_x, B_y and B_z),
- and total plasma energy (E),

where

$$E = \rho e + \rho \frac{\mathbf{u} \cdot \mathbf{u}}{2} + \frac{\mathbf{B} \cdot \mathbf{B}}{2\mu_0} . \tag{1}$$

In addition, the ideal-gas equation of state

$$e = \frac{p}{(\gamma - 1)\,\rho} \tag{2}$$

is used to relate pressure and energy, and Ampère's law is used to relate magnetic field and current density.

The ideal MHD equations, in the form they are used for this work, are given below. Vinokur has carried out a careful derivation, including effects of non-idealities, that goes beyond what is given here.

2.1 Conservation of Mass

The conservation of mass for a plasma is the same as that for a fluid, i.e.

$$\frac{\partial \rho}{\partial t} + \nabla \cdot (\rho \mathbf{u}) = 0 . \tag{3}$$

2.2 Faraday's Law

In a moving medium, the total time rate of change of the magnetic flux across a given surface S bounded by curve ∂S is[13]

$$\frac{d}{dt} \int_S \mathbf{B} \cdot \mathbf{dS} = \int_S \frac{\partial \mathbf{B}}{\partial t} \cdot \mathbf{dS} + \oint_{\partial S} \mathbf{B} \times \mathbf{u} \cdot d\ell + \int_S \nabla \cdot \mathbf{B} \mathbf{u} \cdot \mathbf{dS} \tag{4}$$

where the third term on the right-hand side arises from the passage of the surface S through an inhomogeneous vector field in which flux lines are generated. Using Stokes' theorem, and the fact that E' is zero in the co-moving frame, Faraday's law,

$$-\frac{d}{dt} \int_S \mathbf{B} \cdot \mathbf{dS} = \oint_{\partial S} \mathbf{E}' \cdot d\ell \tag{5}$$

becomes

$$\frac{\partial \mathbf{B}}{\partial t} + \nabla \cdot (\mathbf{uB} - \mathbf{Bu}) = -\mathbf{u} \nabla \cdot \mathbf{B} . \tag{6}$$

The term $\mathbf{u}\nabla \cdot \mathbf{B}$, which is typically dropped in the derivation due to the absence of magnetic monopoles, is kept here for reasons to be discussed in Section 2.8.

2.3 Conservation of Momentum

Conservation of momentum in differential form is

$$\frac{\partial (\rho \mathbf{u})}{\partial t} + \nabla \cdot (\rho \mathbf{u}\mathbf{u} + p\mathbf{I}) = \mathbf{j} \times \mathbf{B} \ . \tag{7}$$

Under the assumptions of ideal MHD, Ampère's law is

$$\mathbf{j} = \frac{1}{\mu_0} \nabla \times \mathbf{B} \ , \tag{8}$$

where μ_0 is the permeability of vacuum. Thus, conservation of momentum for ideal MHD can be written

$$\frac{\partial (\rho \mathbf{u})}{\partial t} + \nabla \cdot (\rho \mathbf{u}\mathbf{u} + p\mathbf{I}) = \frac{1}{\mu_0} (\nabla \times \mathbf{B}) \times \mathbf{B} \ . \tag{9}$$

Rewriting Equation 9, using a vector identity for $(\nabla \times \mathbf{B}) \times \mathbf{B}$, gives

$$\frac{\partial (\rho \mathbf{u})}{\partial t} + \nabla \cdot \left(\rho \mathbf{u}\mathbf{u} + \left(p + \frac{\mathbf{B} \cdot \mathbf{B}}{2\mu_0} \right) \mathbf{I} - \frac{\mathbf{B}\mathbf{B}}{\mu_0} \right) = -\frac{1}{\mu_0} \mathbf{B} \nabla \cdot \mathbf{B} \ . \tag{10}$$

As with Faraday's law, a term proportional to $\nabla \cdot \mathbf{B}$ is retained for reasons discussed in Section 2.8.

2.4 Conservation of Energy

Conservation of hydrodynamic energy density,

$$E_{hd} = \rho e + \rho \frac{\mathbf{u} \cdot \mathbf{u}}{2} \tag{11}$$

$$= \frac{p}{\gamma - 1} + \rho \frac{\mathbf{u} \cdot \mathbf{u}}{2} \tag{12}$$

for a fixed control volume of conducting fluid is given by

$$\frac{\partial E_{hd}}{\partial t} + \nabla \cdot (\mathbf{u} (E_{hd} + p)) = \mathbf{j} \cdot \mathbf{E} \ . \tag{13}$$

Using Ampère's law and the identity

$$\mathbf{E} \times \mathbf{B} = (\mathbf{B} \cdot \mathbf{B}) \mathbf{u} - (\mathbf{u} \cdot \mathbf{B}) \mathbf{B} \ ,$$

the $\mathbf{j} \cdot \mathbf{E}$ term can be expressed in terms of $\mathbf{u}$ and $\mathbf{B}$ as

$$\mathbf{j} \cdot \mathbf{E} = \frac{1}{\mu_0} \left[\mathbf{B} \cdot \frac{\partial \mathbf{B}}{\partial t} - (\mathbf{u} \cdot \mathbf{B}) \nabla \cdot \mathbf{B} - \nabla \cdot ((\mathbf{B} \cdot \mathbf{B}) \mathbf{u} - (\mathbf{u} \cdot \mathbf{B}) \mathbf{B}) \right] \ .$$

Finally, defining the total energy density of the plasma

$$E = E_{hd} + \frac{\mathbf{B}\cdot\mathbf{B}}{2\mu_0} \tag{14}$$

$$= \frac{p}{\gamma - 1} + \rho\frac{\mathbf{u}\cdot\mathbf{u}}{2} + \frac{\mathbf{B}\cdot\mathbf{B}}{2\mu_0} \tag{15}$$

the energy equation becomes

$$\frac{\partial E}{\partial t} + \nabla\cdot\left[\left(E + p + \frac{\mathbf{B}\cdot\mathbf{B}}{2\mu_0}\right)\mathbf{u} - \frac{1}{\mu_0}(\mathbf{u}\cdot\mathbf{B})\,\mathbf{B}\right] = -\frac{1}{\mu_0}(\mathbf{u}\cdot\mathbf{B})\,\nabla\cdot\mathbf{B}\,. \tag{16}$$

2.5 Non-Dimesionalization

It is usual to non-dimensionalize the ideal MHD equations, using, for example, L (a reference length) , a_∞ (the free-stream ion-acoustic speed), and ρ_∞ (the free-stream density). In addition, the current and magnetic field are scaled with $\sqrt{\mu_0}$, which results in the removal of μ_0 from the equations. This non-dimensional scaled form of the equations is used from this point on in this paper.

2.6 Quasilinear Form of Equations

For the eigensystem analysis necessary to develop the Riemann solver, it is convenient to write the governing equations as a quasilinear system in the primitive variables,

$$\mathbf{W} = (\rho, u, v, w, B_x, B_y, B_z, p)^T\ . \tag{17}$$

The primitive variables can be related to the vector of conserved variables

$$\mathbf{U} = (\rho, \rho u, \rho v, \rho w, B_x, B_y, B_z, E)^T \tag{18}$$

by the Jacobian matrices

$$\frac{\partial \mathbf{U}}{\partial \mathbf{W}} = \begin{bmatrix} 1 & 0 & 0 & 0 & 0 & 0 & 0 & 0 \\ u & \rho & 0 & 0 & 0 & 0 & 0 & 0 \\ v & 0 & \rho & 0 & 0 & 0 & 0 & 0 \\ w & 0 & 0 & \rho & 0 & 0 & 0 & 0 \\ 0 & 0 & 0 & 0 & 1 & 0 & 0 & 0 \\ 0 & 0 & 0 & 0 & 0 & 1 & 0 & 0 \\ 0 & 0 & 0 & 0 & 0 & 0 & 1 & 0 \\ \frac{\mathbf{u}\cdot\mathbf{u}}{2} & \rho u & \rho v & \rho w & B_x & B_y & B_z & \frac{1}{\gamma-1} \end{bmatrix} \tag{19}$$

$$\frac{\partial \mathbf{W}}{\partial \mathbf{U}} = \begin{bmatrix} 1 & 0 & 0 & 0 & 0 & 0 & 0 & 0 \\ -\frac{u}{\rho} & \frac{1}{\rho} & 0 & 0 & 0 & 0 & 0 & 0 \\ -\frac{v}{\rho} & 0 & \frac{1}{\rho} & 0 & 0 & 0 & 0 & 0 \\ -\frac{w}{\rho} & 0 & 0 & \frac{1}{\rho} & 0 & 0 & 0 & 0 \\ 0 & 0 & 0 & 0 & 1 & 0 & 0 & 0 \\ 0 & 0 & 0 & 0 & 0 & 1 & 0 & 0 \\ 0 & 0 & 0 & 0 & 0 & 0 & 1 & 0 \\ \frac{(\gamma-1)}{2}\mathbf{u}\cdot\mathbf{u} & ku & kv & kw & kB_x & kB_y & kB_z & (\gamma-1) \end{bmatrix} \tag{20}$$

where $k = (1-\gamma)$.

Collecting Equations 3, 6, 9 and 13, performing the non-dimensionalization and expressing them in terms of primitive variables gives

$$\frac{\partial \mathbf{W}}{\partial t} + (\mathbf{A_x}, \mathbf{A_y}, \mathbf{A_z}) \cdot \nabla \mathbf{W} = 0 \ , \tag{21}$$

where

$$\mathbf{A_x} = \begin{bmatrix} u & \rho & 0 & 0 & 0 & 0 & 0 & 0 \\ 0 & u & 0 & 0 & 0 & \frac{B_y}{\rho} & \frac{B_z}{\rho} & \frac{1}{\rho} \\ 0 & 0 & u & 0 & 0 & -\frac{B_x}{\rho} & 0 & 0 \\ 0 & 0 & 0 & u & 0 & 0 & -\frac{Bx}{\rho} & 0 \\ 0 & 0 & 0 & 0 & u & 0 & 0 & 0 \\ 0 & B_y & -B_x & 0 & 0 & u & 0 & 0 \\ 0 & B_z & 0 & -B_x & 0 & 0 & u & 0 \\ 0 & \gamma p & 0 & 0 & 0 & 0 & 0 & u \end{bmatrix} \tag{22}$$

$$
\mathbf{A_y} = \begin{bmatrix}
v & 0 & \rho & 0 & 0 & 0 & 0 & 0 \\
0 & v & 0 & 0 & -\frac{B_y}{\rho} & 0 & 0 & 0 \\
0 & 0 & v & 0 & \frac{B_x}{\rho} & 0 & \frac{B_z}{\rho} & \frac{1}{\rho} \\
0 & 0 & 0 & v & 0 & 0 & -\frac{B_y}{\rho} & 0 \\
0 & -B_y & B_x & 0 & v & 0 & 0 & 0 \\
0 & 0 & 0 & 0 & 0 & v & 0 & 0 \\
0 & 0 & B_z & -B_y & 0 & 0 & v & 0 \\
0 & 0 & \gamma p & 0 & 0 & 0 & 0 & v
\end{bmatrix}
$$

$$
\mathbf{A_z} = \begin{bmatrix}
w & 0 & 0 & \rho & 0 & 0 & 0 & 0 \\
0 & w & 0 & 0 & -\frac{B_z}{\rho} & 0 & 0 & 0 \\
0 & 0 & w & 0 & 0 & -\frac{B_z}{\rho} & 0 & 0 \\
0 & 0 & 0 & w & \frac{B_x}{\rho} & \frac{B_y}{\rho} & 0 & \frac{1}{\rho} \\
0 & -B_z & 0 & B_x & w & 0 & 0 & 0 \\
0 & 0 & -B_z & B_y & 0 & w & 0 & 0 \\
0 & 0 & 0 & 0 & 0 & 0 & w & 0 \\
0 & 0 & 0 & \gamma p & 0 & 0 & 0 & w
\end{bmatrix} .
$$

2.7 Divergence Form of Equations

Collecting the Equations 3, 6 . 10, and 16, and applying the non-dimensionalization. the normalized divergence form

$$
\frac{\partial \mathbf{U}}{\partial t} + (\nabla \cdot \mathbf{F})^T = \mathbf{S} , \tag{23}
$$

may be written, where **U** is the vector of conserved quantities defined by Equation 18, **F** is a flux tensor,

$$
\mathbf{F} = \begin{pmatrix}
\rho \mathbf{u} \\
\rho \mathbf{u}\mathbf{u} + \left(p + \frac{\mathbf{B}\cdot\mathbf{B}}{2}\right) \mathbf{I} - \mathbf{B}\mathbf{B} \\
\mathbf{u}\mathbf{B} - \mathbf{B}\mathbf{u} \\
\mathbf{u} \left(E + p + \frac{\mathbf{B}\cdot\mathbf{B}}{2}\right) - (\mathbf{u}\cdot\mathbf{B})\,\mathbf{B}
\end{pmatrix}^T \tag{24}
$$

and **S** is a "source" vector, containing the terms that can not be expressed in divergence form:

$$\mathbf{S} = -\nabla\cdot\mathbf{B}\begin{pmatrix} 0 \\ \mathbf{B} \\ \mathbf{u} \\ \mathbf{u}\cdot\mathbf{B} \end{pmatrix} \tag{25}$$

2.8 A Note on the $\nabla\cdot\mathbf{B}$ Source Term in the Divergence Form

The terms proportional to $\nabla\cdot\mathbf{B}$ in Equation 23 arise solely from rewriting the magnetic-field terms in the governing equations in divergence form. Equation 23 (*with* the source term) is exactly equivalent to Equation 21. Although for physical fields there are no magnetic monopoles, and the source term is therefore zero, dropping the source term from the analysis changes the character of the equations. This has been pointed out previously by Godunov[14]. He found that the ideal MHD equations written in pure divergence form (i.e. Equation 23 *without* the source term) are not symmetrizable. He further found that the system could be rendered symmetrizable only by adding a factor of the constraint $\nabla\cdot\mathbf{B}=0$ to each of the equations, and that the resulting symmetrizable form was that of Equation 23 *with* the source term.

Symmetrizable systems of conservation laws have been studied by Godunov [14] and Harten [15], among others. One property of the symmetrizable form of a system of conservation laws is that an added conservation law

$$\frac{\partial(\rho s)}{\partial t}+\frac{\partial(\rho u s)}{\partial x}+\frac{\partial(\rho v s)}{\partial y}+\frac{\partial(\rho w s)}{\partial z}=0$$

for the entropy s can be derived by multiplying each equation in the system by a factor and adding the resulting equations. For the ideal MHD equations, as for the gasdynamic equations, the entropy is $s=\log(p/\rho^{\gamma})$. Another property is that the system is Gallilean invariant; all waves in the system propagate at speeds $u\pm c$ (for MHD, the possible values of c are the Alfvén, magnetofast and magentoslow speeds, described below). Neither of these properties holds for the MHD system if the source term is ignored.

Equation 21, or Equation 23 *with* the source term, yields the following evolution equation for $\nabla\cdot\mathbf{B}$:

$$\frac{\partial}{\partial t}(\nabla\cdot\mathbf{B})+\nabla\cdot(\mathbf{u}\nabla\cdot\mathbf{B})=0\ . \tag{26}$$

This is a statement that the quantity $\nabla\cdot\mathbf{B}/\rho$ satisfies the equation for a passively convected scalar ϕ, i.e.

$$\frac{\partial}{\partial t}(\rho\phi)+\nabla\cdot(\rho\mathbf{u}\phi)=0\ . \tag{27}$$

Thus, for a solution of this system, the quantity $\nabla \cdot \mathbf{B}/\rho$ is constant along particle paths and therefore, since the initial and boundary conditions satisfy $\nabla \cdot \mathbf{B} = 0$, the same will be true for all later times throughout the flow. The only ambiguity arises in regions which are cut off from the boundaries; i.e. isolated regions of recirculating flow. These can occur in three-dimensional flow fields, and do in some of the cases that have been run. In practice, these regions do not lead to numerical difficulties. This may be due to the fact that, in a numerical calculation, these regions are not truly isolated from the outer flow, due to numerical dissipation. Thus, although not connected to the outer flow via a streamline, the magnetic field inside the recirculating region must be compatible with that of the outer flow. This remains to be proven, however.

The downside of the solving the equations in the form given in Equation 23 is, of course, that they are not strictly conservative. Terms of order $\nabla \cdot \mathbf{B}$ are added to what would otherwise be a divergence form. The danger of this of is that shock jump conditions may not be correctly met, unless the added terms are small, and/or they alternate in sign in such a way that the errors are local, and in a global sense cancel in some way with neighboring terms. This downside, however, has to be weighed against the alternative; a system (i.e. the one without the source term) that, while conservative, is not Gallilean invariant, has a zero eigenvalue in the Jacobian matrix, and is not symmetrizable.

The approach taken in this paper is therefore to solve the equations in their symmetrizable form, i.e. the form of Equation 23. As shown previously [11, 12], this form of the equations allows the derivation of an eight-wave approximate Riemann solver that can be used to construct an upwind solution scheme for multi-dimensional flows. The elements of the solution scheme are described in the following section.

3 Elements of Solution Scheme

3.1 Overview of Scheme

The scheme described here is an explicit, solution-adaptive, high-resolution, upwind finite-volume scheme. In a finite-volume approach, the governing equations in the form of Equation 23 are integrated over a cell in the grid, giving

$$\int_{cell\ i} \frac{\partial \mathbf{U}}{\partial t}\, dV + \int_{cell\ i} \nabla \cdot \mathbf{F}\, dV = \int_{cell\ i} \mathbf{S}\, dV \tag{28}$$

$$\frac{d\mathbf{U}_i}{dt} V_i + \oint_{\partial(cell\ i)} \mathbf{F} \cdot \hat{\mathbf{n}} dS = \mathbf{S}_i V_i\ . \tag{29}$$

where $\mathbf{U}_i$ and $\mathbf{S}_i$ are the cell-averaged conserved variables and source terms, respectively, V_i is the cell volume, and $\hat{\mathbf{n}}$ is a unit normal vector, pointing outward from the boundary of the cell. In order to evaluate the integral, a quadrature scheme must be chosen; a simple midpoint rule is used here, giving

$$\frac{d\mathbf{U}_i}{dt}V_i + \sum_{faces} \mathbf{F}\cdot\hat{\mathbf{n}}dS = \mathbf{S}_i V_i \ , \tag{30}$$

where the $\mathbf{F}\cdot\hat{\mathbf{n}}$ terms are evaluated at the midpoints of the faces of the cell. The source term $\mathbf{S}_i$ is proportional to the volume average of $\nabla\cdot\mathbf{B}$ for a cell. That average is computed by

$$\nabla\cdot\mathbf{B}_{cell\ i} = \frac{1}{V_i}\sum_{faces}\mathbf{B}\cdot\hat{\mathbf{n}}dS \ ;$$

the equation to be integrated in time is therefore

$$\frac{d\mathbf{U}_i}{dt}V_i + \sum_{faces}\mathbf{F}\cdot\hat{\mathbf{n}}dS = -\begin{pmatrix} 0 \\ \mathbf{B} \\ \mathbf{u} \\ \mathbf{u}\cdot\mathbf{B} \end{pmatrix}_i \sum_{faces}\mathbf{B}\cdot\hat{\mathbf{n}}dS \ . \tag{31}$$

The evaluation of $\mathbf{F}\cdot\hat{\mathbf{n}}$ at the interface is done by a Roe scheme for MHD, as described in Section 3.5. Other approximate Riemann solvers have been used in the code described here, including an MHD version of the HLLE scheme [6]. These solvers are all based on the eigensystem of the symmetric equations, described in Section 3.5. The time-integration scheme for Equation 30, the solution-adaptive technique and the limited reconstruction technique that makes the scheme second order in space are also described in the following sections.

3.2 Grid and Data Structure

The grid used in this work is an adaptive Cartesian one, with an underlying tree data structure. The basic underlying unit is a block of structured grid of arbitrary size. In the limit, the patch could be $1 \times 1 \times 1$, i.e. a single cell; more typically, blocks of anywhere from $4 \times 4 \times 4$ cells to $10 \times 10 \times 10$ cells are used. Each grid block corresponds to a node of the tree: the root of the tree is a single coarse block of structured grid covering the entire solution domain. In regions flagged for refinement, a block is divided into eight octants; in each octant, Δx, Δy and Δz are each halved from their value on the "parent" block. Two neighboring blocks, one of which has been refined and one of which has not, are shown in Figure 1. Any of these blocks can in turn be refined, and so on, building up a tree of successively finer blocks. The data structure is described more fully elsewhere [16, 17]. The approach

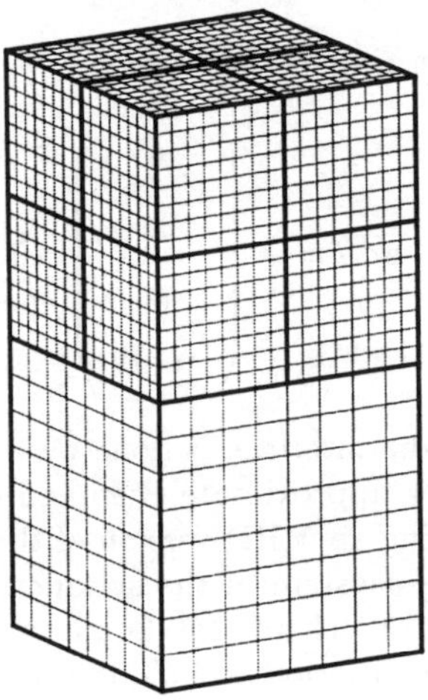

Figure 1: Example of Neighboring Refined and Unrefined Blocks

closely follows that first developed for two-dimensional gas dynamics calculations by Berger [18, 19, 20].

This block-based tree data structure is advantageous for two primary reasons. One is the ease with which the grid can be adapted. If, at some point in the calculation, a particular region of the flow is deemed to be sufficiently interesting, better resolution of that region can be attained by refining a block, and inserting the eight finer blocks that result from this refinement into the data structure. Removing refinement in a region is equally easy. Decisions as to where to refine and coarsen are made based on comparison of local flow quantities to threshold values. Refinement criteria used in this work are local values of

$$\begin{aligned} \epsilon_c &= |\nabla \cdot \mathbf{u}| \sqrt{V} \qquad (32) \\ \epsilon_r &= |\nabla \times \mathbf{u}| \sqrt{V} \\ \epsilon_t &= |\nabla \times \mathbf{B}| \sqrt{V} \end{aligned}$$

These represent local measures of compressibility, rotationality and current density. V is the cell volume; a scaling of this type is necessary to allow the scheme to resolve smooth regions of the flow as well as discontinous ones [21].

Another advantage of this approach is ease of parallelization: blocks of grid can easily be farmed out to separate processors, with communication limited to the boundary between a block and its parent [22, 16, 17]. The number of cells in the refinement blocks can be chosen so as to facilitate load balancing; in particular, an octant of a block is typically refined, so that each block of cells in the grid has the same number of cells.

3.3 Limited Linear Reconstruction

In order for the scheme to be more than first-order accurate, a local reconstruction must be done; in order for the scheme to yield oscillation-free results, the reconstruction must be limited. The limited linear reconstruction described here is due to Barth [23]. A least-squares gradient is calculated, using the cell-centered values in neighboring cells, by locally solving the following non-square system for the gradient of the kth component of the primitive variable vector W by a least-squares approach

$$\mathcal{L} \nabla W^{(k)} = f \qquad (33)$$

$$\mathcal{L} = \begin{pmatrix} \Delta x_1 & \Delta y_1 & \Delta z_1 \\ \vdots & \vdots & \vdots \\ \Delta x_N & \Delta y_N & \Delta z_N \end{pmatrix} \qquad f = \begin{pmatrix} \Delta W_1^{(k)} \\ \vdots \\ \Delta W_N^{(k)} \end{pmatrix} \tag{34}$$

where

$$\begin{aligned} \Delta x_i &= x_i - x_0 \\ \Delta y_i &= y_i - y_0 \\ \Delta z_i &= z_i - z_0 \\ \Delta W_i^{(k)} &= \bar{W}_i^{(k)} - \bar{W}_0^{(k)} \end{aligned}$$

and the points are numbered so that 0 is the cell in which the gradient is being calculated, and i is one of N neighboring cells used in the reconstruction.

The gradients calculated in this manner must be limited in order to avoid overshoots. A typical choice is a limiter due to Barth and Jesperson [24]. The reconstructed values are limited by a quantity $\phi^{(k)}$ in the following way

$$W^{(k)}(\mathbf{x}) = \bar{W}^{(k)} + \phi^{(k)}(\mathbf{x} - \bar{\mathbf{x}}) \cdot \nabla W^{(k)}, \tag{35}$$

where $\phi^{(k)}$ is given by

$$\phi^{(k)} = \min\left(1, \frac{|\bar{W}^{(k)} - \max_{neighbors}(\bar{W}^{(k)})|}{|\bar{W}^{(k)} - \max_{cell}(W^{(k)})|}, \frac{|\bar{W}^{(k)} - \min_{neighbors}(\bar{W}^{(k)})|}{|\bar{W}^{(k)} - \min_{cell}(W^{(k)})|}\right). \tag{36}$$

In the above, $\bar{W}^{(k)}$ is the value of the kth component of $\mathbf{W}$ at a cell center $\bar{\mathbf{x}}$, the subscript *neighbors* denotes the neighboring cells used in the gradient reconstruction, and the subscript *cell* denotes the unlimited ($\phi = 1$) reconstruction to the centroids of the faces of the cell.

At the interfaces of blocks that are at different refinement levels, states are constructed in two layers of "ghost cells" so that the interface is transparent to the reconstruction described above. Since refinement level differences of greater than one are not allowed, there are only two types of ghost cells: those created for a coarse block from values on a neighboring finer block; and those created for a fine block from values on a neighboring coarser block. A simple trilinear interpolation is used to construct the values in the ghost cells.

3.4 Multi-Stage Time Stepping

The time-stepping scheme used is one of the optimally-smoothing multi-stage schemes developed by Van Leer et al [25]. The general m-stage scheme for integrating Equation 30 from time-level n to time-level $n+1$ is

$$\mathbf{U}^{(0)} = \mathbf{U}^n \tag{37}$$

$$\mathbf{U}^{(k)} = \mathbf{U}^{(0)} + \alpha_k \Delta t \mathbf{R}\left(\mathbf{U}^{(k-1)}\right) \qquad k = 1 \ldots m \tag{38}$$

$$\mathbf{U}^{(n+1)} = \mathbf{U}^{(m)} \tag{39}$$

where

$$\mathbf{R} = \mathbf{S}_i - \frac{1}{V_i} \sum_{faces} \mathbf{F} \cdot \hat{\mathbf{n}} dS \, .$$

The multi-stage coefficients α_k and the associated time-step constraint are those that give optimal smoothing of high-frequency error modes in the solution, thereby accelerating convergence to a steady state. [25]. Typically, the two-stage optimal second-order scheme is used. For this scheme, $\alpha_1 = 0.4242$, α_2, and the corresponding CFL number used to choose Δt is 0.4693.

3.5 Approximate Riemann Solver

An approximate Riemann solver is used to compute the interface fluxes needed for the finite-volume scheme of Equation 30. A Roe scheme is used here: it is based on the eigensystem of the matrix

$$\mathbf{A_n} = (\mathbf{A_x}, \mathbf{A_y}, \mathbf{A_z}) \cdot \hat{\mathbf{n}} \, , \tag{40}$$

where $\mathbf{A_x}$, $\mathbf{A_y}$ and $\mathbf{A_z}$ are the matrices in the quasilinear form of the equations (Equation 21) and $\hat{\mathbf{n}}$ is the normal to the face for which the flux is being computed. For simplicity, the derivation is done here for $\hat{\mathbf{n}} = \hat{\mathbf{x}}$; results for an arbitrarily aligned face can be obtained by use of a simple rotation matrix.

3.5.1 Eigensystem of the Governing Equations

For the matrix $\mathbf{A} \cdot \hat{\mathbf{x}}$, there are eight waves, with their corresponding eigenvalues λ, left eigenvectors ℓ and right eigenvectors r. The eigenvalues are:

- $\lambda_e = u$, corresponding to an entropy wave;
- $\lambda_d = u$, corresponding to a magnetic-flux wave;
- $\lambda_a = u \pm B_x/\sqrt{\rho}$, corresponding to a pair of Alfvén waves; and
- $\lambda_{f,s} = u \pm c_{f,s}$, corresponding to two pairs of magneto-acoustic waves.

The magneto-acoustic speeds are given by

$$c_{f,s} = \sqrt{\frac{1}{2}\left(\frac{\gamma p + \mathbf{B}\cdot\mathbf{B}}{\rho} \pm \sqrt{\left(\frac{\gamma p + \mathbf{B}\cdot\mathbf{B}}{\rho}\right)^2 - 4\frac{\gamma p B_x^2}{\rho^2}}\right)}\,.$$

The eigenvectors corresponding to these waves are unique only up to a scaling factor. A suitable choice of scaling is given by Roe and Balsara [3]; that choice was used in the current work. (Recently, Barth [26] has introduced a scaling that is slightly better conditioned.) The scaled version of the eigenvectors comes from defining

$$\alpha_f^2 = \frac{a^2 - c_s^2}{c_f^2 - c_s^2} \qquad \alpha_s^2 = \frac{c_f^2 - a^2}{c_f^2 - c_s^2} \tag{41}$$

and

$$\beta_y = \frac{B_y}{\sqrt{B_y^2 + B_z^2}} \qquad \beta_z = \frac{B_z}{\sqrt{B_y^2 + B_z^2}}\,. \tag{42}$$

The scaled eigenvectors are:

Entropy

$$\begin{aligned} \lambda_e &= u \\ \ell_e &= \left(1, 0, 0, 0, 0, 0, 0, -\frac{1}{a^2}\right) \\ r_e &= (1, 0, 0, 0, 0, 0, 0, 0)^T \end{aligned} \tag{43}$$

Magnetic Flux

$$\begin{aligned} \lambda_d &= u \\ \ell_d &= (0, 0, 0, 0, 1, 0, 0, 0) \\ r_d &= (0, 0, 0, 0, 1, 0, 0, 0)^T \end{aligned} \tag{44}$$

Alfvén

$$\lambda_a = u \pm \frac{B_x}{\rho}$$

$$\ell_a = \left(0, 0, -\frac{\beta_z}{\sqrt{2}}, \frac{\beta_y}{\sqrt{2}}, 0, \pm\frac{\beta_z}{\sqrt{2\rho}}, \mp\frac{\beta_y}{\sqrt{2\rho}}, 0\right)$$

$$r_a = \left(0, 0, -\frac{\beta_z}{\sqrt{2}}, \frac{\beta_y}{\sqrt{2}}, 0, \pm\sqrt{\frac{\rho}{2}}\beta_z, \mp\sqrt{\frac{\rho}{2}}\beta_y, 0\right)^T \tag{45}$$

Fast

$$\lambda_f = u \pm c_f$$

$$\ell_f = \left(0, \pm\frac{\alpha_f c_f}{2a^2}, \mp\frac{\alpha_s}{2a^2}c_s\beta_y \operatorname{sgn} B_x, \mp\frac{\alpha_s}{2a^2}c_s\beta_z \operatorname{sgn} B_x, 0, \frac{\alpha_s}{2\sqrt{\rho}a}\beta_y, \frac{\alpha_s}{2\sqrt{\rho}a}\beta_z, \frac{\alpha_f}{2\rho a^2}\right)$$

$$r_f = \left(\rho\alpha_f, \pm\alpha_f c_f, \mp\alpha_s c_s \beta_y \operatorname{sgn} B_x, \mp\alpha_s c_s \beta_z \operatorname{sgn} B_x, 0, \alpha_s\sqrt{\rho}a\beta_y, \alpha_s\sqrt{\rho}a\beta_z, \alpha_f\gamma p\right)^T \tag{46}$$

Slow

$$\lambda_s = u \pm c_s$$

$$\ell_s = \left(0, \pm\frac{\alpha_s c_s}{2a^2}, \pm\frac{\alpha_f}{2a^2}c_f\beta_y \operatorname{sgn} B_x, \pm\frac{\alpha_f}{2a^2}c_f\beta_z \operatorname{sgn} B_x, 0, -\frac{\alpha_f}{2\sqrt{\rho}a}\beta_y, -\frac{\alpha_f}{2\sqrt{\rho}a}\beta_z, \frac{\alpha_s}{2\rho a^2}\right)$$

$$r_s = \left(\rho\alpha_s, \pm\alpha_s c_s, \pm\alpha_f c_f \beta_y \operatorname{sgn} B_x, \pm\alpha_f c_f \beta_z \operatorname{sgn} B_x, 0, -\alpha_f\sqrt{\rho}a\beta_y, -\alpha_f\sqrt{\rho}a\beta_z, \alpha_s\gamma p\right)^T \tag{47}$$

The eigenvectors given above are orthonormal, and, since α_f, α_s, β_y and β_z all lie between zero and one, the eigenvectors are all well-formed, once these four parameters are defined. The only difficulties in defining these occur when $B_y^2 + B_z^2 = 0$, in which case β_y and β_z are ill-defined, and when $B_y^2 + B_z^2 = 0$ *and* $B_x^2 = \rho a^2$, in which case α_s and α_f are ill-defined. The first case is fairly trivial; β_y and β_z represent direction cosines for the tangential component of the B-field, and in the case of a zero component, it is only important to choose so that $\beta_y^2 + \beta_z^2 = 1$. The choice used here is the same as that proposed by Brio and Wu [1],

$$\beta_y = \frac{1}{\sqrt{2}} \qquad \beta_z = \frac{1}{\sqrt{2}}\,. \tag{48}$$

An approach for the case in which α_s and α_f are ill-defined is outlined by Roe and Balsara [3]. No special treatment of this type was needed for the cases shown in this paper. Indeed, it is shown in [3] that although the linearized Riemann problem has multiple solutions in this case, they all give the same value for the interface flux.

3.5.2 Construction of the Flux Function

The flux function used in this work is defined in the manner of Roe [27] as

$$\mathbf{F}\cdot\hat{\mathbf{n}}(\mathbf{U}_L,\mathbf{U}_R) = \frac{1}{2}\left(\mathbf{F}\cdot\hat{\mathbf{n}}(\mathbf{U}_L) + \mathbf{F}\cdot\hat{\mathbf{n}}(\mathbf{U}_R)\right) - \sum_{k=1}^{8} \mathbf{L}_k(\mathbf{U}_R - \mathbf{U}_L)\,|\lambda_k|\,\mathbf{R}_k \quad (49)$$

where k is an index for the loop over the entropy, divergence, Alvén, magneto-acoustic waves. The conservative eigenvectors are

$$\mathbf{L}_k = \ell_k \frac{\partial \mathbf{W}}{\partial \mathbf{U}} \quad (50)$$

$$\mathbf{R}_k = \frac{\partial \mathbf{U}}{\partial \mathbf{W}} r_k \quad (51)$$

In Equation 49, the terms denoted with subscripts L and R are evaluated from the face-midpoint states just to the left and right of the interface, as determined by the limited linear reconstruction procedure described above. The eigenvalues and eigenvectors are evaluated at an "interface" state that is some combination of the L and R states. For gas dynamics, there is a unique interface state (the "Roe-average state") that Roe has shown exhibits certain desired properties [27]. For MHD, while some interesting work has been done on finding an analogous state for MHD (see, for example, [28]), a unique, efficiently computable Roe average is still elusive. In this paper, a simple arithmetic averaging of the primitive variables is done to compute the interface state.

If a so-called "entropy fix" is not applied to Roe's scheme, expansion shocks can be permitted [29]. The entropy fix is applied to the magnetosonic waves to bound those eigenvalues away from zero when the flow is expanding. This is done by replacing $|\lambda_k|$ in Equation 49 with $|\lambda_k^*|$ (for the values of k corresponding to the magnetoacoustic waves only) where $|\lambda_k^*|$ is given by

$$|\lambda_k^*| = \begin{cases} |\lambda_k| & |\lambda_k| \geq \frac{\delta\lambda_k}{2} \\ \frac{\lambda_k^2}{\delta\lambda_k} + \frac{\delta\lambda_k}{4} & |\lambda_k| < \frac{\delta\lambda_k}{2} \end{cases}, \quad (52)$$

where

$$\delta\lambda_k = \max\left(4\left(\lambda_{kR} - \lambda_{kL}\right), 0\right).$$

4 Validation of Scheme

For the purposes of validation and accuracy assessment, smooth and non-smooth problems with exact solutions were simulated with the method presented in this paper, and the computed solutions for several grids were compared with the exact solutions. The results of the validation runs are presented here.

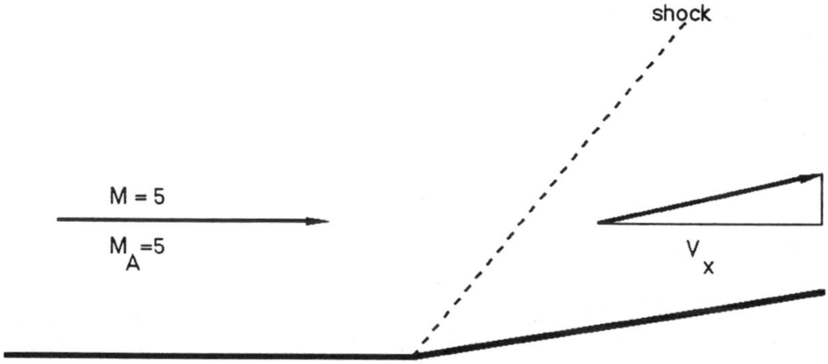

Figure 2: Setup of Validation Case

4.1 Attached Oblique Shocks

Two oblique shock cases were studied: in one, the magnetic field and velocity vectors upstream of the shock are taken to be parallel; in the other, they are perpendicular to each other. For both cases, the acoustic Mach number $M = 5$, the Alfvén number $M_A = 5$, and $\gamma = 5/3$ were taken as the upstream conditions. For both cases, flow past a wedge was computed by the method presented in this paper. The problem is depicted in Figure 2. Shock polars (i.e. plots of post-shock vertical versus post-shock horizontal velocity components) were constructed by varying the wedge angle, and plotting the downstream V_x versus downstream V_y for several wedge angles with the two upstream conditions. Exact shock polars were computed by iteratively solving the appropriate MHD Rankine-Hugoniot relations. Figure 3 is a plot of the exact (solid lines) and computed (symbols) shock polars for the two cases. As is clear from the plot, the agreement is excellent.

In order to assess order of accuracy of the method for non-smooth flows, a single case ($M = 5$, $M_A = 5$, $10°$ wedge, upstream magnetic field and velocity parallel) was run on a sequence of successively finer uniform grids. Limited reconstruction was turned off, so the expectation is of first-order accuracy. Relative errors were calculated in an L_1 norm defined as

$$L_1^\eta = \frac{1}{N}\sum_{i=1}^{N} |\delta_i^\eta|$$

where δ_i^n is the relative error in cell i of some quantity η. For example, relative

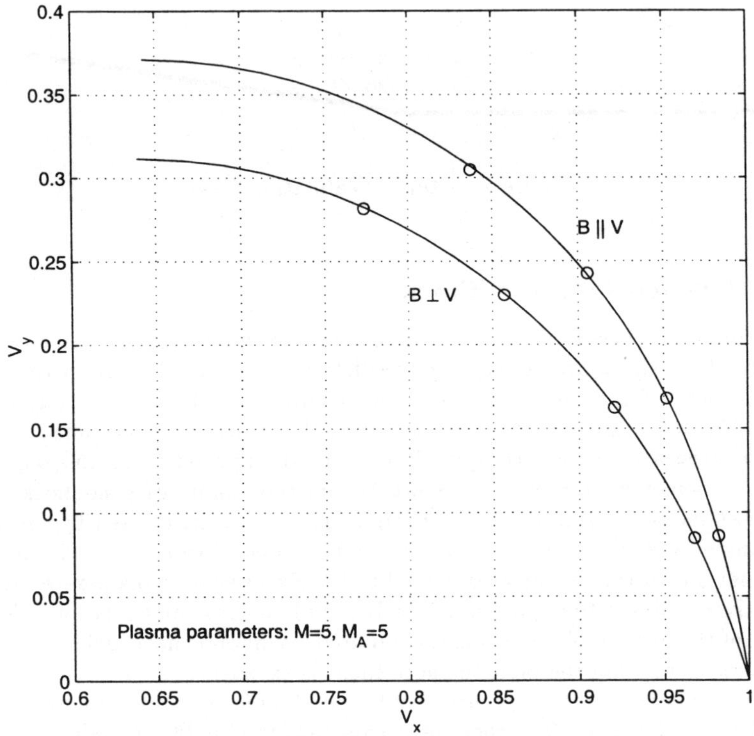

Figure 3: Computed and Exact Shock Polars

L_1^p	L_1^B	$L_1^{h\nabla\cdot\mathbf{B}}$	Resolution
0.2022690	0.1072600	0.00301172	1/16
0.130427	0.0700573	0.00143521	1/32
0.0789827	0.0422129	0.000676634	1/64
0.0449624	0.0239818	0.00032158	1/128
0.0242786	0.0131832	0.000155886	1/256
0.0127462	0.00727291	0.0000766793	1/512

Table 1: Grid Convergence for Oblique-Shock Test Case

errors of pressure and magnetic field magnitude are

$$\delta_i^p = \frac{p_i - p_{exact}}{p_{exact}} \tag{53}$$

$$\delta_i^B = \frac{B_i - B_{exact}}{B_{exact}} \tag{54}$$

To assess the ability of the scheme to maintain $\nabla \cdot \mathbf{B} = 0$, the relative error

$$\delta_i^{h\nabla\cdot\mathbf{B}} = \frac{\iint_{cell\ i} B_n ds}{\iint_{cell\ i} |B_n|\, ds}$$

was calculated, where B_n is the component of the magnetic field normal to a cell face, computed by averaging the values at the cell centroids to the "left" and "right" of the face centroid. This error is denoted as $h\nabla \cdot \mathbf{B}$ because it scales as

$$\left|\delta_i^{h\nabla\cdot\mathbf{B}}\right| \propto \frac{V\,|\nabla \cdot \mathbf{B}|}{A\,|B|}$$

where V is the cell volume and A is the cell surface area; the ratio V/A goes as the mesh spacing h.

Figure 4 shows grid-convergence results for pressure, magnetic-field magnitude, and divergence of magnetic field. The tabulated values are shown in Table 1. Both the plot and the table show an imputed order of accuracy of one, as expected. In addition, it is interesting to note that the error in $h\nabla \cdot \mathbf{B}$ not only converges at the same rate as the error in other variables, it is on each grid more than an order of magnitude lower than the error in the magnetic field. The bad news here is that, since $h\nabla \cdot \mathbf{B}$ is first order, $\nabla \cdot \mathbf{B}$ itself is constant with grid refinement.

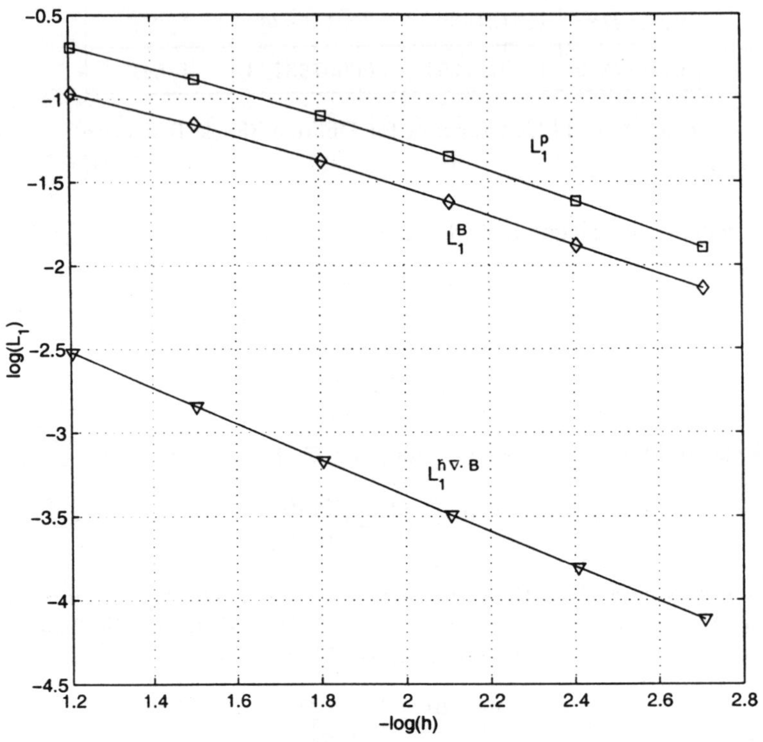

Figure 4: Grid Convergence for Oblique-Shock Test Case

Level	9	8	7	6	5	4	3	2	1
$\Sigma_{\nabla\cdot\mathbf{B}} \times 10^5$	1.670	1.569	1.220	0.846	0.543	0.373	0.234	0.028	0.011

Table 2: Telescoping of magnetic field divergence on a set of consecutively coarsened grids.

However, this is, perhaps, to be expected. For any oblique discontinuity, the three terms comprising $\nabla \cdot \mathbf{B}$ will each be non-zero and of order $1/h$, and will not cancel perfectly. Since, as can be seen from examining the multi-stage scheme (Section 3.4), the term added in updating the conserved variables is proportional to $\Delta t \nabla \cdot \mathbf{B}$, and $\Delta t \approx h$ (from the CFL condition), comparing the $h\nabla \cdot \mathbf{B}$ term to the relative error in the magnetic field itself is appropriate.

It is also interesting to note the structure of the $\nabla \cdot \mathbf{B}$ errors. The only non-zero values are in the vicinity of the shock. Figure 5 shows contours of $\nabla \cdot \mathbf{B}$ in the vicinity of the shock; positive values are denoted by solid countours; negative values are denoted by dashed contours. The extent of the contours of non-zero divergence is less than five cells across, typical of numerical oblique shock structures.

As can be seen, the $\nabla \cdot \mathbf{B}$ that is created numerically does not appear as isolated magnetic monopoles; any positive $\nabla \cdot \mathbf{B}$ that is created is paired with a negative contribution. This plot, and the fact that the far-field boundary conditions are divergence-free, suggest a "telescoping" property: integration of $\nabla \cdot \mathbf{B}$ over successively larger control volumes should lead to successively smaller values. Defining

$$\Sigma_{\nabla\cdot\mathbf{B}} = \sum_{i=1}^{N} \left| \iint_{control\ volume\ i} B_n \, ds \right|$$

where N is the number of control volumes into which the grid is divided. This telescoping property can be studied by taking succesively larger control volumes for the same solution. In Table 2, the quantity $\Sigma_{\nabla\cdot\mathbf{B}}$ is reported for successively larger control volumes: level 9 corresponds to taking each cell in the grid as a control volume, level 8 to a control volume consisting of eight control volumes from level 9, and so on up to level 1, where the control volume is the entire computational domain.

4.2 Weber-Davis Flow

Weber-Davis flow is a smooth solution to the ideal MHD equations approximating the solar wind in the equatorial plane of the interplanetary medium [9]. While a complete analytic solution for this flow does not exist, certain quantities, including

$$\Phi_M = \rho v_r r^2 \tag{55}$$

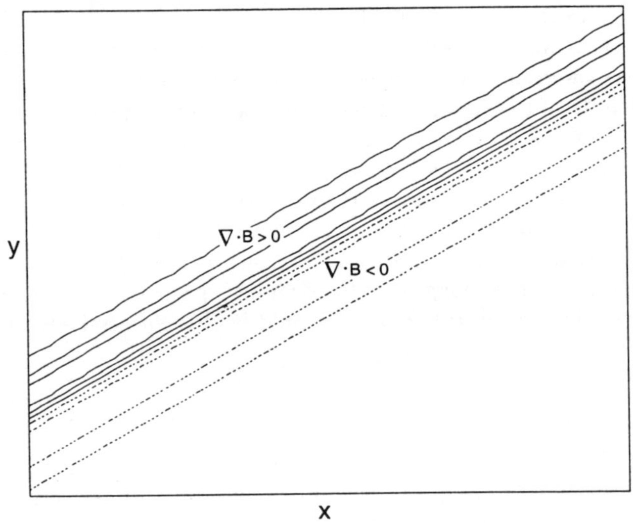

Figure 5: Structure of $\nabla \cdot \mathbf{B}$ Truncation Error — Magnified View of a Portion of a Captured Shock

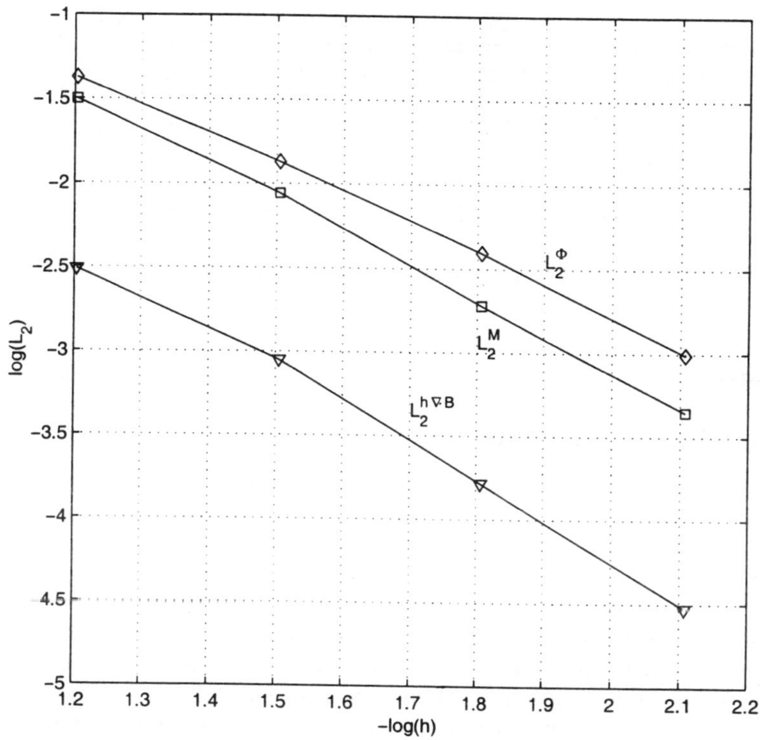

Figure 6: Grid Convergence for Weber-Davis Test Case

$$\Phi_B = B_r r^2 \tag{56}$$

are invariant. Thus, the method presented in this paper can be validated by calculating the degree to which Φ_M and Φ_B remain constant. The L_2 norms of the relative errors in Φ_B, Φ_M and magnetic-field divergence are plotted in Figure 6 and Table 3 for various levels of grid resolution; the results show second-order accuracy. Again, as in the non-smooth flow, the divergence error is more than an order of magnitude smaller than the errors in other variables.

$L_2^{\Phi_M}$	$L_2^{\Phi_B}$	$L_2^{\nabla\cdot\mathbf{B}}$	Resolution
0.0314475	0.042268	0.0030893	1/16
0.0087872	0.0134876	0.000889703	1/32
0.00190635	0.0039304	0.000164449	1/64
0.000444465	0.00098265	0.000029798	1/128

Table 3: Grid Convergence for Weber-Davis Test Case

5 Results

5.1 Parallel Performance

BATS-R-US was designed from the ground up with a view to achieving very high performance on massively parallel architectures. The underlying upwind finite-volume solution algorithm, with explicit time stepping, has a very compact stencil and is therefore highly local in nature. This results in low inter-processor communication overhead. It also permits the more efficient use of memory and cache. The hierarchical data structure and self-similar blocks make domain decomposition of the problem almost trivial and readily enable good load-balancing, a crucial element for truly scalable computing. A natural load balancing is accomplished by simply distributing the blocks equally amongst the processors and for 10 blocks per processor the load imbalance is less that 10% (the load imbalance is less than 1% for 100 blocks per node). The self-similar nature of the solution blocks also means that serial performance enhancements apply to all blocks and that fine grain parallelization of the algorithm is possible. The parallel implementation of the algorithm has been carried out to such an extent, that even the grid adaptation is performed in parallel.

Other features of the parallel implementation include the use of FORTRAN 90 as the programming language and the message passing interface (MPI) library for performing the message passing. Use of these standards greatly enhances the portability of the code and led to very good serial and parallel performance. The message passing is performed in an asynchronous fashion with gathered wait states and message consolidation such that it typically accounts for less than 3-5% of processor time. The serial performance of the algorithm, and hence the overall parallel performance of the method, has also been greatly enhanced by: 1) avoiding the use of indirect addressing and allocatable arrays when defining memory for primary solution variables; 2) strip mining computationally intensive routines to achieve desired strides through memory for more efficient use of cache (i.e., high

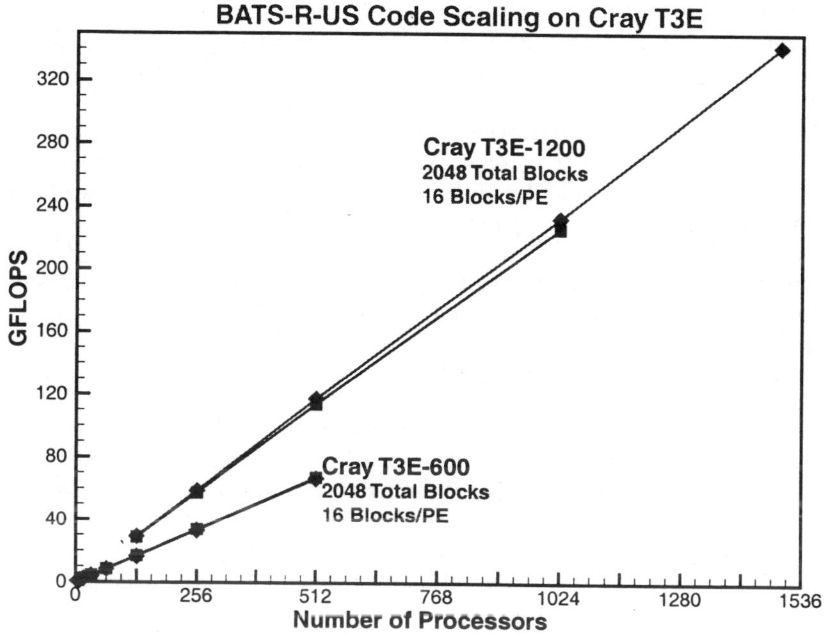

Figure 7: Parallelscaling of BATS-R-US on Cray-T3E-600 and Cray-T3E-1200 processors. Dashed lines show timings for a problem that has a fixed size per processor; solid lines show timings for a problem that has a fixed total size.

cache reuse is desirable); and 3) performing loop optimizations, such as unrolling and code inlining.

BATS-R-US has been developed on Cray T3E parallel computers. Implementation of the algorithm has also been carried out on SGI and Sun workstations, on SGI shared-memory machines. on a Cray T3D, and on several IBM SP2s. BATS-R-US nearly perfectly scales to 1,500 processors and a maximum of 344.5 GFlops has been attained on a Cray T3E-1200 using 1,490 PEs. The scalability of BATS-R-US is shown in Figure 7. It shows two pairs of curves obtained for Cray-T3E-600 and Cray-T3E-1200 processors. Dashed lines show timings for a problem that has a fixed size per processor; solid lines show timings for a problem that has a fixed total size. Of the two, the second is the harder — as a fixed-size problem is distributed across more and more processors, the ratio of communications overhead to computing cost rises. As can be seen from the figure, for both problems, the scaling is nearly perfect. Nearly identical results have been obtained on an IBM SP2. The code is written in Fortran 90 with message-passing via MPI, and hence is portable to a wide range of machines, from integrated shared-memory systems to networks of workstations. For each target architecture, simple single-processor measurements are used to tune the size of the adaptive blocks. In Figure 8 we show the performance obtained on a number of parallel architectures without any particular tuning.

The BATS-R-US code has been successfully applied to a broad range of space plasmas ranging from solar coronal expansion [30, 31], to the interaction of the heliosphere with the interstellar medium [32], to the magnetospheres of Mercury [33], Venus [34], Earth [35, 36] and Saturn [37, 33]. In addition, we successfully simulated the interaction of comets with the solar wind [38] including the emission of cometary x–rays [39], the interaction of Io [40], Europa [41] and Titan [42] with the high speed magnetospheric plasma.

5.2 Simulation of a CME

Coronal mass ejections (CMEs) are highly transient solar events involving the expulsion of mass and magnetic field from the solar surface. On the order of 10^{12} kg of plasma may be expelled from the solar surface during a typical event. These dynamic events originate in closed magnetic field regions of the corona. They produce large-scale reconfiguration of the coronal magnetic field and generate large solar wind disturbances that, as mentioned above, appear to be the primary cause of major geomagnetic storms at Earth.

The physical mechanisms involved in the initiation of CMEs are not well understood. Many scenarios have been put forth for their release. Early on it was suggested that thermally driven pressure pulses from solar flares drive the release [43], yet more recently it is felt that it is the large-scale destabilization of the coronal magnetic field that initiates CMEs. [44] and [45] have recently considered the release of CMEs as a two-step process: first, there is a CME which opens up an initially closed coronal magnetic field; this is followed by a flare resulting from the reconnective-closing of field lines trailing the ejecta [46, 47, 48]. Another scenario put forth for the formation and release of CMEs involves the buildup of magnetic energy and subsequent destabilization of the field due to the quasi-static shearing of the footprints of closed magnetic field lines on the solar surface [49, 50, 51, 52]. This effect is likely to be a trigger mechanism but is not, by itself, sufficient to explain the CME phenomenon [53, 54]. Finally, it has been suggested that the onset of CMEs may be produced by the emergence of magnetic flux ropes that gain energy as they are continually stressed and deformed by chromospheric and photospheric motions [55]. Prior to eruption, the flux ropes are confined by the large mass in the flux tubes, but when confinement fails CMEs are initiated due to the magnetic buoyancy of the ropes [56, 57, 58, 59].

After release, CMEs accelerate and become part of the outward flow of the solar wind. They are either accelerated by the solar wind so as to come into equilibrium with the ambient wind or act as drivers moving faster than the background solar wind. Close to the earth, the typical dimension of a CME is less than a solar radius. As the CMEs propagate outward from the corona, they expand dramatically and may extend over tenths of an AU by the time Earth's orbit is reached at 1 AU. Moreover, many, if not all, CMEs are associated with magnetic clouds and the plasma properties within these clouds can differ substantially from those of the ambient solar wind.

Global computational models based on first principles mathematical descriptions of the physics represent a very important component of efforts to understand the initiation, structure, and evolution of CMEs. Recent examples of the application of MHD models to the study of coronal and solar wind plasma flows include the studies by [60, 52, 53, 61, 62, 63, 64, 65, 66], and [67, 68, 69].

Here we show our numerical results for a CME driven by local plasma density enhancement. In this calculation, the background solar wind solution described above was used as an initial solution and then a localized pressure and density enhancement was introduced at the solar surface just above the equatorial plane. The CME was initiated by introducing a localized isothermal density enhancement at the solar surface. In this enhancement the density and pressure are locally increased by a factor f_{CME} given by

$$f_{\mathrm{CME}} = \begin{cases} 1 + 134 \exp\left(-\frac{(\mathbf{r}-\mathbf{r}_{\mathrm{CME}})^2}{\Delta_r^2} - \frac{(t-t_1)^2}{\Delta_t^2}\right) & \text{for } t < t_1 \,, \\ 1 + 134 \exp\left(-\frac{(\mathbf{r}-\mathbf{r}_{\mathrm{CME}})^2}{\Delta_r^2}\right) & \text{for } t_1 \le t \le t_2 \,, \\ 1 + 134 \exp\left(-\frac{(\mathbf{r}-\mathbf{r}_{\mathrm{CME}})^2}{\Delta_r^2} - \frac{(t-t_2)^2}{\Delta_t^2}\right) & \text{for } t > t_2 \,, \end{cases} \tag{57}$$

where $t_1 = 2$ hrs, $t_2 = 10$ hrs, $\Delta_t = 1$ hrs, $\Delta_r = 0.13$ R_s and $\mathbf{r}_{\mathrm{CME}}$ is located 11.5° above the solar equator. This produces a localized maximum density increase by a factor of 135 in a small region just above the solar equator for a duration of about 12 hours.

6 Concluding Remarks

A scheme for solving the compressible MHD equations in their symmetrizable form has been presented in this paper. The scheme is solution-adaptive, and based on an approximate solution to the MHD Riemann problem. Grid-convergence studies were carried out on smooth and non-smooth problems, validating the accuracy of the scheme. In addition, a method for splitting off known steady magnetic fields from the solution was presented, and applied in solving for the interaction of the solar wind with a magnetized planet. The combination of a robust solution method and the solution-adaptive capability yields a method that is very useful for space physics applications, which are characterized by disparate scales.

References

[1] M. Brio and C. C. Wu. An upwind differencing scheme for the equations of ideal magnetohydrodynamics. *J. Comput. Phys.*, 75:400–422, 1988.

[2] A. L. Zachary and P. Colella. A higher-order Godunov method for the equations of ideal magnetohydrodynamics. *J. Comput. Phys.*, 99:341–347, 1992.

[3] P. L. Roe and D. S. Balsara. Notes on the eigensystem of magnetohydrodynamics. *SIAM J. Appl. Math.*, 56(1):57–67, February 1996.

[4] W. Dai and P. R. Woodward. Extension of the piecewise parabolic method (PPM) to multidimensional magnetohydrodynamics. *J. Comput. Phys.*, 111:354, 1994.

[5] J.-P. Croisille, R. Khanfir, and G. Chanteur. Numerical simulation of the MHD equations by a kinetic-type method. *J. Sci. Comput.*, 10(1):81–92, 1995.

[6] T. J. Linde. *A Three-Dimensional Adaptive Multifluid MHD Model of the Heliosphere.* PhD thesis, Univ. of Mich., Ann Arbor, May 1998.

[7] G. Tóth and D. Odstrčil. Comparison of some flux corrected transport and total variation diminishing numerical schemes for hydrodynamic and megnetohydrodynamic problems. *J. Comput. Phys.*, 128(1):82–100, October 1996.

[8] J.U. Brackbill and D.C. Barnes. The effect of nonzero $\nabla \cdot \mathbf{B}$ on the numerical solution of the magnetohydrodynamic equations. *J. Comput. Phys.*, 35:426–430, 1980.

[9] J. M. Stone, J. F. Hawley, C. R. Evans, and M. L. Norman. A test suite for magnetohydrodynamical simulations. *Astrophys. J.*, 388(2):415–437, April 1992.

[10] S. K. Godunov. Symmetric form of the equations of magnetohydrodynamics (in russian). In *Numerical Methods for Mechanics of Continuum Medium*, volume 1, pages 26–34. Siberian Branch of USSR Acad. of Sci., 1972.

[11] K. G. Powell. An approximate Riemann solver for magnetohydrodynamics (that works in more than one dimension). Technical Report 94-24, Inst. for Comput. Appl. in Sci. and Eng., NASA Langley Space Flight Center, Hampton, Va., 1994.

[12] K. G. Powell, P. L. Roe, T. J. Linde, T. I. Gombosi, and D. L. De Zeeuw. A solution-adaptive upwind scheme for ideal magnetohydrodynamics. *J. Comput. Phys.*, 154(2):284–309, September 1999.

[13] W. K. H. Panofsky and M. Phillips. *Classical Electricity and Magnetism*. Addison Wesley, Reading, Massachusetts, 1955.

[14] S. K. Godunov. An interesting class of quasilinear systems. *Sov. Math. Dokl.*, 2:947–949, 1961.

[15] A. Harten. On the symmetric form of systems of conservation laws with entropy. *J. Comput. Phys.*, 49:151–164, 1983.

[16] Q. F. Stout, D. L. De Zeeuw, T. I. Gombosi, C. P. T. Groth, H. G. Marshall, and K. G. Powell. Adaptive blocks: A high-performance data structure. In *Proc. Supercomputing'97*, 1997.

[17] Q.F. Stout, D. L. De Zeeuw, T. I. Gombosi, C. P. T. Groth, H. G. Marshall, and K. G. Powell. Adaptive parallel computation of a grand-challenge problem: Prediction of the path of a solar coronal mass ejection. In *Proc. Supercomputing'98*, 1998.

[18] M. J. Berger and P. Colella. Local adaptive mesh refinement for shock hydrodynamics. *J. Comput. Phys.*, 82:67–84, 1989.

[19] M. J. Berger. Adaptive mesh refinement for hyperbolic partial differential equations. *J. Comput. Phys.*, 53:484–512, 1984.

[20] M. J. Berger and A. Jameson. Automatic adaptive grid refinement for the Euler equations. *AIAA Journal*, 32(4):561–568, 1985.

[21] G. Warren, W. K. Anderson, J. Thomas, and S. Krist. Grid convergence for adaptive methods. In *AIAA 10th Computational Fluid Dynamics Conference*, 1991.

[22] J. J. Quirk and U. R. Hanebutte. A parallel adaptive mesh refinement algorithm. Technical Report 93-63, ICASE, August 1993.

[23] T. J. Barth. On unstructured grids and solvers. In H. Deconinck, editor, *Computational Fluid Dynamics*. von Kármán Lecture Series 1990-03. VKI, Rhode-St.-Genese, Belgium, March 1990.

[24] T. J. Barth. Recent developments in high-order k-exact reconstruction on unstructured meshes. In *Proc. 11th AIAA Computational Fluid Dynamics Conference*. AIAA Paper No. 93-3273, Orlando, FL, 1993.

[25] B. van Leer, C. H. Tai, and K. G. Powell. Design of optimally-smoothing multi-stage schemes for the Euler equations. In *AIAA 9th Computational Fluid Dynamics Conference*, Buffalo, New York, 1989. AIA-89-1933-CP.

[26] T. J. Barth. Numerical methods for gasdynamic systems. In D. Kröner, M. Ohlberger, and C. Rohde, editors, *An Introduction to Recent Developments in Theory and Numerics for Conservation Laws.* Springer, 1999.

[27] P. L. Roe. Approximate Riemann solvers, parameter vectors, and difference schemes. *J. Comput. Phys.*, 43:357–372, 1981.

[28] G. Gallice. Resolution numerique des equations de la magnetohydrodynamique ideale bidimmensionale. In *Méthodes numériques pour la MHD*, 1995.

[29] A. Harten, J. M. Hyman, and P. D. Lax. On finite-difference approximations and entropy conditions for shocks. *Commun. Pure Appl. Math.*, 29:297–322, 1976.

[30] C. P. T. Groth, D. L. De Zeeuw, T. I. Gombosi, and K. G. Powell. Three-dimensional MHD simulation of coronal mass ejections. *Adv. Space Res.*, 26(5):793–800, 2000.

[31] C. P. T. Groth, D. L. De Zeeuw, T. I. Gombosi, and K. G. Powell. A parallel adaptive 3D MHD scheme for modeling coronal and solar wind plasma flows. In *Proceedings of SOHO 7 Workshop*, 1998.

[32] T. J. Linde, T. I. Gombosi, P. L Roe, K. G. Powell, and D. L. De Zeeuw. The heliosphere in the magnetized local interstellar medium: Results of a 3D MHD simulation. *J. Geophys. Res.*, 103(A2):1889–1904, February 1998.

[33] T. I. Gombosi, D. L. De Zeeuw, C. P. T. Groth, K. C. Hansen, K. Kabin, and K. G. Powell. MHD simulations of current systems in planetary magnetospheres: Mercury and Saturn. In R. Fujii, M. Hesse, R. Lysak, and S. Ohtani, editors, *Magnetospheric Current Systems*, volume 118 of *AGU Geophys. Monog. er.*, pages 363–370. AGU, 2000.

[34] R. Bauske, A. F. Nagy, T. I. Gombosi, D. L. De Zeeuw, K. G. Powell, and J. G. Luhmann. A three-dimensional MHD study of solar wind mass loading processes at Venus: Effects of photoionization, electron impact ionization, and charge exchange. *J. Geophys. Res.*, 103(A10):23,625–23,638, October 1998.

[35] T. I. Gombosi, D. L. De Zeeuw, C. P. T. Groth, K. G. Powell, and P. Song. The length of the magnetotail for northward IMF: Results of 3D MHD simulations. In T. Chang and J. R. Jasperse, editors, *Physics of Space Plasmas*, volume 15, pages 121–128. MIT Press, Cambridge, Mass., 1998.

[36] T. I. Gombosi, D. L. De Zeeuw, C. P. T. Groth, and K. G. Powell. Magnetospheric configuration for Parker-spiral IMF conditions: Results of a 3D AMR MHD simulation. *Adv. Space Res.*, 26(1):139–149, 2000.

[37] K. C. Hansen, T. I. Gombosi, D. L. DeZeeuw, C. P. T. Groth, and K. G. Powell. A 3D global MHD simulation of Saturn's magnetosphere. *Adv. Space Res.*, 1999. in press.

[38] T. I. Gombosi, D. L. De Zeeuw, R. M. Häberli, and K. G. Powell. Three-dimensional multiscale MHD model of cometary plasma environments. *J. Geophys. Res.*, 101(A7):15233–15253, July 1996.

[39] R. M. Häberli, T. I. Gombosi, D. L. DeZeuuw, M. R. Combi, and K. G. Powell. Modeling of cometary x-rays caused by solar wind minor ions. *Science*, 276:939–942, May 1997.

[40] M. R. Combi, K. Kabin, T.I. Gombosi, D.L. De Zeeuw, and K.G. Powell. Io's plasma environment during the Galileo flyby: Global three-dimensional MHD modeling with adaptive mesh refinement. *J. Geophys. Res.*, 103(A5):9071–9081, May 1998.

[41] K. Kabin, M. R. Combi, T. I. Gombosi, A. F. Nagy, D. L. De Zeeuw, and K. G. Powell. On Europa's magnetospheric interaction: An MHD simulation of the E4 flyby. *J. Geophys. Res.*, 104(A9):19,983–19,992, September 1999.

[42] K. Kabin, T. I. Gombosi, D. L. DeZeeuw, K. G. Powell, and P. L. Israelevich. Interaction of the Saturnian magnetosphere with Titan: Results of a 3D MHD simulation. *J. Geophys. Res.*, 104(A2):2451–2458, February 1999.

[43] M. Dryer, S. T. Wu, R. S. Steinolfson, and R. M. Wilson. Magnetohydrodynamic models of coronal transients in the meridional plane, II, Simulation of the coronal transient of 1973 August 21. *Astrophys. J.*, 227:1059–1079, 1979.

[44] B. C. Low. Equilibrium and dynamics of coronal magnetic fields. *Annu. Rev. Astron. Astrophys.*, 28:491–524, 1990.

[45] A. J. Hundhausen. Coronal mass ejections: A summary of SMM observations from 1980 and 1984-1989. In K. T. Strong, J. L. R. Soba, and B. M. Haisch, editors, *The Many Faces of the Sun*, pages 143–200. Springer-Verlag, New York, 1999.

[46] T. Hirayama. Theoretical models of flares and prominences, I, evaporating flare model. *Sol. Phys.*, 34:323–338, 1974.

[47] R. Kopp and G. W. Pneuman. Magnetic reconnection on the corona and the loop prominencce phenomenon. *Sol. Phys.*, 50:85–98, 1976.

[48] B. C. Low. Magnetohydrodynamic processes in the solar corona: Flares, coronal mass ejections, and magentic helicity. *Plasma Phys.*, 1:1684–1690, 1994.

[49] B. C. Low. Evolving force-free magnetic fields, I, The development of the preflare stage. *Astrophys. J.*, 212:234–242, 1977.

[50] S. T. Wu, Y. Q. Hu, Y. Nakagawa, and E. Tandberg-Hanssen. Induced mass and wave motions in the lower solar atmosphere, I, Effects of shear motion on flux tubes. *Astrophys. J.*, 266:866–881, 1983.

[51] Z. Mikić, D. Barnes, and D. D. Schnack. Dynamical evolution of a solar coronal magnetic field arcade. *Astrophys. J.*, 328:830–847, 1988.

[52] Z. Mikić and J. A. Linker. Disruption of coronal magnetic field arcades. *Astrophys. J.*, 430:898–912, August 1994.

[53] J. A. Linker and Z. Mikić. Disruption of a helmet streamer by photospheric shear. *Astrophys. J.*, 438:L45–L48, January 1995.

[54] B. C. Low. Solar activity and the corona. *Sol. Phys.*, 167:217–265, 1996.

[55] A. A. van Ballegooijen and P. C. H. Martens. Magnetic fields in quiescent prominences. *Astrophys. J.*, 361:283–289, 1990.

[56] B. C. Low. Eruptive magnetic fields. *Astrophys. J.*, 251:352–363, 1981.

[57] R. R. Fisher and A. I. Poland. Coronal activity below 2 $R_\odot$: February 15-17. *Astrophys. J.*, 246:1004–1009, 1981.

[58] B. C. Low, R. H. Munro, and R. R. Fisher. The initiation of a coronal transient. *Astrophys. J.*, 254:335–342, 1982.

[59] R. M. E. Illing and A. J. Hundhausen. Disruption of a coronal streamer by an eruptive prominence and coronal mass ejection. *J. Geophys. Res.*, 91:10,951–10,960, 1986.

[60] J. A. Linker, Z. Mikić, and D. D. Schnack. Modeling coronal evolution. In *Proceedings of the Third SOHO Workshop*, pages 249–252. European Space Agency, Estes Park, Colo., 1994.

[61] S. T. Suess, A.-H. Wang, and S. T. Wu. Volumetric heating in coronal streamers. *J. Geophys. Res.*, 101(A9):19,957–19,966, September 1996.

[62] A.-H. Wang, S. T. Wu, S. T. Suess, and G. Poletto. Global model of the corona with heat and momentum addition. *J. Geophys. Res.*, 103:1913–1922, 1998.

[63] S. T. Wu and W. P. Guo. A self-consistent numerical magnetohydrodynamic (MHD) model of helmet streamer and flux rope interactions: Initiation and propagation of coronal mass ejections (CMEs). In N. Crooker, J. A Joselyn, and J. Feynman, editors, *Coronal Mass Ejections*, volume 99 of *Geophys. Monogr. Ser.*, pages 83–89. Amer. Geophysical Union, 1997.

[64] W. P. Guo and S. T. Wu. A magnetohydrodynamic description of coronal helmet streamers containing a cavity. *Astrophys. J.*, 494:419–429, 1998.

[65] R. Lionello, Z. Mikić, and D. D. Schnack. Magnetohydrodynamics of solar coronal plasmas in cylindrical geometry. *J. Comput. Phys.*, 140:172–201, 1998.

[66] M. Dryer. Multidimensional, magnetohydrodynamic simulation of solar-generated disturbances: Space weather forecasting of geomagnetic storms. *AIAA Journal*, 3:365–370, 1998.

[67] D. Odstrčil and V. J. Pizzo. Distortion of the interplanetary magnetic field by three-dimensional propagation of coronal mass ejections in a structured solar wind. *J. Geophys. Res.*, 104:28,225–28,239, 1999.

[68] D. Odstrčil and V. J. Pizzo. Three-dimensional propagation of CMEs in a structured solar wind flow, 1, CME launched within the streamer belt. *J. Geophys. Res.*, 104:483–492, January 1999.

[69] D. Odstrčil and V. J. Pizzo. Three-dimensional propagation of coronal mass ejections in a structured solar wind flow, 2, CME launched adjacent to the streamer belt. *J. Geophys. Res.*, 104:493–503, January 1999.

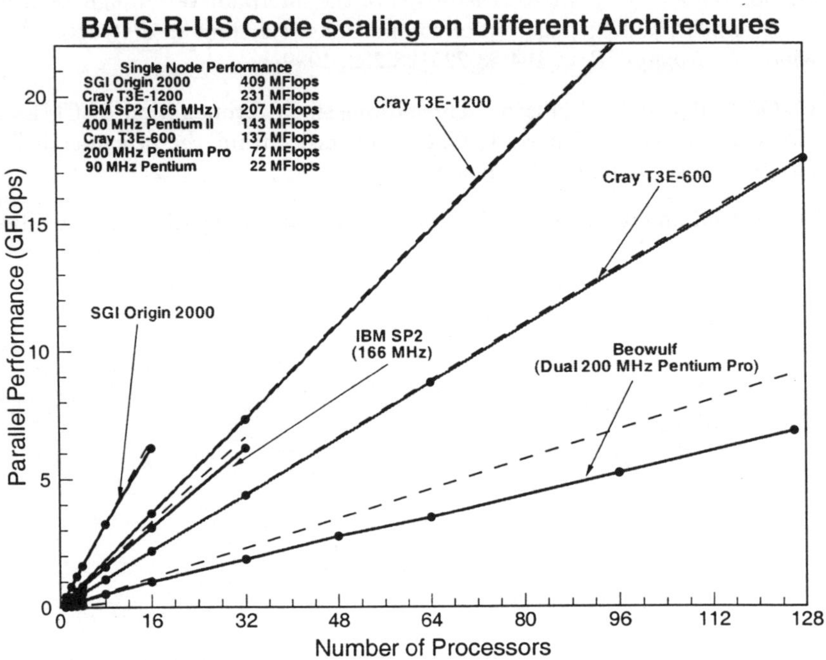

Figure 8: Parallel performance of BATS-R-US for a variety of parallel architectures. The dashed line indicates ideal scale-up performance based on single node performance and solid lines indicate actual performance achieved on each of the machines.

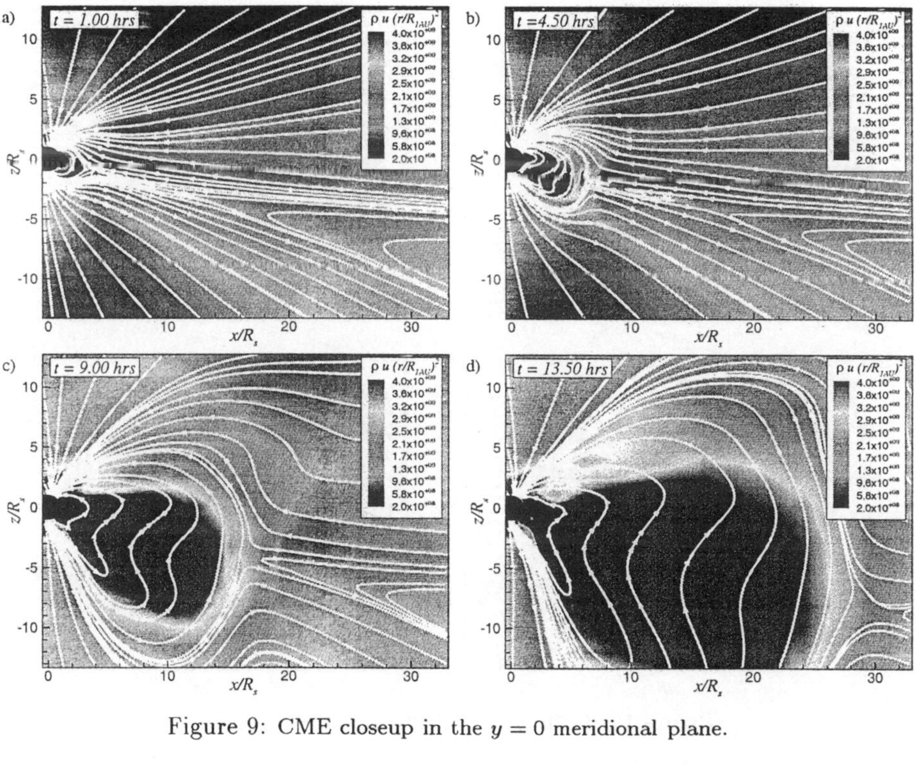

Figure 9: CME closeup in the $y = 0$ meridional plane.

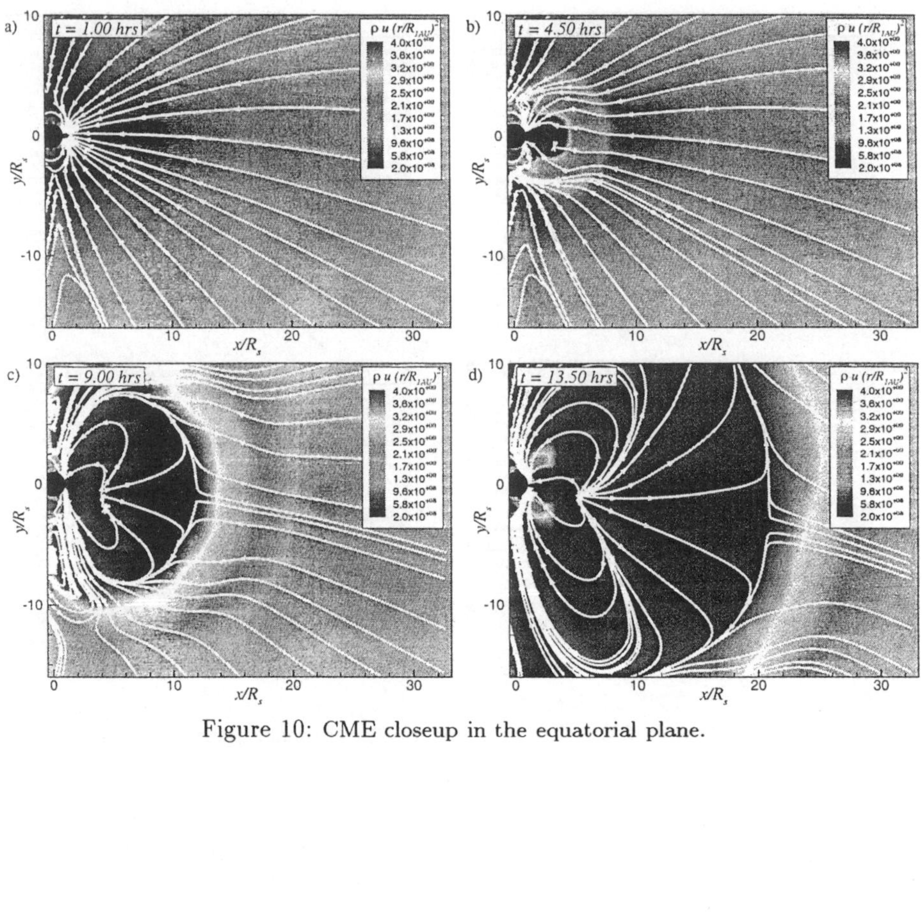

Figure 10: CME closeup in the equatorial plane.

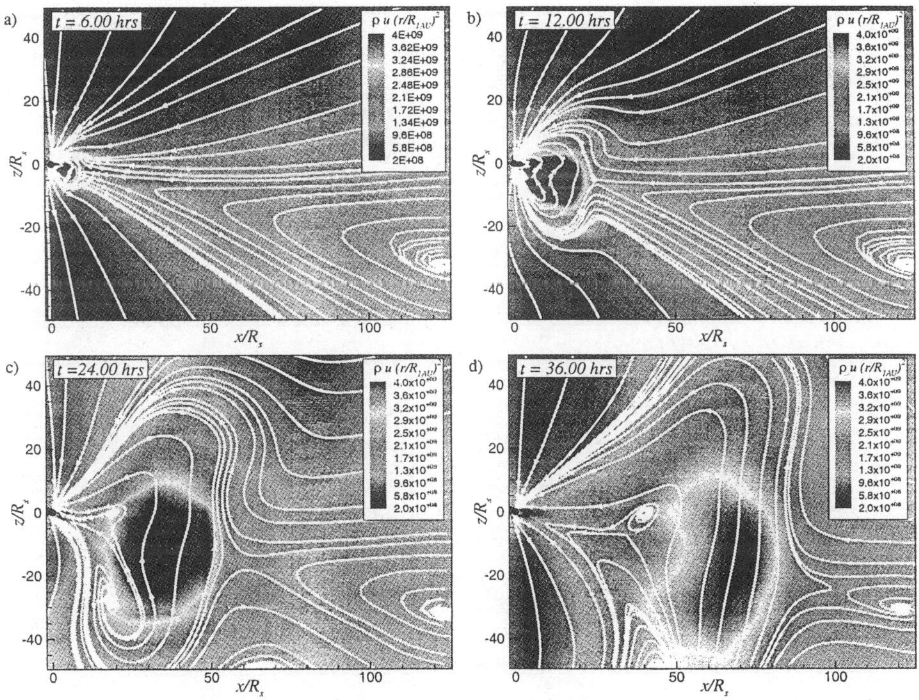

Figure 11: CME in the $y = 0$ meridional plane.

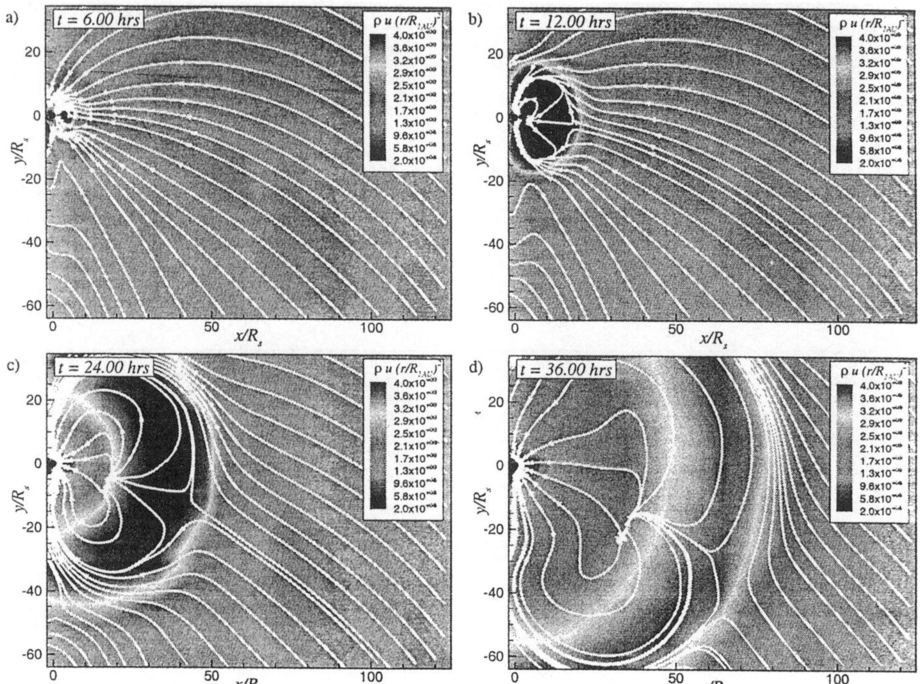

Figure 12: CME in the equatorial plane.

The Penultimate Scheme for Systems of Conservation Laws: Finite Difference ENO with Marquina's Flux Splitting[1]

Ronald P. Fedkiw

Computer Science Department
Stanford University, Stanford, California 94305
Email:fedkiw@cs.stanford.edu

Barry Merriman

Department of Mathematics
University of California Los Angeles, Los Angeles, California, 90095
Email:barry@math.ucla.edu

Rosa Donat

Departament de Matemàtiques Aplicades
Universitat de València, Spain
Email:donat@uv.es

Stanley Osher

Department of Mathematics
University of California Los Angeles, Los Angeles, California, 90095
Email:sjo@math.ucla.edu

Keywords: Euler Equations, Compressible Flow, Finite Difference Methods, ENO, Flux Splitting

Abstract: This paper provides a users' guide to a new, general finite difference method for the numerical solution of systems of convection dominated conservation laws. We include both extensive motivation for the method design, as well as a detailed formulation suitable for direct implementation.

Essentially Non-Oscillatory (ENO) methods are a class of high accuracy, shock capturing numerical methods for hyperbolic systems of conservation laws, based on upwind biased differencing in local characteristic fields. The earliest ENO methods used control volume discretizations, but subsequent work [12] has produced a simpler finite difference form of the ENO method. While this method has achieved excellent results in a great variety of compressible flow problems, there are still special situations where noticeable spurious oscillations develop. Why this occurs is not always understood, and there has been no elegant way to eliminate these problems.

Based on the extensive work of Donat and Marquina [1], it appears that these difficulties arise from using a *single* transformation to local characteristic

[1]This paper was presented in "Solutions of PDE" Conference in honour of Prof. Roe on the occassion of his 60th birthday, July 1998, Arachaon, France

variables at cell walls in the course of computing wall fluxes. In concrete terms this is the practice of evaluating the flux Jacobian matrix at cell walls using an average of adjacent cell states, such as the Roe average or linear average. When the states differ greatly across the cell wall, using such an intermediate state in the transformation adds subtle spurious features to the solution. As an alternative, Donat and Marquina recommend obtaining the wall flux from a splitting procedure based on fluxes computed separately from the left and right sides. This approach avoids introducing artificial intermediate states, and seems to improve the robustness of many characteristic based methods.

Applying their splitting in the ENO framework, the left and right sided fluxes are evaluated by the ENO interpolation technique, i.e. using the smoothest high order interpolations from each side. In the resulting method, the spurious oscillations are eliminated without sacrificing high resolution. Thus this seems to be an ideal scheme for general hyperbolic systems: it provides high accuracy and shock capturing without numerical artifacts, problem dependent "fixes", or free parameters that must be "tuned". (Of course, for scalar equations this "fix" is unnecessary and nonexistent.)

This paper is intended as a self-contained guide to this new approach, in the context of solving general systems of convection-diffusion-reaction conservation laws. We provide all the conceptual background needed to understand the design of numerical methods for systems of hyperbolic conservation laws in general, and the finite difference ENO method and Marquina's flux splitting procedure in particular. We then give a detailed presentation of the preferred form of ENO with Marquina's splitting. We conclude with one example where this eliminates a severe, non-physical oscillation in a complicated ENO based calculation.

1 Introduction

Essentially Non-Oscillatory (ENO) methods were developed to address the special difficulties that arise in the numerical solution of systems of nonlinear conservation laws, such as those arising in high speed gas dynamics and other convective transport problems. Numerical methods for these problems must be able to handle steep gradients—shocks and contact discontinuities—that may develop spontaneously and then persist in these flows. Classical numerical schemes had a tendency to either produce large spurious oscillations near steep gradients, or to greatly smear out both these gradients and the fine details of the flow. An excellent introductory discussion of these difficulties and the methods developed to deal with them can be found in Leveque's book [8].

The primary goal of the ENO effort has been to develop a general purpose numerical method for systems of conservation laws that has high accuracy (at least third order) in smooth regions and captures the motion of unresolved steep gradients in the flow, without creating spurious oscillations and without the use of problem dependent fixes or tunable parameters. An additional priority has been to formulate the methods within a systematic mathematical framework.

The philosophy underlying the ENO methods is simple: when reconstructing a profile for use in a convective flux term, one should not use high order polynomial interpolation across a steep gradient in the data. Such an interpolant would be highly oscillatory and ultimately corrupt the computed solution. ENO methods use an adaptive polynomial interpolation constructed to avoid steep gradients in the data. The polynomial is also biased to extrapolate from data from the direction of information propagation—"upwind"— for physical consistency and stability. In the case of a system, this interpolation must be done in the local characteristic fields, since it is these quantities—not the primitive conserved variables such as mass, momentum and energy—that are properly thought of as propagating in various directions. The ENO approach is completed by combining this interpolation method with a discrete conservation form for the equations. This form insures that shocks and other steep gradients in the flow are "captured", i.e. move at the right speed even if they are not fully resolved.

The original ENO schemes were based on the conservative control volume discretization of the equations, which yields discrete evolution equations for grid cell averages of the conserved quantities, e.g. mass, momentum and energy. This formulation has the disadvantage of requiring complicated transfers between cell averages and cell center nodal values in the algorithm. In particular, the transfer process becomes progressively more complicated in one, two and three spatial dimensions. The formulation also results in space and time discretizations that are coupled in a way that becomes complicated for higher order accurate versions.

To eliminate these complications, Shu and Osher [12] developed a conservative finite difference form of the ENO method, which uses only nodal values of the conserved variables. Their method is faster and easier to implement than the cell averaged formulation. In addition, the finite difference ENO method extends to higher dimensions in a "dimension by dimension" fashion, so that the 1D method applies unchanged to higher dimensional problems. They also use the method of lines for time integration, which decouples the time and space discretizations. To complete the scheme, Shu and Osher developed a special family of Runge-Kutta time integration schemes that are easy to implement, have good stability properties, and also have a "Total Variation Diminishing" (TVD) property. The TVD property prevents the time stepping scheme from introducing spurious spatial oscillations into upwind-biased spatial discretizations. We emphasize that this is *not* dimensional splitting in time, which has accuracy limitations unlike the "dimension by dimension" approach.

While both the cell averaged and finite difference formulations of ENO perform well on a great variety of compressible flow calculations, there are still special circumstances under which they produce spurious oscillatory results. Some of these situations are well known, such as the case of a slow moving shock. In this case, the cause of the oscillations is largely understood, but this has not resulted in a general, elegant way to eliminate the problem. In other cases, such as the examples provided in section 8, the cause of the oscillations is not understood due to the complexity of the physical problem.

It is apparent now—based mainly on the work of Donat and Marquina [1], as well as a model problem described in [2]—that the manner in which the transformation to local characteristic variables is evaluated within the cell wall flux calculation is responsible for these occasional spurious oscillations. In particular, the problem is due to evaluating the transformation—or, equivalently, the flux Jacobian matrix—at a cell wall that separates two very different states. The common approach in all characteristic-based methods is to evaluate this transformation at some reasonable average of the adjacent states. However, there is clearly a great deal of ambiguity in choosing this average, and any particular choice seems to introduce subtle spurious features into the solution. To avoid this ambiguity, Marquina introduced a flux splitting technique based on the unambiguous data on the left and right sides of the cell wall. There, the transformations to characteristic variables and subsequent flux calculations are well defined, Marquina combines the results in an upwind fashion to determine the cell wall flux. The details of these old and new approaches are described in section 2.7.

When the Marquina's splitting technique is applied to the standard ENO flux calculation, it fixes all known problematic cases. Thus the resulting finite difference ENO method with Marquina's splitting seems to meet the original goal of an elegant, general, accurate, robust, parameter-free method for hyperbolic systems of conservation laws. If this turns out to be the case, it may be the ultimate conservative difference scheme [3, 4, 5, 6, 7]. Since only further experience can determine its limitations, for now we propose it as the penultimate method. In any case, it is an important enhancement of the original ENO method, and should replace it for future applications.

Our primary goal here is to present—in a self-contained, accessible form—this new hybrid method consisting of Shu-Osher finite difference ENO with Marquina's flux splitting technique. We hope this will encourage widespread application of this technique.

This paper divides naturally into two parts. The first part is a tutorial on scheme design for hyperbolic systems of conservation laws, and is directed mainly at those not familiar with this field. The goal is to motivate the many details that go into the final scheme design described in the second part. The second part of the paper is a users' guide for the preferred form of the new method.

The first part provides the conceptual background needed to appreciate characteristic based methods for systems of convective conservation laws. This includes both the basic ingredients that go into the numerical method design, such as CFL restriction and shock capturing, as well as the advanced issue of conservative finite difference discretization, the upwind biased ENO interpolation technique, and the Jacobian evaluation problem that motivates Marquina's splitting and distinguishes it from previous practice.

In the second part, we start by showing how this method fits in as part of a comprehensive space and time discretization that can handle general systems of conservation laws that arise in physical problems. Next, we present the preferred form of the new method in a concise, detailed fashion suitable for direct application in numerical calculations. We finish with a few new examples illustrating the effectiveness of this approach.

2 Background and Motivation

In order to make this presentation self-contained, we provide some conceptual motivation and background for the various ingredients used in the new scheme. We will motivate the use of characteristic based schemes, discuss the form of the finite difference discretization used in the ENO methods, and discuss ENO interpolation. We then contrast Marquina's splitting with the traditional approach used in characteristic based schemes.

A general introduction to the properties of systems of conservation laws and their associated numerical methods can be found in LeVeque's book [8]. Comments on the original motivation and development of finite difference ENO can be found in the first paper of Shu and Osher, [11]. Marquina's splitting is motivated and introduced in the papers of Donat and Marquina, [1, 9].

Our goal here is to introduce convective—or hyperbolic—systems of conservation laws, and understand how their fundamental features impact the design of appropriate numerical methods.

2.1 General Conservation Laws

A continuum physical system is described by the laws of conservation of mass, momentum, and energy. That is, for each conserved quantity, the rate of change of the total amount in some region is given by its flux (convective or diffusive) through the region boundary, plus whatever internal sources exist. The integral form of this conservation law is

$$\frac{d}{dt}\int_R U dV + \int_{\partial R} \vec{F}(U) \cdot dA = \int_R S(U) dV \tag{1}$$

where U is the density of the conserved quantity, $\vec{F}(U)$ is the flux, and $S(U)$ is the source rate, and the volume and surface integrals indicated are over the region R and its boundary ∂R. By taking R to be an infinitesimal volume and applying the divergence theorem, we get the differential form of the conservation law,

$$\frac{\partial U}{\partial t} + \nabla \cdot \vec{F}(U) = S(U) \tag{2}$$

which is the basis for the numerical modeling of all continuum systems. Any physical system will be described by a system of such equations, i.e. a system of conservation laws. These also form the basis for their numerical modeling.

We will write most equations in one dimension, in which case our notation for the differential conservation equation 2 takes the more compact form:

$$U_t + F(U)_x = S(U) \tag{3}$$

2.2 Convective (Hyperbolic) Conservation Laws

A conserved quantity, such as mass, can be transported by convective or diffusive fluxes. The distinction is that diffusive fluxes are driven by gradients in density, while convective fluxes persist even in the absence of gradients. Here we will concentrate on the convective transport, ignoring diffusion (mass diffusion, viscosity and thermal conductivity) and also the source terms (such as chemical reactions, atomic excitations, and ionization processes). We take this simplified approach because the convective transport requires specialized numerical treatment. If present, diffusive and reactive effects can be treated by standard numerical methods that are independent of those for the convective terms. Stiff reactions, however, do present numerical difficulties.

Conservation laws with only convective fluxes are known as "hyperbolic" conservation laws (a more careful definition is given is section 2.5). A vast array of physical phenomena are modeled by such systems. The physics of explosives and high speed aircraft were two major driving forces in the development of these models. They also provide the basis for modeling astrophysical and fusion reactor plasma, mixed phase flow in fission reactor cooling systems, and combustion in jet engines, to mention a few of the important technological applications.

2.3 Convective Phenomena, Models and Numerical Implications

Our goal here is to mention the most universal aspects of the physics of hyperbolic systems, and relate it to the design of appropriate numerical methods.

The important physical phenomena exhibited by convective conservation laws are bulk convection, waves, contact discontinuities, shocks, and rarefactions. We will briefly describe the physical features and mathematical model equations for each effect, and most importantly, note the implications they have for numerical method design.

Bulk Convection and Waves Bulk convection is simply the bulk movement of matter—carrying it from one spot to another, like water streaming from a hose. Waves are small amplitude smooth rippling disturbances that transmit through the system without any bulk transport—like ripples on a water surface or sound waves through air. Whereas convective transport occurs at the gross velocity of the material, waves propagate at the "speed of sound" in the system (relative to the bulk convective motion of the system). Waves interact by superposition, so that they can either cancel out (interfere) or enhance each other.

The simplest model equation that describes bulk convective transport is the linear convection equation

$$\rho_t + v\rho_x = 0, \tag{4}$$

where v is a constant, equal to the convection velocity. This has the equivalent, but less often used, conservation form

$$\rho_t + (v\rho)_x = 0, \tag{5}$$

The solution to this is simply that ρ translates at the constant speed v. This same equation can also be taken as a simple model of wave motion, if ρ is a sine wave and v is interpreted as the sound speed.

The linear convection equation is also an important model for understanding smooth transport in any conservation law: as long as U has no jumps in it, and F is smooth, the general law $U_t + F(U)_x = 0$ can be expanded to

$$U_t + vU_x = 0, \tag{6}$$

where $v = F'(U)$. Thus, locally in smooth parts of the flow, any conservation law behaves like bulk convection with convective velocity $F'(U)$. This is called the characteristic velocity of the flow.

Bulk convection and waves are important because they imply that signals propagate in definite directions at definite speeds. This in contrast to a phenomena like diffusion which propagates signals in all directions at arbitrarily large speeds depending on the severity of the driving gradients. Thus we anticipate suitable numerical methods for hyperbolic systems will also have directional biases in space—which leads to the idea of upwind differencing, see section 2.4—and a definite relation between the space and time step (discrete propagation speed)—which will roughly be that the discrete propagation speed $\Delta x/\Delta t$ must be the same as the physical propagation speeds (characteristic speeds) in the problem. The general form of this relation is called the Courant-Friedrichs-Lewy (CFL) restriction, and it says the discrete speed must be at least as large as any characteristic speed in the problem.

Also, note that wave motion and bulk convection don't create any new sharp features in the flow. The other remaining phenomena are all special because they involve discontinuous jumps in the transported quantities. Because smooth features can be accurately represented by a polynomial interpolation, we expect to be able to develop extremely high accuracy numerical methods for the wave and convective effects. Conversely, since jump functions are poorly represented by polynomials, we expect little accuracy and perhaps great difficulty in numerically approximating the discontinuous phenomena.

The linear convection model also has an important implication for the time integration numerical method, i.e. the numerical method used to discretize

$\partial U/\partial t$. If we Fourier transform the linear advection equation we end up with an ordinary differential equation needing only time integration:

$$\hat{\rho}_t - ikv\hat{\rho} = 0, \tag{7}$$

where $\hat{\rho}(k)$ is the Fourier transform of $\rho(x)$. The important thing to note is that this is an ODE of the form $y_t = \lambda y$, where the growth rate λ is purely imaginary. Thus we must use an ODE integration method that is stable for imaginary growth rates. This is true for standard third order and fourth order Runge-Kutta methods, for example. But it is not true for common first order ("explicit Euler") or second order ("Heun's method") Runge-Kutta schemes. *Use of these common low order schemes is not compatible with an accurate spatial discretization of convection.* That is, using these methods with standard hyperbolic spatial discretizations would lead to the development of severe grid point to grid point oscillations, due solely to the poor choice of time stepping procedure. For explicit time stepping stability, a third or fourth order Runge-Kutta method should be used.

Contacts A contact discontinuity is a persistent, discontinuous jump in mass density moving by bulk convection through the system. Since there is negligible mass diffusion, such a jump persists. These jumps usually appear at the point of contact of different materials, for example, a contact discontinuity separates oil from water. Contacts move at the local bulk convection speed, or more generally, the characteristic speed, and can be modeled by using step-function initial data in the bulk convection equation 4. Since contacts are simply a bulk convection effect, they retain any perturbations they receive. Thus we expect contacts to be especially sensitive to numerical methods—any spurious alteration of the contact will tend to persist and accumulate.

Shocks A shock is a spatial jump in material properties—like pressure and temperature—that develops spontaneously from smooth distributions and then persists. That is, the shock jump is self-forming and also self-maintaining. This is unlike a contact, which must be put in the system initially, and will not re-sharpen itself if it is smeared out by some other process. Shocks develop through a feedback mechanism in which strong impulses move faster than weak ones, and thus tend to steepen themselves up into a "step" profile as they travel through the system. Familiar examples are the "sonic boom" of a jet aircraft, or the "bang" from a gun. These sounds are our perceptions of a sudden jump in air pressure.

The simplest model equation that describes shock formation is Burgers' equation

$$u_t + (\frac{u^2}{2})_x = 0. \tag{8}$$

Formally, this looks like the convection equation 4, with a non-constant convective speed of $v = u$. Thus larger u values move faster, and they will overtake smaller values, ultimately resulting in the development of a right-going shock if the initial data for u is any positive, decreasing function, e.g. $1 - \tanh(x)$.

Shocks move at a speed that is not simply related to the bulk flow speed or characteristic speed, and is not immediately evident from examining the flux, in contrast to contacts. Shock speed is controlled simply by the difference between influx and outflux of conserved quantity into the region. Specifically, suppose a conserved quantity U with conservation law 3 has a step function profile with one constant value extending to the left, U_L, and a lower constant value to the right, U_R, with a single shock jump transition between these two, and this jump location is moving with speed s to the right. Then the integral form of the conservation law 1, applied to any interval containing the shock, gives the relation

$$s(U_R - U_L) = F_R - F_L, \tag{9}$$

which is, of course, just another statement that the rate at which U appears, $s(U_R - U_L)$, in the interval of interest is given by the difference in fluxes across the interval. However, it also determines the shock speed s in terms of densities and fluxes well away from the shock itself.

Thus we see that the proper speed of the shock is directly determined by—and only by—conservation of U via the flux F. This has an important implication for numerical method design: namely, a numerical method will only "capture" the correct shock speeds if it has "conservation form", i.e. if the rate of change of U at some node is the difference of fluxes which are accurate approximations of the real flux F.

The self-sharpening feature of shocks has two implications for numerical methods. First, it means that even if the initial data is smooth, steep gradients and jumps will form spontaneously; thus our numerical method must be prepared to deal with shocks even if none are present in the initial data. Second, there is a beneficial effect from self-sharpening, because modest numerical errors introduced near a shock (smearing or small oscillations) will tend to be eliminated, and will not accumulate. The shock is naturally driven towards its proper shape. Because of this, computing strong shocks is mostly a matter of having a conservative scheme in order to get their speed correct—the basic jump itself will be preserved by the physical self-sharpening.

Rarefactions A rarefaction is a discontinuous jump or steep gradient in properties that dissipates as a smooth expansion. A common example is the jump in air pressure from outside to inside a balloon, which dissipates as soon as the balloon is burst and the high pressure gas inside is allowed to expand. Such an expansion also occurs when the piston in an engine is rapidly pulled outward from the cylinder. The expansion (density drop) associated with a rarefaction propagates outward at the sound speed of the system, relative to the underlying bulk convection speed.

A rarefaction can be modeled by Burgers' equation 8, with initial data that starts out as a steep increasing step, for example $u(x) = \tanh(\frac{x}{\epsilon})$, where ϵ is a small (perhaps 0) width for the step. This step will broaden and smooth out during the evolution.

A rarefaction tends to smooth out local features, which is somewhat good for numerical modeling. It tends to diminish numerical errors over time and make the solution easier to represent by polynomials, which form the basis for our numerical representation. However, a rarefaction often connects to a smooth (e.g. constant) solution region and this results in a "corner" which is notoriously difficult to capture accurately.

The main numerical problem posed by rarefactions is that of initiating the expansion. If the initial data is a perfect, symmetrical step, such as $u(x) = \text{sign}(x)$, it may be "stuck" in this form, since the steady state Burgers' equation is satisfied identically (i.e. the flux $\frac{u^2}{2}$ is constant everywhere, and similarly in any numerical discretization). However, local analysis can identify this stuck expansion, because the characteristic speed u on either side points away from the jump, suggesting its potential to expand. In order to get the initial data unstuck, some small amount of smoothing must be applied to introduce some intermediate state values and thus have a non-constant flux to drive expansion. In numerical methods, this smoothing applied at a jump where the effective local velocity indicates expansion should occur is called an "entropy fix", since it allows the system to evolve from the artificial low entropy (i.e. very symmetrical) initial state to the proper increased entropy state of a free expansion.

Systems of Equations In general, a hyperbolic system will simultaneously contain all these processes: smooth processes of bulk convection and wave motion, and discontinuous processes involving contacts, shocks and rarefactions. For example, if a gas in a tube is initially prepared with a jump in the states (density, velocity and temperature) across some surface, as the evolution proceeds in time these jumps will break up into a combination of shocks, rarefactions and contacts, in addition to any bulk motion and sound waves that may exist or develop.

Based on these considerations, in a general system we expect that the density of mass, momentum and energy will be smooth in large regions, separated by these discontinuous jumps in properties. Further, these jumps are moving through the system, interacting in complex ways. What we can generally hope for is numerical methods that are high accuracy in the smooth regions, don't distort the jumps too much (smear them or add oscillations), and move the jumps at the correct speeds, which are usually not known a priori. We want high accuracy methods to effectively model smooth convection and wave motion. We expect contacts to be the most sensitive indicators of numerical errors on discontinuities. We expect the shocks to be robust features, and we expect rarefactions to not be a problem as long as their initial expansion from a jump can be made to occur.

The simplest system of *physically realistic* model equations for convective transport is the Euler equations for gas dynamics, which describe the conservative transport of mass, momentum and energy in a gas in one spatial dimension (e.g. in a long tube):

$$\rho_t + (\rho v)_x = 0 \tag{10}$$
$$(\rho v)_t + (\rho v^2 + p)_x = 0 \tag{11}$$
$$E_t + ((E + p)v)_x = 0 \tag{12}$$

where ρ is the mass density, v is the flow velocity, p is the pressure and E is the total energy (kinetic plus internal) density. To be completely specified, these equations require an "equation of state" for the pressure, i.e. a relation $p = p(\rho, E)$. One of the simplest reasonable forms is the gamma law gas relation, $p = p_0(\rho/\rho_0)^\gamma$, where $\gamma \geq 1$ is a constant and p_0, ρ_0 are the reference pressure and density.

Despite their simple form—looking like linear convection and Burgers' equations—the Euler equations support extremely complex dynamic behavior which can be difficult to understand and predict, due to the nonlinear, coupled form of the equations. However, it is true that any isolated jump discontinuity in the state variables will, as time goes on, break up into some "combination" of a shock, a contact and a rarefaction. This justifies a simple intuitive model for the structure of this system: a "toy" version of the Euler equations consists of three independent scalar equations: one convection (for representing the effects of bulk convection, waves and contacts), one Burgers' equation (for shock formation) and another independent Burgers' equation (for independent rarefaction formation):

$$(u_1)_t + (u_1)_x = 0 \tag{13}$$
$$(u_2)_t + \left(\frac{u_2^2}{2}\right)_x = 0 \tag{14}$$
$$(u_3)_t + \left(\frac{u_3^2}{2}\right)_x = 0 \tag{15}$$

where the initial data are step functions with jumps at $x = 0$ as follows: an arbitrary jump in u_1 (a contact, moving right at speed 1), a decreasing jump in u_2 (a shock, or shock precursor if smoothed out slightly), and an increasing jump in u_3 (a rarefaction). Since these three equations are independent, the subsequent evolution is obvious. However, let us form a new, equivalent system by multiplying this system by a 3×3 invertible matrix R. If the original system in vector form is written as

$$\vec{U}_t + [\vec{F}(\vec{U})]_x = 0 \tag{16}$$

then the new system can be written as

$$\vec{V}_t + [\vec{G}(\vec{V})]_x = 0 \tag{17}$$

where $\vec{V} = R\vec{U}, \vec{G} = R\vec{F}$. When considered in terms of the "mixed" variables $\vec{V} = (v_1, v_2, v_3)$, the behavior of this system is not at all obvious, and the simple contact, shock and rarefaction present in the system will cause a rather complicated evolution of $\vec{V}$. The intuitive point to understand is that the real Euler equations, as well as other hyperbolic systems we encounter in physical problems, are written in what are effectively the mixed variables, where the apparent behavior is quite complicated. It requires some transformation to decouple them back into unmixed fields that exhibit the pure contact, shock and rarefaction phenomena (as well as bulk convection and waves). In this toy model here, there is a single linear transformation that perfectly decouples the mixed equations, namely the inverse R^{-1}. In a real system, this perfect decoupling is not possible because the mixing is nonlinear, but it can be achieved approximately—over a small space and time region—and this provides the basis for the theoretical understanding of the structure of general hyperbolic systems of conservation laws. This is called a transformation to characteristic variables, and we will present it in detail in section 2.5. As we shall see, this transformation also provides the basis for designing appropriate numerical methods.

Summary of Numerical Implications As we have examined the properties of hyperbolic systems, we have compiled a list of associated implications for numerical method design. For clarity, we will summarize these here.

- **CFL Condition** to correctly model the propagation of information, the space and time grids must satisfy $\Delta x/\Delta t \geq s_{\max}$, where $s_{\max}$ is the largest propagation speed (characteristic speed) in the problem.
- **Upwind Biasing** the directed propagation of signals implies there will be directional biases in the choice of spatial nodes—in the "upwind direction"—used to discretize the equations.
- **High Accuracy** modeling of weak wave propagation and smooth convection benefits greatly from numerical methods based on high order accurate polynomial interpolation.
- **Time Integration** A third or fourth order accurate Runge-Kutta method (or other method stable for imaginary growth rates) must be used for the numerical time integration, to avoid instability.
- **Entropy Fix** proper numerical modeling of rarefactions require a small amount of smoothing of a jump where the nearby characteristic speeds indicates potential for expansion.
- **Sharp Contacts** the most sensitive indicator of how well a numerical method handles jumps is the treatment of contacts discontinuities (in a linear convection equation).

- **Conservation Form** in order to capture shock speeds, the numerical method must have conservation form, i.e. be written in terms of a discrete difference in fluxes.
- **Characteristic Decomposition** systems of conservation laws can be best understood by transforming to local characteristic variables that display largely decoupled scalar behavior.

By addressing all these points, we can design methods that accurately and efficiently compute the behavior of hyperbolic systems of conservations laws, or the hyperbolic parts of general systems of conservation laws.

2.4 Upwind Biased ENO Interpolation

Here we give a more detailed motivation for upwind biased discretization, and the Essentially Non-Oscillatory (ENO) interpolation technique that forms the basis for our numerical methods.

As noted in the summary of section 2.2, to assess the quality of our numerical method we can focus on the treatment of contact discontinuities in the linear convection equation 4. Since the time discretization can be handled by a high accuracy Runge-Kutta method, we will focus on the spatial discretization and assume the time evolution takes place exactly—i.e. at each time step Δt, the spatial profile just translates rigidly by the amount $v\Delta t$.

Spatially, the contact is initially represented by a discrete step function, i.e. nodal values that are constant at one value ρ_L on nodes $x_1, \dots, x_J$, and then constant at a different value, ρ_R, at all remaining nodes $x_{J+1}, \dots, x_N$.

To update in time the value ρ_i at a given node x_i, we first reconstruct the graph of a function $\rho(x)$ near x_i by interpolating nearby nodal ρ values, shift that $\rho(x)$ graph spatially by $v\Delta t$ (the exact time evolution), and then reevaluate it at the node x_i to obtain the updated ρ_i. We require our local interpolant be smooth at the point x_i, since in actual practice we are going to use it to evaluate the derivative term $(v\rho)_x$ there.

The simplest symmetrical approach to smooth interpolation near a node x_i is to run a parabola through the nodal data at x_{i-1}, x_i, x_{i+1}. This interpolation is an accurate reconstruction of $\rho(x)$ in smooth regions, and this approach will work well there. However, near the jump, at x_J and x_{J+1}, the parabola will greatly overshoot the nodal ρ data itself, by an amount comparable to the jump $\rho_L - \rho_R$, and this overshoot will show up in the nodal values once the shift is performed. Successive time steps will further enhance these spurious oscillations. In this way, repeated parabolic interpolation and shifting introduces severe oscillations that totally destroy the structure of the contact. This approach corresponds to standard central differencing applied to the convection equation 4.

To avoid the oscillations from parabolic interpolation, we could instead try to use a smooth linear interpolation near x_i. However, there are two to choose from, namely the line through data at nodes x_i and x_{i-1}, or through data at x_i

and x_{i+1}. The direction of information propagation determines which one will result in a non-oscillatory reconstruction. Assuming the convection speed v is positive, the data is moving from the left to the right. Thus, for a short time, the only ρ values that will arrive at node x_i in the exact solution are those over the interval (x_{i-1}, x_i). If we use a linear interpolation based on these two upwind nodes, when we shift it right by $v\Delta t$ we will not introduce any new extrema in ρ at x_i, since the result will lie between ρ_{i-1} and ρ_i (as long as the shift $v\Delta t$ is less than the width of the interval $\Delta x = x_i - x_{i-1}$, which is exactly the CFL restriction on the time step). In contrast, if the linear interpolation were based on the "downwind" nodes (x_i, x_{i+1}), a shift right would cause a part of this line *not* between ρ_i and ρ_{i+1} appear as the new ρ value at x_i, and this can and will introduce new, spurious extreme values and oscillations into the nodal data. In this way, we see clearly why it necessary to base the linear interpolant on the upwind point x_{i-1}: interpolating from that direction represents the data that is supposed to arrive at the point of interest, so that no spurious values are introduced.

The main problem with the linear upwind biased interpolant is that it has low accuracy. Each interpolate and shift step will smear out the contact jump over more nodes. If we naively go to higher accuracy by using a higher order upwind biased interpolant, such as running a parabola through x_i, x_{i-1}, x_{i-2} to advance ρ_i, we will run into the spurious oscillation problem again—at nodes x_{J+1} and x_{J+2} this upwind parabola will interpolate across the jump and thus have large overshoots just as for the centrally interpolated parabolas. By forcing the parabola to cross a jump, it no longer reflects the data on the interval (x_i, x_{i-1}) that will be arriving at x_i during the next time step.

A solution to the problem of achieving more accuracy while avoiding spurious overshoots in the interpolant is to use the upwind biased, Essentially Non-Oscillatory (ENO) interpolation technique [11, 12]. The motivation for this approach is that we must use a higher degree polynomial interpolant to achieve more accuracy, and it must involve the immediate upwind node to properly represent the propagation of data. But, as we saw, we must also avoid polluting this upwind data with spurious oscillations that come from interpolating across jumps in data. Thus the remaining interpolation nodes are chosen based on smoothness considerations. Specifically, to update ρ_i using a degree k interpolant requires $k+1$ interpolation nodes. We will choose $k+1$ consecutive nodes that include the immediate upwind node from x_i, which is x_{i-1} if $v > 0$. Still, that leaves k different lists of nodes—"stencils"—to choose from. Of these, we will use the one for which the resulting interpolating polynomial is the smoothest, by some measure. (For example, we can measure smoothness by the size of the kth derivative, or the total variation, or by any other convenient means. For more detailed considerations see [10].) In particular, this approach will—if at all possible—not run an interpolant across a jump in the data. Thus, it avoids introducing large, spurious overshoots. However very small interpolation overshoots do occur near extrema in the nodal data, as they must, since any smooth function will slightly overshoot its values as sampled at discrete points near extrema. This is the sense in which the method is only *Essentially* Non-Oscillatory (ENO)—it is not a failing; it simply reflects the real relation between smooth functions and their discretely sampled values.

In practice, there are simple, efficient ways to generate the upwind biased ENO interpolant of any desired order, based on the divided difference table of the nodal data.

2.5 Characteristic Based Schemes for Hyperbolic Systems

In this section we describe the use of characteristic decomposition for designing suitable upwind biased numerical schemes.

Consider a system of N convective conservation laws in one spatial dimension,

$$\vec{U}_t + [\vec{\mathcal{F}}(\vec{U})]_x = 0 \tag{18}$$

The basic idea of characteristic numerical schemes is to transform this nonlinear system to a system of (nearly) independent scalar equations of the form

$$u_t + vu_x = 0 \tag{19}$$

discretize each scalar equation independently in an v-upwind biased fashion, and then transform the discretized system back into the original variables.

In a smooth region of the flow, we can get a better understanding of the structure of the system by expanding out the derivative as

$$\vec{U}_t + J\vec{U}_x = 0 \tag{20}$$

where $J = \frac{\partial \vec{\mathcal{F}}}{\partial \vec{U}}$ is the Jacobian matrix of the convective flux function. Note that if J were a diagonal matrix, with real diagonal elements, this system would be decoupled into N independent scalar equations as desired.

In general J is not of this form, but we can hope to transform this system to that form by multiplying through by a matrix that diagonalizes J. If this is possible, we call the system *hyperbolic*. The fortunate thing is that most physical convective transport equations turn out to be "hyperbolic". In this case, the necessary matrices turn out to be the matrices of left-multiplying and right-multiplying eigenvectors of J. Specifically, for a hyperbolic system we require following properties (which allow our strategy to work): first, we require that J have N real eigenvalues $\lambda^p, p = 1, \ldots, N$, and that there be N eigenvectors for multiplying against J from the right. If we use these as columns of a matrix R, this is expressed by the matrix equation

$$JR = RDiag(\lambda^p) \tag{21}$$

where $Diag(\lambda^p)$ denotes a diagonal matrix with the elements $\lambda^p, p = 1, \ldots, N$ on the diagonal. Similarly, we also require that there be N eigenvectors for multiplying against J from the left; when these are used as the rows of a matrix L, this is expressed by the matrix equation

$$LJ = Diag(\lambda^p)L \tag{22}$$

We finally require that these matrices L and R can be chosen to be inverses

$$LR = RL = I \tag{23}$$

These matrices transform to a system of coordinates in which J is diagonalized as desired:

$$LJR = Diag(\lambda^p) \tag{24}$$

Suppose we want to discretize our equation at the node x_0, where L and R have values L_0 and R_0. To get a locally diagonalized form, we multiply our system equation by the *constant* matrix L_0 which nearly diagonalizes J over the region near x_0 (we require a constant matrix so that we can move it inside all derivatives):

$$[L_0\vec{U}]_t + L_0 J R_0 [L_0\vec{U}]_x = 0 \tag{25}$$

We have inserted $I = R_0 L_0$ to put the equation in a more recognizable form. The spatially varying matrix $L_0 J R_0$ is exactly diagonalized at the point x_0, with eigenvalues λ_0^p, and it is nearly diagonalized at nearby points. Thus the equations are sufficiently decoupled for us to apply upwind biased discretizations independently to each component, with λ_0^p determining the upwind biased direction for the p-th component equation. Once this system is fully discretized, we multiply the entire system by $L_0^{-1} = R_0$ to return to the original variables.

In terms of our original equation 18, our procedure for discretizing at a point x_0 is simply to multiply the entire system by the left eigenvector matrix L_0,

$$[L_0\vec{U}]_t + [L_0\vec{\mathcal{F}}(\vec{U})]_x = 0 \tag{26}$$

and discretize the $p = 1, \ldots, N$ scalar components of this system

$$[(L_0\vec{U})_p]_t + [(L_0\vec{\mathcal{F}}(\vec{U}))_p]_x = 0 \tag{27}$$

independently, using upwind biased differencing with the upwind direction for the p-th equation determined by the sign of λ^p. We then multiply the resulting spatially discretized system of equations by R_0 to recover the spatially discretized fluxes for the the original variables:

$$\vec{U}_t + R_0\Delta(L_0\vec{\mathcal{F}}(\vec{U})) = 0 \tag{28}$$

where Δ stands for the upwind biased discretization operator.

We call λ^p the p-th characteristic velocity or speed, $(L_0\vec{U})_p = \vec{L}_0^p \cdot \vec{U}$ the p-th characteristic state or field (here L^p denotes the p-th row of L, i.e. the p-th left eigenvector of J), and $(L_0\vec{\mathcal{F}}(\vec{U}))_p = \vec{L}_0^p \cdot \vec{\mathcal{F}}(\vec{U})$ the p-th characteristic flux. According to the local linearization, it is approximately true the p-th characteristic field rigidly translates in space at the p-th characteristic velocity. Thus this decomposition corresponds to the local physical propagation of independent "waves" or "signals".

2.6 The Conservative Finite Difference Form

To ensure that shocks and other steep gradients are captured by the scheme—i.e. they move at the right speed even if they are unresolved—we must write the equation in a discrete conservation form. That is, a form in which the rate of change of conserved quantities is equal to a difference of fluxes. This form guarantees that we conserve the total amount of the states $\vec{U}$ (e.g. mass, momentum and energy) present, in analogy with the integral form given by equation 1. More importantly, this can be shown to imply that steep gradients or jumps in the discrete profiles must propagate at the physically correct speeds [8] as discussed in section 2.3.

Usually, conservation form is derived for *control volume* methods, that is methods that evolve cell average values in time rather than nodal values. In this approach, a grid node x_i is assumed to be the center of a grid cell $(x_{i-\frac{1}{2}}, x_{i+\frac{1}{2}})$, and we integrate the conservation law 3 across this control volume to obtain (set the source to 0 for simplicity)

$$\bar{U}_t + (F(U_{i+\frac{1}{2}}) - F(U_{i-\frac{1}{2}})) = 0 \tag{29}$$

where $\bar{U}$ is the integral of U over the cell, and $U_{i\pm\frac{1}{2}}$ are the (unknown) values of U at the cell walls. This has the desired conservation form, in that the rate of change of the cell average is a difference of fluxes. The difficulty with this formulation is that it requires transforming between cell averages of U (which are directly evolved in time by the scheme) and cell wall values of U (which must be reconstructed) to evaluate the needed fluxes. While this is manageable in 1-D, in higher dimensional problems the series of transformations necessary to convert the cell averages to cell wall quantities becomes increasingly complicated. The distinction between cell average and midpoint values is usually ignored for schemes whose accuracy is no higher than second order (e.g. TVD schemes). This is because the cell average and the midpoint value differ by $O(\Delta x^2)$.

Instead, we seek a fully finite difference scheme—i.e. a scheme that directly evolves nodal values in time. For the finite difference approach, the derivation of conservation form is less obvious. We define the "numerical flux function", $\vec{F}$, by the property that the real flux divergence is a finite difference of numerical fluxes:

$$\vec{\mathcal{F}}(\vec{U})_x = \frac{\vec{F}(x + \frac{\Delta x}{2}) - \vec{F}(x - \frac{\Delta x}{2})}{\Delta x} \tag{30}$$

at every x (here Δx is some constant spacing). We call it the numerical flux since we require it in our numerical scheme, and also to distinguish it from the closely related "physical flux", $\vec{\mathcal{F}}(\vec{U})$. It is not obvious that the numerical flux function exists, but from relationship 30 one can solve for its Taylor expansion (or, using a Fourier transform gives a quick derivation). The result is

$$\vec{F} = \vec{\mathcal{F}}(\vec{U}) - \frac{(\Delta x)^2}{24}\vec{\mathcal{F}}(\vec{U})_{xx} + \frac{7(\Delta x)^4}{5760}\vec{\mathcal{F}}(\vec{U})_{xxxx} - \cdots \tag{31}$$

Note that to second order accuracy in Δx the physical and numerical flux functions are the same. As described in section 6.2, direct use of the Taylor series is not the most convenient way to compute the numerical flux in the ENO algorithm. The series is simply useful for understanding the relation between physical and numerical fluxes.

The finite difference discretization is not based directly on the differential form of the conservation law 18; rather, it is based on the equivalent conservative finite difference form

$$\vec{U}_t + \frac{\vec{F}(x + \frac{\Delta x}{2}) - \vec{F}(x - \frac{\Delta x}{2})}{\Delta x} = 0 \tag{32}$$

The discretization is based on characteristic upwind differencing, but now it is the numerical flux $\vec{F}$ that must be discretely approximated, rather than the physical flux derivative, $\vec{\mathcal{F}}(\vec{U})_x$. To accomplish this, we generalize the conclusions of the previous characteristic variables analysis to the following procedure: to determine $\vec{F}$ at a point x_0, we should multiply all the local nodal physical flux vectors by L_0, and then use these, component by component, to construct scalar characteristic numerical fluxes in an upwind biased fashion. We then project these back to original variable numerical fluxes by multiplying by R_0.

2.7 Comments on Jacobian Evaluation

We will briefly outline the significance of the Jacobian evaluation for characteristic based methods, and how Marquina's procedure differs from the evaluation commonly used in the ENO method [11, 12].

The Jacobian matrix of the convective flux vector is quite important to any characteristic based scheme, as it defines the local linearization of the nonlinear problem. As previously described, it determines the transformation to the local characteristic fields, and thus what the upwind directions are as well as what quantities are to be upwind differenced.

In finite volume methods it is natural that the fluxes—and thus the transformation to characteristic fields needed to evaluate them in an upwind way—be evaluated at cell walls. The analogous situation occurs in the conservative finite difference formulation as well. There we want to discretely approximate the conservation equation, 18, at grid nodes, x_i. Thus, by the numerical flux relation, 30, we require values of the *numerical flux* at the midpoint between nodes, $x_0 = x_{i+\frac{1}{2}}$. We refer to these midpoints as "cell walls", in analogy with the finite volume case.

Thus, in order to transform to characteristic fields to evaluate the numerical fluxes, we require values of the Jacobian (and its eigensystem) at cell walls. In the finite difference setting, we only know values for $\vec{U}$ at the nodes, so evaluation of a Jacobian at the cell walls requires some form of interpolation. In standard ENO schemes it was thought that the precise form of this interpolation was not so important. But recent developments show that in fact it can make a great deal of difference in causing or eliminating spurious oscillations.

The standard ENO method uses a single Jacobian evaluated at the linear average of the states at nodes adjacent to the midpoint,

$$J_{i+\frac{1}{2}} = J\left(\frac{\vec{U}_i + \vec{U}_{i+1}}{2}\right) \tag{33}$$

In smooth regions, this centered linear approximation is second order accurate. Moreover, in a smooth region it makes little difference whether the derivatives are computed in an upwind biased fashion or in some combination of upwind and downwind. Thus the precise determination of the Jacobian (and the transformation to characteristic fields)—in addition to having little uncertainty anyway—is not so important. It is between nodes in an unresolved steep gradient that the centrally averaged Jacobian might cause problems. It can differ significantly from the left and right Jacobians interpolated from left and right nodal state values, and there is no clear reason why this central Jacobian value is the proper choice for a midpoint Jacobian. The only justification for its use is that in practice it seems to work well for many problems. However, based on the following considerations we can see that it has the *potential* to allow spurious oscillations under special circumstances.

In the Separating Box Problem [2], we showed that small perturbations to the Jacobian matrix can lead to large oscillations in an ENO numerical solution. The intuition developed there was that small errors in the Jacobian would cause one to transform into the wrong characteristic variables, i.e. ones which were mixtures of the true characteristics. Upwinding on these slightly mixed fields amounts to a small amount of downwind differencing on the "true" characteristic fields, combined with the desired upwind differencing. This small amount of downwind differencing can create noticeable oscillations near unresolved steep gradients in the flow. We used this example to argue that one should use the exact formulas for the Jacobian and associated transformation to characteristic variables, rather than simplified approximate expressions that often seem attractive in complex problems.

Donat and Marquina [1, 9] independently took this idea much further. They realized that near an unresolved steep gradient in the flow, in which the states vary by a large amount from one node to the next, there is no clear way to determine "the" value of the Jacobian midway between the nodes (where it is required for ENO and other methods). There may be unambiguous values of the Jacobian when extrapolating from nodal data from the left or from the right of the midpoint, but these left sided and right sided Jacobians can differ substantially. They propose to make use of these two Jacobians separately, in an upwind fashion, rather than attempt to define a single representative midpoint Jacobian. In doing so, they seem to have avoided the substantial uncertainty in the value of the Jacobian that results from trying to choose a single one from the large range of possible values "between" the left and right Jacobians. This uncertainty insures that any single choice of Jacobian will be a large perturbation from the true Jacobian of the exact solution of the underlying flow problem, and, as before, the result is an inaccurate transformation to characteristic fields. This allows for a mix of upwind and downwind differencing with the associated potential for oscillations.

Donat and Marquina make use of the left and right Jacobian in a consistent upwind way to compute the discretized convective flux terms. In the context of ENO methods, they propose to evaluate the left Jacobian with the left side biased interpolation of the conserved variables, and the right Jacobian with the right side biased interpolation of the conserved variables. Each of these interpolations is done in a high order accurate ENO fashion, e.g. the left interpolant is chosen as the smoothest possible polynomial interpolant of the desired degree that includes the left node in its stencil. Using each of these two Jacobian matrices separately, we are to compute convective flux derivatives in each characteristic field using the ENO method. Of these, only the right moving fluxes from the left and left moving fluxes from the right are actually used, the rest are discarded. The fluxes are then taken out of the characteristic fields, yielding two vector flux functions for the conserved variables. Adding these two vector fluxes together gives a consistent, high order accurate numerical flux function.

In [1, 9], the authors show many examples illustrating the advantages of using Marquina's Jacobian. In section 8 we present one more, since it is of special interest to us and also illustrates how their technique can be of value in more complicated applications.

We note a special case that could occur when using a second order accurate approximation to the conserved variables. If the smoothest possible approximation to the conserved variables from the left and from the right both happen to be the central linear average, then the resulting scheme is equivalent that using the standard ENO Jacobian evaluation.

It is tempting to think that some other way of constructing a Jacobian—or a $\vec{U}$ value—at the midpoint would yield a more appropriate value for the ENO scheme. Some form of interpolation must be used, since only nodal variable values are available; the linear average is merely a symmetrical choice. One might suppose that instead an upwind biased interpolation should be used to determine a midpoint value of the variables, since this would better reflect what information actually reaches the midpoint during the course of a time step. However, this idea is complicated by the fact that the upwind directions are only defined for the characteristic fields, while the conserved variables to be interpolated are mixtures of those fields. Thus, it is not possible to pick an upwind direction for a single conserved variable.

Still, this idea can be developed to determine a more physically reasonable midpoint Jacobian evaluation: based on the value of the Jacobian at the left and right adjacent nodes, one can transform to characteristic fields on the left side and the right side using these respective Jacobians. Then, one can interpolate all right moving characteristic fields from the left side to the midpoint, and all left moving characteristic fields from the right side to the midpoint, combine them into a single characteristic state vector, and transform this vector back to primitive variables to obtain a properly "upwind interpolated" midpoint state vector. The Jacobian evaluated at this state provides a single Jacobian for use with the ENO method. Our experiments show that this is indeed a superior choice over the standard linear average, in that it does greatly reduce spurious oscillations in those special cases when they occur. However, it does not perform as well as Marquina's procedure for making use of separate left and right Jacobians. It seems that no single midpoint Jacobian adequately represents the situation when the left and right nodal Jacobians differ greatly.

3 Model Equations and Discretizations

The numerical method we will present can be applied to a general system of convection-diffusion-reaction conservation equations in any number of spatial dimensions. For example, in two spatial dimensions (x, y) the vector form of the equations is

$$\vec{U}_t + [\vec{\mathcal{F}}(\vec{U})]_x + [\vec{\mathcal{G}}(\vec{U})]_y = [\vec{\mathcal{F}}_d(\vec{\nabla}\vec{U})]_x + [\vec{\mathcal{G}}_d(\vec{\nabla}\vec{U})]_y + \vec{\mathcal{S}}(\vec{U}) \tag{34}$$

where $\vec{U} = \vec{U}(x, y, t)$ is the vector of conserved variables, $\vec{\mathcal{F}}(\vec{U})$ and $\vec{\mathcal{G}}(\vec{U})$ are the vectors of convective fluxes, $\vec{\mathcal{F}}_d(\vec{\nabla}\vec{U})$ and $\vec{\mathcal{G}}_d(\vec{\nabla}\vec{U})$ are the vectors of diffusive fluxes, and $\vec{\mathcal{S}}(\vec{U})$ is the vector of reaction terms. Subscripts t, x, y denote the corresponding time and space partial derivatives. Note that our techniques can also be extended to apply to systems that have convective or diffusive terms that are not in conservation form. For example, non-conservative convective terms arise as the "thermal forces" in the Braginskii equations describing transport in a multi-species plasma.

4 Spatial Discretization

The diffusion terms can be evaluated with standard second or fourth order conservative central differencing. The reaction terms, which involve no derivatives, are simply evaluated at point values.

The finite difference ENO method is used to evaluate convection terms. It is applied independently to $[\vec{\mathcal{F}}(\vec{U})]_x$ and to $[\vec{\mathcal{G}}(\vec{U})]_y$—a "dimension by dimension" discretization. On a rectangular 2-D grid, we sweep through the grid from bottom to top performing ENO on 1-D horizontal rows of grid points to evaluate the $[\vec{\mathcal{F}}(\vec{U})]_x$ term. The $[\vec{\mathcal{G}}(\vec{U})]_y$ term is evaluated in a similar way by sweeping through the grid from left to right performing ENO on 1-D vertical rows of grid points. The ENO method will be fully described in subsequent sections.

The spatial discretization can be extended to cover equations that include first and second derivative terms that are not expressible in conservation form. Second derivative terms can still be treated with standard central differences, second or fourth order. For non-conservative convective terms, the ENO procedure must be based on a linearized convective matrix of the entire first order system, not just the part that is in conservation form. We will describe this extension in detail in a future report.

5 Time Discretization

Once we have a numerical approximation to each of the spatial terms in equation 34, we can write it abstractly as a system of Ordinary Differential Equations (ODEs)

$$\vec{U}_t = \vec{f}(\vec{U}) \tag{35}$$

This equation could be discretized in time by any ODE integrating method that has suitable accuracy and stability properties. If the spatial reaction or diffusion terms are particularly strong, to the point where their time step restrictions are much more limiting than that of the Courant-Friedrich-Lewy (CFL) restriction for the convective terms, it is better to handle them separately via a time splitting procedure and a stiff ODE integrator as described in [2].

For the general time integration of this ODE, the Total Variation Diminishing (TVD) Runge-Kutta methods of Shu and Osher [12] are particularly well suited. In addition to the simplicity of Runge-Kutta methods, they are specially designed for time integrating spatially discretized convection equations in a way that will not create spurious oscillations in the solution.

First order TVD Runge-Kutta is simply the forward Euler method,

$$\vec{U}^{n+1} = \vec{U}^n + \Delta t \vec{f}(\vec{U}^n) \tag{36}$$

Second order TVD Runge-Kutta is Heun's predictor-corrector method,

$$\vec{U}^\star = \vec{U}^n + \Delta t \vec{f}(\vec{U}^n) \tag{37}$$

$$\vec{U}^{n+1} = \vec{U}^n + \Delta t \left(\frac{1}{2}\vec{f}(\vec{U}^n) + \frac{1}{2}\vec{f}(\vec{U}^\star) \right) \tag{38}$$

A third order TVD Runge-Kutta method is given by,

$$\vec{U}^\star = \vec{U}^n + \Delta t \vec{f}(\vec{U}^n) \tag{39}$$

$$\vec{U}^{\star\star} = \vec{U}^n + \Delta t\left(\frac{1}{4}\vec{f}(\vec{U}^n) + \frac{1}{4}\vec{f}(\vec{U}^\star)\right) \tag{40}$$

$$\vec{U}^{n+1} = \vec{U}^n + \Delta t\left(\frac{1}{6}\vec{f}(\vec{U}^n) + \frac{1}{6}\vec{f}(\vec{U}^\star) + \frac{2}{3}\vec{f}(\vec{U}^{\star\star})\right) \tag{41}$$

There are no convenient fourth order or higher TVD Runge-Kutta methods; they do exist, but they only maintain the TVD property when used with special, more complicated spatial discretizations. The standard fourth order accurate Runge-Kutta method can be used, but it is not TVD. This means it could cause spurious spatial oscillations, though in practice this has not been a problem.

The third order TVD method is generally recommended, since it has the greatest accuracy and largest time step stability region of the TVD schemes. Due to its large stability region (which includes a segment of purely imaginary linear growth rates), for a sufficiently small time step it is guaranteed to be linearly stable for the entire class of problems considered here. In contrast, the first and second order methods both require some spatial diffusion terms in order to be stable. Without that, no matter how small the time step is they may be mildly unstable (although it turns out the ENO spatial discretization can prevent this instability to some extent). For this reason, they should not be used unless there is substantial spatial diffusion in the problem.

6 The Finite Difference ENO Scheme

We are now ready to proceed with the precise presentation of the ENO finite difference discretization.

6.1 Reducing a System to Independent Scalar Equations

First, we show how the discretization for a system is reduced to that of independent scalar problems.

Consider the Jacobian matrix J of the $\vec{\mathcal{F}}(\vec{U})$ term in equation 18. We assume that this $N \times N$ Jacobian matrix has a complete eigensystem consisting of eigenvalues , $\lambda^p(\vec{U})$, left eigenvectors, $\vec{L}^p(\vec{U})$, and right eigenvectors, $\vec{R}^p(\vec{U})$, for $p = 1, \ldots, N$ that satisfy inversion and diagonalization relations 23 and 24.

At a specific point $x_{i_0+\frac{1}{2}}$ midway between two grid nodes, we wish to find the numerical flux function $\vec{F}_{i_0+\frac{1}{2}}$. We evaluate the eigensystem at $\vec{U}_{i_0+\frac{1}{2}}$. Our method for approximating the value of $\vec{U}_{i_0+\frac{1}{2}}$ for use in the left sided and right sided Jacobian evaluations is explained in section 7.1.

In the p-th characteristic field we have an eigenvalue $\lambda^p(\vec{U}_{i_0+\frac{1}{2}})$, left eigenvector $\vec{L}^p(\vec{U}_{i_0+\frac{1}{2}})$, and right eigenvector $\vec{R}^p(\vec{U}_{i_0+\frac{1}{2}})$. We put $\vec{U}$ values and $\vec{\mathcal{F}}(\vec{U})$ values into the p-th characteristic field by taking the dot product with the left eigenvector,

$$u = \vec{L}^p(\vec{U}_{i_0+\frac{1}{2}}) \cdot \vec{U} \tag{42}$$

$$f(u) = \vec{L}^p(\vec{U}_{i_0+\frac{1}{2}}) \cdot \vec{\mathcal{F}}(\vec{U}) \tag{43}$$

where u and $f(u)$ are scalars. Once in the characteristic field we perform a scalar version of ENO, obtaining a scalar numerical flux function $F_{i_0+\frac{1}{2}}$ in the scalar field. We take this flux out of the characteristic field by multiplying with the right eigenvector,

$$\vec{F}^p_{i_0+\frac{1}{2}} = F_{i_0+\frac{1}{2}} \vec{R}^p(\vec{U}_{i_0+\frac{1}{2}}) \tag{44}$$

where $\vec{F}^p_{i_0+\frac{1}{2}}$ is the portion of the numerical flux function $\vec{F}_{i_0+\frac{1}{2}}$ from the p-th field.

Once we have evaluated the contribution to the numerical flux function from each field, we get the total numerical flux from summing the contributions from each field,

$$\vec{F}_{i_0+\frac{1}{2}} = \sum_p \vec{F}^p_{i_0+\frac{1}{2}} \tag{45}$$

6.2 Finite Difference ENO for Scalar Equations

Once the system has been reduced to independent scalar conservation equations, we need only develop ENO in this simple setting.

ENO is performed in each scalar characteristic field, on the scalar equation

$$u_t + f(u)_x = 0 \tag{46}$$

where u and $f(u)$ come from equations 42 and 43 respectively.

We define the numerical flux function F through the relation

$$f(u_i)_x = \frac{F_{i+\frac{1}{2}} - F_{i-\frac{1}{2}}}{\Delta x} \tag{47}$$

where the $F_{i\pm\frac{1}{2}}$ are the values of the numerical flux function at the cell walls. To obtain a convenient algorithm for computing this numerical flux function, we proceed as follows: define $h(x)$ implicitly through the following equation,

$$f(u(x)) = \frac{1}{\Delta x} \int_{x-\frac{\Delta x}{2}}^{x+\frac{\Delta x}{2}} h(y)dy \tag{48}$$

taking a derivative on both sides of equation 48 yields,

$$f(u(x))_x = \frac{h(x+\frac{\Delta x}{2}) - h(x-\frac{\Delta x}{2})}{\Delta x} \tag{49}$$

which shows that h is identical to the numerical flux function F at the cell walls. That is $F_{i\pm\frac{1}{2}} = h(x_{i\pm\frac{1}{2}})$ for all i. We can in turn calculate h by finding its primitive

$$H(x) = \int_{x_{-\frac{1}{2}}}^{x} h(y)dy \tag{50}$$

and then taking a derivative. From relation 48, it turns out that $f(u(x_i))$ provides values for the first divided differences of H on the grid, which allows us to accurately and efficiently interpolate the derivative of H to any other necessary points.

We will calculate H at the cell walls with polynomial interpolation. Our goal is to calculate $h = H'$, so we do not need the zeroth order divided differences of H that vanish with the derivative. The zeroth order divided differences, $D^0_{i+\frac{1}{2}}$, and all higher order *even* divided differences of H exist at the cell walls and will have the subscript $i \pm \frac{1}{2}$. The first order divided differences D^1_i and all higher order *odd* divided differences of H exist at the grid points and will have the subscript i.

The first order divided differences of H are,

$$D^1_i H = \frac{H(x_{i+\frac{1}{2}}) - H(x_{i-\frac{1}{2}})}{\Delta x} = f(u(x_i)) \tag{51}$$

where the second equality sign comes from

$$H(x_{i+\frac{1}{2}}) = \int_{x_{-\frac{1}{2}}}^{x_{i+\frac{1}{2}}} h(y)dy = \sum_{j=0}^{i} \left(\int_{x_{j-\frac{1}{2}}}^{x_{j+\frac{1}{2}}} h(y)dy \right) = \Delta x \sum_{j=0}^{i} f(u(x_j)) \tag{52}$$

which follows from equations 50 and 48. The higher divided differences are,

$$D^2_{i+\frac{1}{2}} H = \frac{f(u(x_{i+1})) - f(u(x_i))}{2\Delta x} = \frac{1}{2} D^1_{i+\frac{1}{2}} f \tag{53}$$

$$D^3_i H = \frac{1}{3} D^2_i f \tag{54}$$

and they continue in that manner.

According to the rules of polynomial interpolation, we can take any path along the divided difference table to construct H, although they do not all give good results. ENO reconstruction consists of two important features:

1. Choose $D^1_i H$ in the upwind direction.
2. Choose higher order divided differences by taking the smaller in absolute value of the two possible choices.

Once we construct $H(x)$, we evaluate $H'(x_{i+\frac{1}{2}})$ to get the numerical flux $F_{i+\frac{1}{2}}$.

It is important to note that there are other ways to choose the higher order divided differences. For example, in step 2 one can bias the decision towards the more central divided difference, which lowers the truncation error of the scheme in smooth regions.

6.3 The ENO-Roe Discretization (Third Order Accurate)

For a specific cell wall, located at $x_{i_0+\frac{1}{2}}$, we find the associated numerical flux function $F_{i_0+\frac{1}{2}}$ as follows:

If $\lambda^p(\vec{U}_{i_0+\frac{1}{2}}) > 0$, then $k = i_0$. Otherwise, set $k = i_0 + 1$. Define

$$Q_1(x) = (D_k^1 H)(x - x_{i_0+\frac{1}{2}}) \tag{55}$$

If $|D^2_{k-\frac{1}{2}}H| \le |D^2_{k+\frac{1}{2}}H|$, then $c = D^2_{k-\frac{1}{2}}H$ and $k^\star = k - 1$. Otherwise, $c = D^2_{k+\frac{1}{2}}H$ and $k^\star = k$. Define

$$Q_2(x) = c(x - x_{k-\frac{1}{2}})(x - x_{k+\frac{1}{2}}) \tag{56}$$

If $|D^3_{k^\star}H| \le |D^3_{k^\star+1}H|$, then $c^\star = D^3_{k^\star}H$. Otherwise, $c^\star = D^3_{k^\star+1}H$. Define

$$Q_3(x) = c^\star(x - x_{k^\star-\frac{1}{2}})(x - x_{k^\star+\frac{1}{2}})(x - x_{k^\star+\frac{3}{2}}) \tag{57}$$

Then,

$$F_{i_0+\frac{1}{2}} = H'(x_{i_0+\frac{1}{2}}) = Q_1'(x_{i_0+\frac{1}{2}}) + Q_2'(x_{i_0+\frac{1}{2}}) + Q_3'(x_{i_0+\frac{1}{2}}) \tag{58}$$

which simplifies to

$$F_{i_0+\frac{1}{2}} = D_k^1 H + c\,(2(i_0 - k) + 1)\,\Delta x + c^\star\left(3(i_0 - k^\star)^2 - 1\right)(\Delta x)^2 \tag{59}$$

by using equations 55, 56, and 57.

6.4 The Entropy Fix

The ENO-Roe discretization can admit entropy violating expansion shocks near sonic points. That is, at a place where a characteristic velocity changes sign—a "sonic point"—it is possible to have a stationary "expansion shock" solution with a discontinuous jump in value. If this jump were smoothed out even slightly, it would break up into an expansion "fan" (i.e. rarefaction) and dissipate, which is the desired physical solution. For a specific cell wall, $x_{i_0+\frac{1}{2}}$, if there are no "nearby" sonic points, then we use ENO-Roe. Otherwise, we add *high order* dissipation to our calculation of $F_{i_0+\frac{1}{2}}$ which is extremely small if the solution is locally smooth, but is large enough to break up an expansion shock. We explain when a sonic point is considered "nearby" in the next section. This approach retains a uniformly high order accurate scheme in smooth regions, and eliminates any "entropy violating" expansion shocks.

Consider two primitive functions H^+ and H^-. We compute a divided difference table for each of them. Their first divided differences are,

$$D_i^1 H^\pm = \frac{1}{2} f(u_i) \pm \frac{1}{2}\alpha_{i_0+\frac{1}{2}} u_i \tag{60}$$

where $\alpha_{i_0+\frac{1}{2}}$ is defined in section 7.4. We define the second divided differences $D^2_{i+\frac{1}{2}}H^{\pm}$ and the third divided differences $D^3_i H^{\pm}$ in the standard way, like those of H.

For H^+, set $k = i_0$. Then, replacing H with H^+ everywhere, define $Q_1(x)$, $Q_2(x)$, $Q_3(x)$, and finally $F^+_{i_0+\frac{1}{2}}$ by using the algorithm above. For H^-, set $k = i_0 + 1$. Then, replacing H with H^- everywhere, define $Q_1(x)$, $Q_2(x)$, $Q_3(x)$, and finally $F^-_{i_0+\frac{1}{2}}$ by using the algorithm above. Then,

$$F_{i_0+\frac{1}{2}} = F^+_{i_0+\frac{1}{2}} + F^-_{i_0+\frac{1}{2}} \tag{61}$$

is the new numerical flux function with added high order dissipation.

7 Marquina's Flux Splitting

7.1 Finding $\vec{U}_{i_0+\frac{1}{2}}$ (Third Order Accurate Algorithm)

We will need two approximations for the value of $\vec{U}$ at the cell wall $x_{i_0+\frac{1}{2}}$. The value from the left, $\vec{U}^L_{i_0+\frac{1}{2}}$, is interpolated from the x_{i_0} side. The value from the right, $\vec{U}^R_{i_0+\frac{1}{2}}$, is interpolated from the x_{i_0+1} side.

We need the divided differences of $\vec{U}$ for polynomial interpolation. This takes no extra work, since we have already computed the divided difference of $\vec{U}$ for use in the entropy fix portion of the ENO algorithm. The interpolation is done for each conserved variable, v, and we phrase the algorithm in terms of v. The divided differences of v are: $D^0_i v = v_i$, $D^1_{i+\frac{1}{2}}v$, and $D^2_i v$. For each conserved variable, v, we choose the zeroth order divided difference in the left or right direction based on whether we are looking for v^L or v^R. Then, we choose the higher order divided differences by taking the smaller in absolute value of the two possible choices.

For a specific cell wall, located at $x_{i_0+\frac{1}{2}}$, we find the approximations to $v^L_{i_0+\frac{1}{2}}$ and $v^R_{i_0+\frac{1}{2}}$ as follows:

If we are looking for $v^L_{i_0+\frac{1}{2}}$, then $k = i_0$. If we are looking for $v^R_{i_0+\frac{1}{2}}$, then $k = i_0 + 1$. Define

$$Q_0(x) = D^0_k v = v_k \tag{62}$$

If $|D^1_{k-\frac{1}{2}}v| \leq |D^1_{k+\frac{1}{2}}v|$, then $c = D^1_{k-\frac{1}{2}}v$ and $k^\star = k - 1$. Otherwise, $c = D^1_{k+\frac{1}{2}}v$ and $k^\star = k$. Define

$$Q_1(x) = c(x - x_k) \tag{63}$$

If $|D^2_{k^\star}v| \leq |D^2_{k^\star+1}v|$, then $c^\star = D^2_{k^\star}v$. Otherwise, $c^\star = D^2_{k^\star+1}v$. Define

$$Q_2(x) = c^\star(x - x_{k^\star})(x - x_{k^\star+1}) \tag{64}$$

Then,

$$v_{i_0+\frac{1}{2}} = Q_0(x_{i_0+\frac{1}{2}}) + Q_1(x_{i_0+\frac{1}{2}}) + Q_2(x_{i_0+\frac{1}{2}}) \tag{65}$$

which simplifies to one of the following

$$v^L_{i_0+\frac{1}{2}} = v_{i_0} + \frac{c\triangle x}{2} + c^\star\left((i_0 - k^\star)^2 - \frac{1}{4}\right)(\triangle x)^2 \tag{66}$$

$$v^R_{i_0+\frac{1}{2}} = v_{i_0+1} - \frac{c\triangle x}{2} + c^\star\left((i_0 - k^\star)^2 - \frac{1}{4}\right)(\triangle x)^2 \tag{67}$$

depending on which one was being computed.

7.2 Constructing Marquina's Left and Right Jacobians

Consider a cell wall, $x_{i_0+\frac{1}{2}}$, where we wish to calculate the numerical flux function $\vec{F}_{i_0+\frac{1}{2}}$. We have two estimates for $\vec{U}_{i_0+\frac{1}{2}}$: one estimate from the left, $\vec{U}^L_{i_0+\frac{1}{2}}$, and one estimate from the right, $\vec{U}^R_{i_0+\frac{1}{2}}$. We use these estimates to compute two Jacobian matrices,

$$J^L = J(\vec{U}^L_{i_0+\frac{1}{2}}), \qquad J^R = J(\vec{U}^R_{i_0+\frac{1}{2}}) \tag{68}$$

and their associated eigensystems.

In the p-th characteristic field, we have an eigenvalue, left eigenvector, and right eigenvector for each of the two Jacobians:

$$(\lambda^p)^L = \lambda^p(\vec{U}^L_{i_0+\frac{1}{2}}), \qquad (\lambda^p)^R = \lambda^p(\vec{U}^R_{i_0+\frac{1}{2}}) \tag{69}$$

$$(\vec{L}^p)^L = \vec{L}^p(\vec{U}^L_{i_0+\frac{1}{2}}), \qquad (\vec{L}^p)^R = \vec{L}^p(\vec{U}^R_{i_0+\frac{1}{2}}) \tag{70}$$

$$(\vec{R}^p)^L = \vec{R}^p(\vec{U}^L_{i_0+\frac{1}{2}}), \qquad (\vec{R}^p)^R = \vec{R}^p(\vec{U}^R_{i_0+\frac{1}{2}}) \tag{71}$$

7.3 Constructing Left and Right Numerical Flux Functions

Consider a cell wall, $x_{i_0+\frac{1}{2}}$, where we wish to find the numerical flux in the p-th characteristic field: $\vec{F}^p_{i_0+\frac{1}{2}}$. For each of the two Jacobians, we find a numerical flux in the p-th characteristic field: $(\vec{F}^p_{i_0+\frac{1}{2}})^L$ and $(\vec{F}^p_{i_0+\frac{1}{2}})^R$. Then we sum them,

$$\vec{F}^p_{i_0+\frac{1}{2}} = (\vec{F}^p_{i_0+\frac{1}{2}})^L + (\vec{F}^p_{i_0+\frac{1}{2}})^R \tag{72}$$

to get the total numerical flux in the p-th field. This is done for each field, and then the total numerical flux is defined by equation 45.

7.4 Constructing the ENO Numerical Scheme

Consider a cell wall, $x_{i_0+\frac{1}{2}}$. If the left and right eigenvalues evaluated at this cell wall agree on the upwind direction then there is no sonic point "nearby", and we use the ENO-Roe discretization. If the eigenvalues disagree at the cell wall, then there is a sonic point "nearby", and we use the version of ENO with the entropy fix. There are 3 cases:

1. If $(\lambda^p)^L > 0$ and $(\lambda^p)^R > 0$, then upwind is from the left. We calculate $(\vec{F}^p_{i_0+\frac{1}{2}})^L$ using ENO-Roe. We set $(\vec{F}^p_{i_0+\frac{1}{2}})^R = 0$.

2. If $(\lambda^p)^L < 0$ and $(\lambda^p)^R < 0$, then upwind is from the right. We calculate $(\vec{F}^p_{i_0+\frac{1}{2}})^R$ using ENO-Roe. We set $(\vec{F}^p_{i_0+\frac{1}{2}})^L = 0$.

3. If $(\lambda^p)^L(\lambda^p)^R \leq 0$, then the eigenvalues disagree. We use the entropy fix version of ENO. For this, we define

$$\alpha_{i_0+\frac{1}{2}} = \max\left(|(\lambda^p)^L|, |(\lambda^p)^R|\right) \qquad (73)$$

as our dissipation coefficient. In the evaluation of $(\vec{F}^p_{i_0+\frac{1}{2}})^L$, we evaluate $F^+_{i_0+\frac{1}{2}}$ normally, but set $F^-_{i_0+\frac{1}{2}} = 0$. Thus, equation 61 becomes $F_{i_0+\frac{1}{2}} = F^+_{i_0+\frac{1}{2}}$. In the evaluation of $(\vec{F}^p_{i_0+\frac{1}{2}})^R$, we evaluate $F^-_{i_0+\frac{1}{2}}$ normally, but set $F^+_{i_0+\frac{1}{2}} = 0$. Thus, equation 61 becomes $F_{i_0+\frac{1}{2}} = F^-_{i_0+\frac{1}{2}}$.

This completes the description of the finite difference ENO discretization using Marquina's Jacobians.

8 Examples

8.1 Example 1: Reflecting Shock in a Thermally Perfect Gas

We are currently developing numerical methods for treating an interface separating a liquid drop and a high speed gas flow. The droplet is an incompressible Navier-Stokes fluid. The gas is a compressible, multi-species, chemically reactive Navier-Stokes fluid. A level set is used for domain decomposition. (This research will be described in detail in a future UCLA CAM report.)

In this example, a 1D "Sod" shock tube was set up in the middle of the domain, with the generated shock moving from left to right. The water droplet is off to the right hand side of the domain. The shock hits the water droplet, reflects off in the opposite direction, and proceeds toward the contact discontinuity. We implement standard 3rd order ENO with the Jacobian matrix evaluated at the linear average of the points adjacent to the flux. This is a second order accurate, central approximation to the Jacobian. Using standard finite difference ENO, there is a great deal of "noise" generated when the shock approaches the contact discontinuity, after reflection off the water droplet. See Figure 1. Note, however, that standard 2nd order ENO (which is a TVD scheme) with the Jacobian matrix evaluated at the linear average does not generate much noise at all.

We run the same problem with 3rd order ENO, but this time we used Marquina's Jacobian evaluated with 3rd order accurate left side and right side biased approximations to the conserved variables. There is no significant noise. See Figure 2.

(Note that the actual values for the density and the pressure of the water droplet are not shown. We use "place holder" values in the figures. However, the values for the velocity and the temperature are unaltered.)

8.2 Example 2: Importance of High Order Accurate Jacobians

We emphasize that it is important to use Marquina's Jacobian with a high order accurate approximation to the conserved variables at the cell walls. To illustrate this, consider the previous problem with 3rd order ENO. The Jacobian is evaluated with 1st, 2nd, and 3rd order accurate approximations to the conserved variables. The results are shown in Figures 3, 4, and 5 respectively. Note that all the ENO algorithms are 3rd order, only the approximations to the conserved variables for the left and right Jacobian evaluations vary in order. Based on these results, we recommend using 3rd order ENO with Marquina's Jacobian also evaluated to 3rd order.

8.3 Further Examples

See [1, 9] for more numerical examples using Marquina's Jacobian to fix a variety of spurious oscillatory effects.

9 Conclusions

ENO methods are a class of high accuracy, shock capturing numerical methods for general hyperbolic systems of conservation laws. They are based on using upwind biased interpolations in the characteristic fields without interpolating across steep gradients in the flow.

The finite difference formulation of the ENO method allows an efficient and convenient implementation that readily applies to any number of spatial dimensions.

This method works well on a great variety of gas dynamics problems, as well as other convective problems, but there are still special circumstances in which spurious oscillations develop.

Based on recent work, we have identified the source of these oscillations as the centered linear average interpolation used to evaluate the Jacobian and eigensystem of the convective flux at the midpoints between nodes, prior to transforming to characteristic fields. This effect can be understood intuitively as well, in terms of unintentionally performing downwind differencing of the true characteristic fields near steep gradients.

Marquina recently devised a way to make use of left side and right side Jacobians at the midpoint, without the need to construct a single Jacobian. The general technique seems to fix all known cases in which serious spurious oscillations have occurred.

We presented a detailed description of the preferred (third order accurate in space and time) finite difference ENO scheme using Marquina's Jacobian evaluation procedure, so that others can readily make use of this (pen)ultimate scheme.

We presented examples demonstrating that this approach fixes a large, non-physical oscillation in a complicated gas dynamics problem. We also showed that it is important to evaluate the Jacobian and eigensystem to high order accuracy from the left and from the right at the midpoint, as this has a large impact on the practical resolution of the scheme. This is contrary to what one would naively expect, since the formal order of accuracy of the scheme is unchanged by the Jacobian evaluation strategy.

More analysis is required to understand why this two sided approach works so well, and why it has such a large effect on resolution without altering the formal order of accuracy. For now, however, it does seem to allow a robust, general, accurate, parameter-free ENO scheme which we expect will have wide application for problems which include a hyperbolic system.

10 Acknowledgements

We dedicate this paper to the memory of Ami Harten whose creativity and personality inspired everyone in the field.

Research of the first, second and fourth authors supported in part by ARPA URI-ONR-N00014-92-J-1890, NSF #DMS 94-04942, and ARO DAAH04-95-1-0155. Research of the third author supported in part by a University of Valencia grant and DGYCIT PB94-0987.

References

[1] Donat, R., and Marquina, A., *Capturing Shock Reflections: An Improved Flux Formula*, J. Comput. Phys. 25, 42-58 (1996).

[2] Fedkiw, R., Merriman, B., and Osher, S., *Numerical Methods for a Mixture of Thermally Perfect and/or Calorically Perfect Gaseous Species with Chemical Reactions*, J. Comput. Phys. 132, 175-190 (1997).

[3] B. van Leer, *Towards the Ultimate Difference Scheme I. The Quest for Monotonicity*, Springer Lecture Notes in Physics 18, 163-168 (1973).

[4] B. van Leer, *Towards the Ultimate Difference Scheme II. Monotonicity and Conservation Combined in a Second Order Scheme*, J. Comput. Phys. 14, 361-370 (1974).

[5] B. van Leer, *Towards the Ultimate Difference Scheme III. Upstream-Centered Finite Difference Schemes for Ideal Compressible Flow*, J. Comput. Phys. 23, 263-275 (1977).

[6] B. van Leer, *Towards the Ultimate Difference Scheme IV. A New Approach to Numerical Convection*, J. Comput. Phys. 23, 276-299 (1977).

[7] B. van Leer, *Towards the Ultimate Difference Scheme V. A Second Order Sequel to Gudonov's Method*, J. Comput. Phys. 32, 101-136 (1979).

[8] Randall J. LeVeque, *Numerical Methods for Conservation Laws*, Birhauser Verlag, Boston, USA. 1992. ISBN 3-8176-2723-5.

[9] Marquina, A., and Donat, R., *Capturing Shock Reflections: A Nonlinear Local Characteristic Approach*, UCLA CAM Report 93-31, April 1993.

[10] Shu, C.W., *Numerical experiments on the accuracy of ENO and modified ENO schemes*, J. Sci. Comput. 5, 127-149 (1990).

[11] Shu, C.W. and Osher, S., *Efficient Implementation of Essentially Non-Oscillatory Shock Capturing Schemes*, J. Comput. Phys. 77, 429-471 (1988).

[12] Shu, C.W. and Osher, S., *Efficient Implementation of Essentially Non-Oscillatory Shock Capturing Schemes II (two)*, J. Comput. Phys. 83, 32-78 (1989).

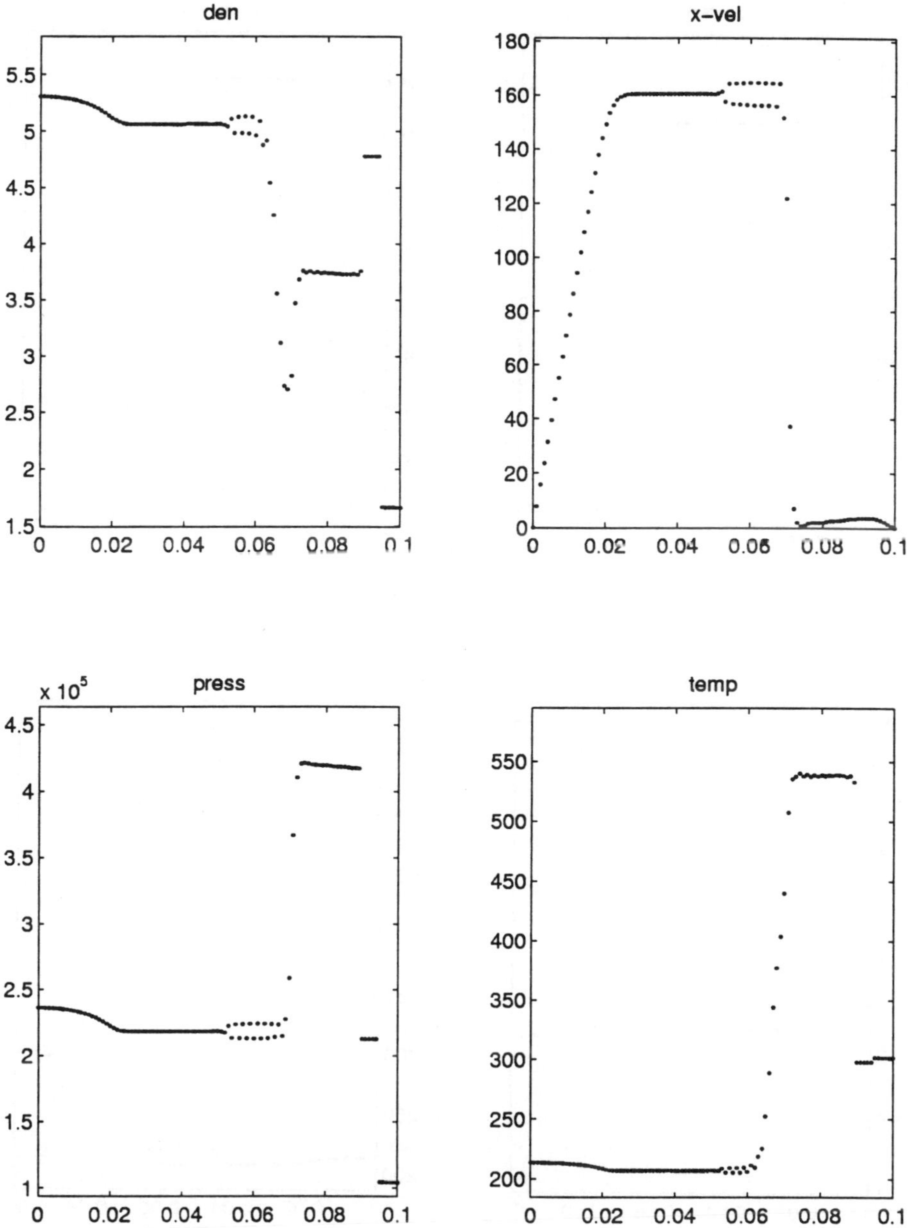

Figure 1: 3rd order ENO, Jacobian matrix evaluated at the linear average. Note the large spurious oscillations near $x = 0.06$.

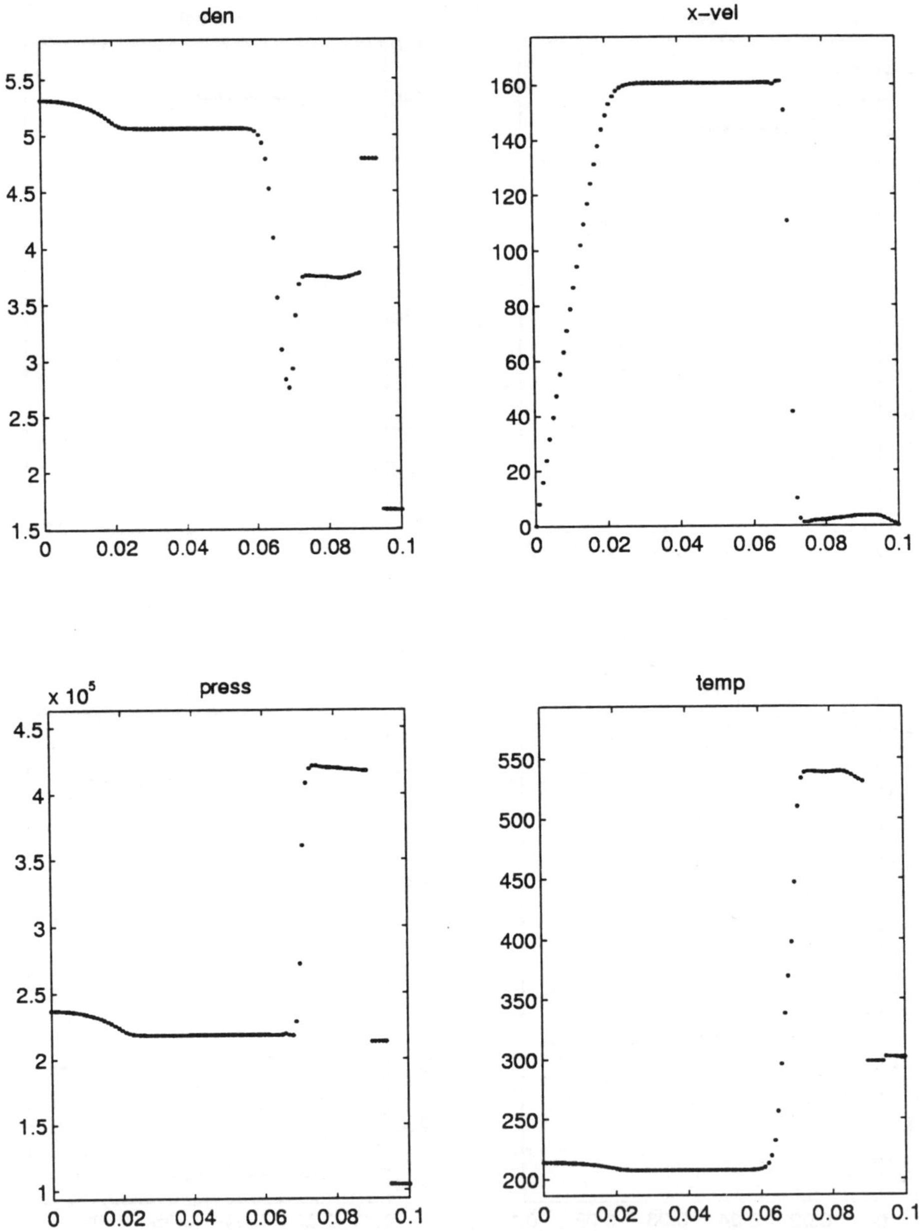

Figure 2: 3rd order ENO, 3rd order Marquina's Jacobian. The spurious oscillations of ENO are eliminated.

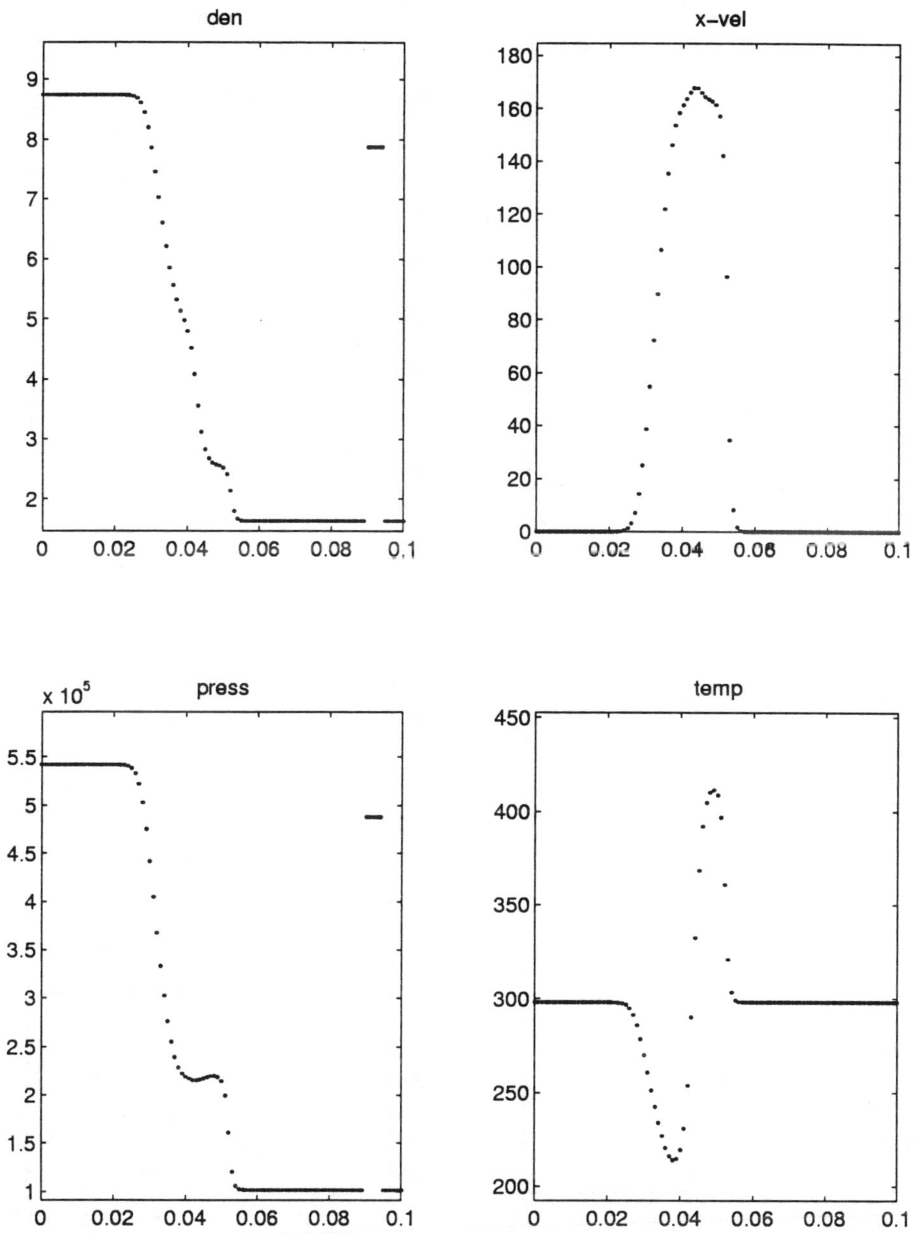

Figure 3: 3rd order ENO, 1st order Marquina's Jacobian. Note the smoothed out features, particularly near $x = 0.04$, due only to the low accuracy of the Jacobian evaluation.

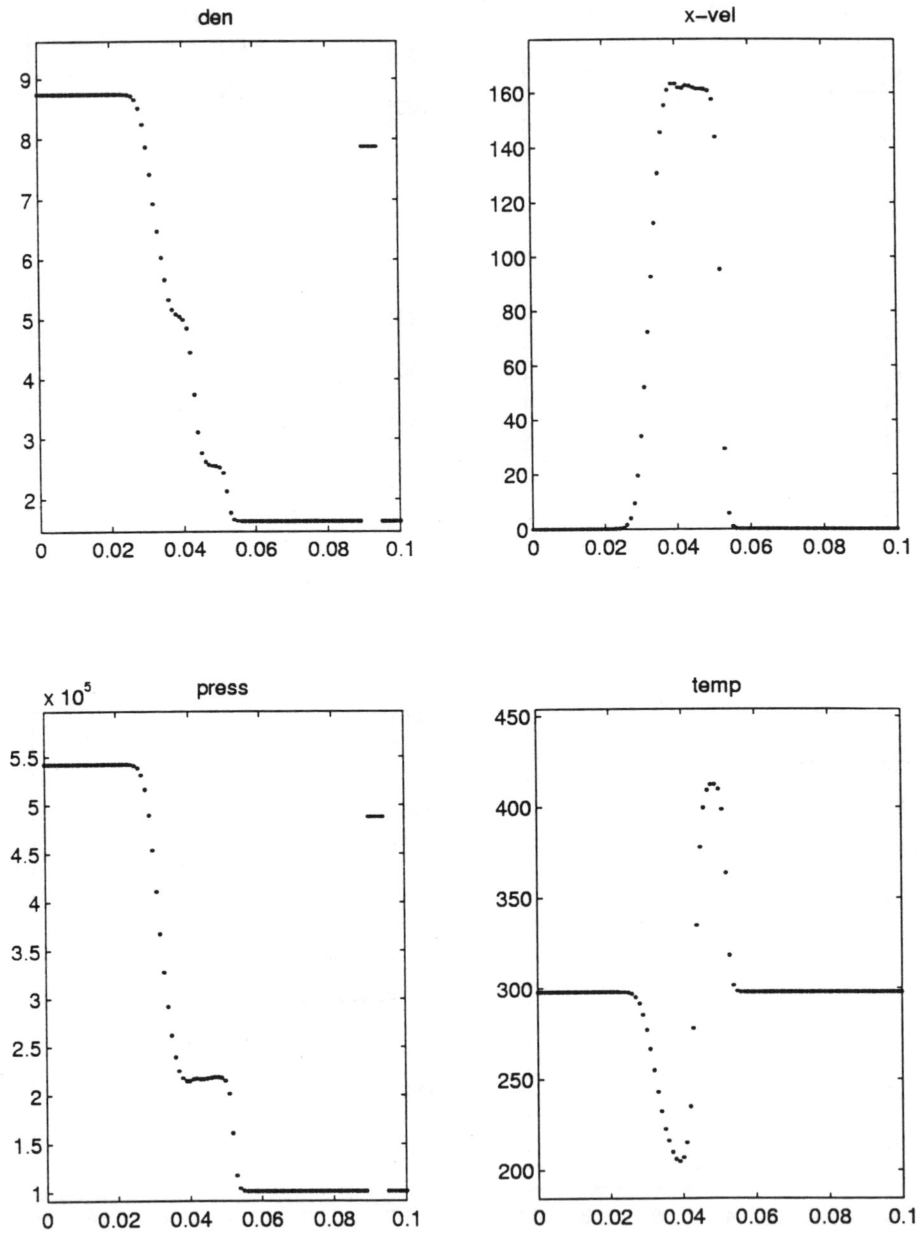

Figure 4: 3rd order ENO, 2nd order Marquina's Jacobian. The features at $x = 0.04$ are sharpened as the Jacobian accuracy is increased.

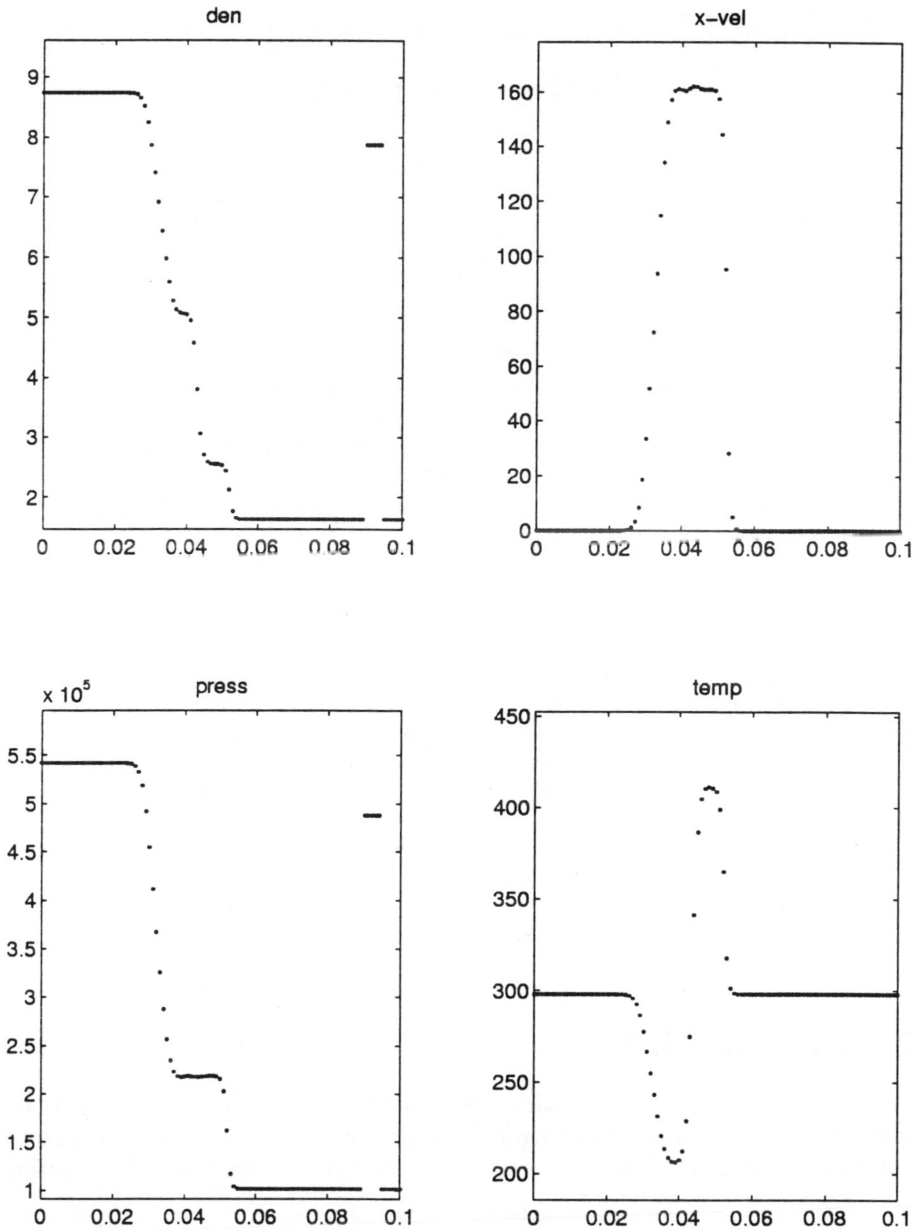

Figure 5: 3rd order ENO, 3rd order Marquina's Jacobian. The features at $x = 0.04$ are now well resolved, due only to the high accuracy of the Jacobian evaluation.

A Finite Element Based Level-Set Method for Multiphase Flows

Björn Engquist * Anna-Karin Tornberg †

April 29, 1999

Abstract

A numerical method based on a level-set formulation for incompressible two-dimensional multiphase flow is presented. A finite element discretization is used, and the method is designed to handle specific features of this problem, such as surface tension forces acting at the interfaces separating two immiscible fluids, as well as the density and viscosity jumps that in general occur across such interfaces. The technique can also be applied to other problems for which methods of shock capturing types are less suitable, as for example, certain simulations of combustions and passive advection.

There are advantages of the finite element method for this problem inherent from the weak formulation of the Navier-Stokes equations. Differentiation of the discontinuous viscosity is avoided and the singular surface tension forces are included in the formulation through the evaluation of an easily approximated line integral along the interface. New methods for handling the discontinuous properties in the finite element integrals are introduced.

Numerical tests are presented. For the case of a rising buoyant bubble the computations are briefly compared to results from a front-tracking method and a new method based on a segment projection technique. Simulations with topology changes, such as merging of bubbles, are presented for the level-set method.

1 Introduction

Very often in fluid flow simulations the fluid properties change rapidly at interfaces. Typical examples are strong shocks, vortex sheets, combustion fronts and multiphase flow. The ideal numerical method is here of shock capturing

*email: Mathematics Department, UCLA, Los Angeles, California 90095-1555 and Department of Numerical Analysis and Computing Science, Royal Institute of Technology, S-100 44 Stockholm, SWEDEN, email: engquist@nada.kth.se; Research supported by the Competence Center PSCI and the NSF grant DMS97-06827

†Department of Numerical Analysis and Computing Science, Royal Institute of Technology, S-100 44 Stockholm, SWEDEN, email: annak@nada.kth.se

type. Nothing special needs to be done at the interface. The algorithm adjusts nonlinearly at the interface and captures the changes in fluid properties over a few grid points or cells. Sometimes it is, however, preferable to locate the interface more precisely. Traditionally this has been done by points following the interface, as in front-tracking.

We shall here consider a variant of the level-set method. The level-set method was introduced by Osher and Sethian in [6] and this method is similar to shock-capturing in many aspects. The physical application we shall consider is multiphase flow.

When designing a method for calculations of multiphase incompressible flow, particular difficulties are present. These difficulties are effects of the internal boundaries, or interfaces, separating two immiscible fluids. The two fluids will in general have different densities and viscosities and these physical quantities will therefore have a jump in value across interfaces. In addition, surface tension forces act at the interfaces, with the strength directly defined by the interface shape. Any method designed to perform multiphase flow calculations must therefore include an accurate description of the moving and deforming interfaces.

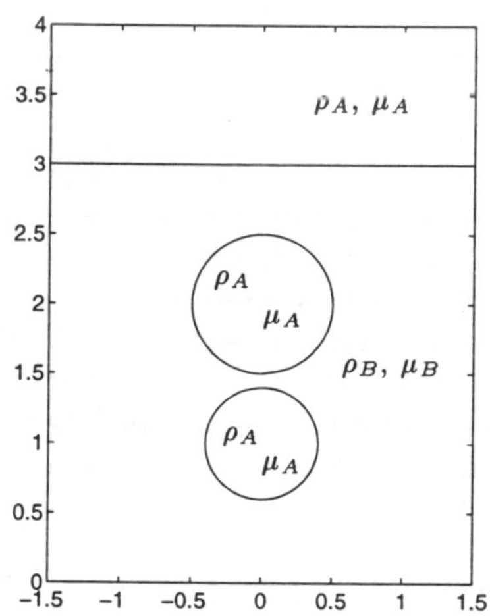

Figure 1: Example of a configuration involving two fluids A and B, with density and viscosity (ρ_A, μ_A) and (ρ_B, μ_B).

1.1 Background

Most methods designed for multiphase flow calculations are based on algorithms where the background mesh/grid is kept fixed, and the internal boundaries are represented by supplying and continuously updating some additional information. An early method of this type was the Marker in Cell (MAC) method [4]. In this method, a fixed number of discrete Lagrangian particles inserted are advected by the local flow. The distribution of these particles identifies the regions occupied by a certain fluid. In volume-tracking or Volume of Fluid (VOF) methods, a fractional volume function is defined to indicate the volume fraction of a certain fluid in each grid cell. An update on such methods, together with a comparison between some of them is presented in [8].

Another method is the front-tracking method, where separate data structures are introduced to represent the interfaces. The basic idea of representing internal boundaries with separate data structures was given by Peskin [7], as he applied his immersed boundary method to calculations of blood flow in the heart. Unverdi and Tryggvason [15] applied this idea in their front-tracking method designed to simulate two-phase flow, or more specifically the motion of bubbles in a surrounding fluid.

The level-set method was first introduced by Osher and Sethian [6]. This method has been further developed for use in many different applications, and its application to multiphase flow calculations has been described by Sussman et al. in [11] and [10]. The basic idea of the level-set method is to represent the interfaces separating two fluids A and B simply as the zero level sets of a continuous function, designed to be of one sign in fluid A, and of opposite sign in fluid B. The level-set function is initialized as a signed distance function, carrying information about the closest distance to an interface. As the level set function is advected by the flow, this property will not be retained, and reinitialization is applied [11].

The main advantage of the level-set methodology compared to the front-tracking methodology is in dealing with simulations in which topology changes occur. When for example merging or splitting of bubbles or drops takes place, the front-tracking method cannot be used without explicit treatment of the connection and splitting of interface data structures. With the level-set method, the distance function can represent an arbitrary (limited by resolution only) number of bubble or drop interfaces, and a topology change is only seen as a change in the values of this function, causing a different pattern for the zero level sets.

1.2 The Present Work

In this paper, a numerical method based on a level-set formulation for incompressible two-dimensional multiphase flow is presented. The discretization is made by a finite element technique. The method handles surface tension forces acting at the interfaces separating the two fluids, as well as the density and viscosity jumps that in general occur across these interfaces.

Both the front-tracking method in [15] and the level-set method in [11] was discretized using finite difference techniques. In both cases, surface tension forces are smoothed by the use of a mollified delta function, as described by Peskin [7]. No such smoothing and explicit discretization of the delta function is needed with our finite element discretization. The weak formulation of the equations includes the singular surface tension forces through the evaluation of a well-defined line integral along the interface (the analogue in three dimensions would be a surface integral). Further, the differential form of the equations includes derivatives of the discontinuous viscosity. In the weak formulation, by using Green's formula, this derivative has been moved over to the test function. It is also advantageous to be able to use variable spatial resolution. The finite element technique is also well suited to perform simulations on domains of various geometrical shapes.

Motivated by the presence of the discontinuous density and viscosity in the integrals in the variational formulation, the errors associated with the evaluation of integrals of discontinuous functions have been analyzed. The discontinuous function is replaced by a smooth approximation before the integrals are evaluated. The error of the integration will consist of two parts, firstly the analytical error made when replacing a discontinuous function by a smooth approximation, and secondly the numerical error from the integration of this smooth function. These errors are analyzed, and we show that vanishing moments of a certain error function are needed to obtain a small analytical error. Compare to the use of vanishing moments in [2], in the construction of approximations to the Dirac δ-function. The regularity of the smooth approximation is shown to be critical for the numerical error.

The outline of the paper is as follows: In Section 2, the problem is formulated, and the discretization of the Navier-Stokes equations is briefly discussed. In the following section, the formulation and discretization of the level-set method are discussed. The errors associated with the evaluation of integrals of discontinuous functions are analyzed in Section 4. In Section 5, a simulation including topology changes is presented. In Section 6, a front-tracking method and the new segment projection method are described, and a brief comparison between the different methods is done.

2 The Navier-Stokes Equations

The equations describing this immiscible multiphase flow are essentially the Navier-Stokes equations for incompressible flow. The contribution of the surface tension forces is in addition to the gravity forces added as a source term.

In this presentation, we assume that we have two different fluids, fluid A and fluid B. The density and viscosity are given by

$$(\rho(\mathbf{x}), \mu(\mathbf{x})) = \begin{cases} (\rho_A, \mu_A) & \text{for } \mathbf{x} \text{ in fluid A,} \\ (\rho_B, \mu_B) & \text{for } \mathbf{x} \text{ in fluid B.} \end{cases} \tag{1}$$

In general $\rho_A \neq \rho_B$ and $\mu_A \neq \mu_B$, so that $\rho(\mathbf{x})$ and $\mu(\mathbf{x})$ are discontinuous at each interface separating fluid A and B. Refer to Figure 1 for an example of a configuration of the two fluids A and B. The Navier-Stokes equations can be written

$$\rho\,(\frac{\partial \mathbf{u}}{\partial t} + \mathbf{u}\cdot\nabla\mathbf{u}) = -\nabla p + \nabla\cdot(\mu\,(\nabla\mathbf{u} + \nabla\mathbf{u}^T)) + \mathbf{f} + \rho\mathbf{g} \qquad \text{in } \Omega \subset \mathbb{R}^2, \tag{2}$$

together with the divergence-free constraint and boundary conditions,

$$\nabla \cdot \mathbf{u} = 0 \qquad \text{in } \Omega \subset \mathbb{R}^2, \quad (3)$$

$$\mathbf{u} = v \qquad \text{on } \partial\Omega, \quad (4)$$

and some appropriate initial condition $\mathbf{u}(\mathbf{x},0) = \mathbf{u}_0(\mathbf{x})$. $\mathbf{u}(\mathbf{x}) : \mathbb{R}^2 \rightarrow \mathbb{R}^2$ denotes the velocity field and $p(\mathbf{x}) : \mathbb{R}^2 \rightarrow \mathbb{R}$ denotes the pressure field. Buoyancy effects arises from the source term $\rho\mathbf{g}$, where the gravitational force $\mathbf{g}$ is multiplied by the discontinuous $\rho(\mathbf{x})$. The source term $\mathbf{f}$ in the right hand side is the surface tension force. At any point along an interface, the direction of this force is towards the local center of curvature. Denote the union of all interfaces separating the two fluids by γ. In general, γ will consist of several separate segments, where each segment can either be closed or emerge from the boundary. For simplicity of description, we often refer to γ as one single segment. The surface tension force is given by

$$\mathbf{f} = \sigma\kappa\hat{\mathbf{n}}\delta_\gamma, \quad (5)$$

where $\sigma \in \mathbb{R}$ is the surface tension coefficient, $\kappa \in \mathbb{R}$ is the (local) curvature and $\hat{\mathbf{n}} \in \mathbb{R}^2$ is a normal vector to the interface γ. The product $\kappa\,\hat{\mathbf{n}}$ yields the direction of the force. Here, δ_γ is a measure of Dirac delta function type with support on γ. Its action on any smooth test function φ is given by

$$\int_\Omega \delta_\gamma\ \varphi\ d\Omega = \int_\gamma \varphi\ d\gamma, \quad (6)$$

where the last quantity denotes the line integral of φ along the interface γ.

In order to give a weak formulation of the equations, introduce the spaces

$$\mathsf{V} = \{\mathbf{v} \in H^1(\Omega)^2 : \mathbf{v} = \{v_i\}_{i=1}^2,\ v_i \in H^1(\Omega),\ \}, \quad (7)$$

$$\Pi = \{q \in L^2(\Omega) : \int_\Omega q\ d\mathbf{x} = 0\}, \quad (8)$$

where

$$H^1(\Omega) = \{v : v \text{ is defined on } \Omega \text{ and } \int_\Omega v^2 + |\nabla v|^2 d\mathbf{x} < \infty\}, \quad (9)$$

$$L^2(\Omega) = \{v : v \text{ is defined on } \Omega \text{ and } \int_\Omega v^2 d\mathbf{x} < \infty\}. \quad (10)$$

We define the subspaces

$$\mathsf{V}_\nu = \{\mathbf{v} \in \mathsf{V} : \mathbf{v} = \nu \text{ on } \partial\Omega\} \quad \text{and} \quad \mathsf{V}_0 = \{\mathbf{v} \in \mathsf{V} : \mathbf{v} = 0 \text{ on } \partial\Omega\}, \quad (11)$$

where ν is given by (4). Multiplying equation (2) by $\mathbf{v} \in \mathsf{V}_0$, (3) by $q \in \Pi$, and integrating over the domain using integration by parts, the following variational formulation of (2)-(4) is obtained:

Find $\mathbf{u}(\mathbf{x},t) \in V_\nu$ and $p(\mathbf{x},t) \in \Pi$ such that $\forall t \in [0,T]$

$$m(\rho, \frac{\partial \mathbf{u}}{\partial t}, \mathbf{v}) + \tilde{a}(\mu, \mathbf{u}, \mathbf{v}) + b(\mathbf{v}, p) + c(\rho, \mathbf{u}, \mathbf{u}, \mathbf{v}) = f_\gamma(\mathbf{v}) + m(\rho, \mathbf{g}, \mathbf{v}) \qquad \forall \mathbf{v} \in V_0, \tag{12}$$

$$b(\mathbf{u}, q) = 0 \qquad \forall q \in \Pi. \tag{13}$$

The form $\tilde{a}(\mu, \cdot, \cdot)$ is defined as

$$\tilde{a}(\mu, \mathbf{u}, \mathbf{v}) = \int_\Omega \mu \left\{ \mathrm{tr}(\nabla \mathbf{u}^T : \nabla \mathbf{v}) + \mathrm{tr}(\nabla \mathbf{u} : \nabla \mathbf{v}) \right\} d\mathbf{x} \tag{14}$$

where $\mathrm{tr}(\nabla \mathbf{u}^T : \nabla \mathbf{v})$ denotes the trace of the dyad product of $\nabla \mathbf{u}^T$ and $\nabla \mathbf{v}$, and similarly for $\mathrm{tr}(\nabla \mathbf{u} : \nabla \mathbf{v})$. Further, the forms $b(\cdot,\cdot)$, $c(\rho,\cdot,\cdot,\cdot)$ and $m(\rho,\cdot,\cdot)$ are defined by

$$b(\mathbf{v}, q) = -\int_\Omega (\nabla \cdot \mathbf{v}) q \, d\mathbf{x}, \tag{15}$$

$$c(\rho, \mathbf{u}, \mathbf{v}, \mathbf{w}) = \int_\Omega \rho (\mathbf{u} \cdot \nabla \mathbf{v}) \cdot \mathbf{w} \, d\mathbf{x}, \tag{16}$$

$$m(\rho, \mathbf{u}, \mathbf{v}) = \int_\Omega \rho \, \mathbf{u} \cdot \mathbf{v} \, d\mathbf{x}. \tag{17}$$

The interfacial force term $f_\gamma(\mathbf{v})$ evaluates as the following line integral along the interface γ,

$$f_\gamma(\mathbf{v}) = \sigma \int_\gamma \kappa \hat{\mathbf{n}} \cdot \mathbf{v} \, d\gamma, \tag{18}$$

according to definition (6). Contrary to the differential equation (2), the weak formulation of the equations does not include δ_γ. Therefore, advantageously, no explicit discretization of this two-dimensional delta function has to be made. The evaluation of the line integral (18) will be discussed in some detail in Section 3.2.

2.1 Discretization of the Navier-Stokes Equations

The time-stepping scheme used to advance the Navier-Stokes equations in time includes a combination of implicit and explicit terms and an iteration over the nonlinearity. The discretization of the time derivative is given by the backward difference formula:

$$\frac{\partial \mathbf{u}}{\partial t} = \frac{3\mathbf{u}^{n+1} - 4\mathbf{u}^n + \mathbf{u}^{n-1}}{2\Delta t} + O((\Delta t)^2), \tag{19}$$

resulting in an implicit approximation. See [12] for details. To resolve the nonlinearities, we introduce a fixed point iteration. After each iteration, the density and viscosity fields are updated by computing a new temporary bubble position. The iteration is terminated when the difference between two consecutive velocity fields is smaller than a chosen tolerance.

The resulting problem to solve in each iteration is a Stokes problem. The divergence-free constraint is enforced using an iterated penalty method (described by Brenner and Scott [3]). With this method, the problem is reformulated as an iterative procedure, where the divergence of the velocity field is removed as the pressure is obtained through iterative updates. The pressure is in this procedure not solved for explicitly, but can be computed from $\mathbf{w} \in V_h$ where $\mathbf{w}$ is obtained from the penalty iterations. The system of equations to be solved will therefore be smaller compared to methods requiring direct solution of the pressure, with the number of unknowns determined solely by the dimension of the approximation space for the velocity. The two velocity components are however coupled.

Piecewise quadratic polynomials have been used for the discretization of this problem. For the evaluation of the integrals in the variational formulation, each element is mapped onto a "reference" element τ, where a 13-point quadrature formula (cf. Strang and Fix [9], Section 4.3) is used.

The linear algebraic system obtained from the discretization is symmetric, and can be solved using a conjugate gradient method. For the system to converge in a feasible number of iterations, a good preconditioner is however needed. The straight-forward way to define the preconditioner used here is to assemble a matrix with $\rho(\mathbf{x}, t)$ and $\mu(\mathbf{x}, t)$ at one instant in time, factor it, and keep it as a preconditioner until the number of iterations needed for convergence exceeds a certain limit. At this point the process is repeated.

3 The Level-Set Method

In the level-set method, the interface is simply represented as the zero level set of a continuous function, designed to be of positive sign in fluid A, and of negative sign in fluid B. The level-set function ϕ is initialized as a signed distance function, carrying information about the closest distance to any interface separating the two fluids.

The level-set function ϕ is simply advected by the flow, when the velocity field $\mathbf{u}$ is known. The advection is given by

$$\frac{\partial \phi}{\partial t} + \mathbf{u} \cdot \nabla \phi = 0. \tag{20}$$

It has been shown [11] that it is critical that ϕ remain a distance function, in order to obtain acceptable conservation of mass. This property is however not preserved during advection. A reinitialization procedure is applied as part of the calculations in order to restore the level-set function as a distance function in regions close to a zero level set.

The following procedure was suggested by Sussman et al. [11]: Let $\psi_0 = \phi(t^*)$ be the level set function obtained in any time step. It will be somewhat distorted from a distance function, and to obtain a true distance function ψ that has the same zero level set as ψ_0, the following problem is solved to steady state:

$$\frac{\partial \psi}{\partial \tau} = S(\psi_0)\ (1 - |\nabla \psi|) \tag{21}$$

$$\psi(\mathbf{x}, 0) = \psi_0(\mathbf{x}), \tag{22}$$

where $S(t)$ is a sign function, given by

$$S(t) = \begin{cases} -1 & \text{for } t < 0, \\ 0 & \text{for } t = 0, \\ 1 & \text{for } t > 0, \end{cases} \tag{23}$$

Alternatively, equation (21) can be written as

$$\frac{\partial \psi}{\partial \tau} + \mathbf{w} \cdot \nabla \psi = S(\psi_0), \quad \mathbf{w} = S(\psi_0) \frac{\nabla \psi}{|\nabla \psi|}. \tag{24}$$

That is, $\mathbf{w}$ is a unit vector, always pointing away from the interface, and information is propagated out from the contour with a speed of one.

The equations (20) and (24) are discretized using piecewise quadratic polynomials, as the Navier-Stokes equations. In order to stabilize the equations, the streamline diffusion method has been used.

Since it is only of interest that we obtain a distance function in the neighborhood of the zero level set, the reinitialization need not to be done until a steady state is reached. After each advection step, we perform M reinitialization "time"-steps, in artificial time τ. Including the streamline diffusion modification, formulating an explicit first order method in time, this reads: Given $\mathbf{u} \in \mathsf{V}_h$ and $\phi^{n-1} = \phi(t_{n-1}) \in W_h$, find $\tilde{\phi}^n \in W_h$ such that

$$(\frac{\tilde{\phi}^n - \phi^{n-1}}{\Delta t}, v) + (\mathbf{u} \cdot \nabla \phi^{n-1}, v + \delta\ \mathbf{u} \cdot \nabla v) = 0 \qquad \forall v \in W_h. \tag{25}$$

Set $\psi^0 = \psi_0 = \tilde{\phi}^n$, and for $m = 1, ..M$, solve

$$\begin{aligned}(\frac{\psi^m - \psi^{m-1}}{\Delta \tau}, v) + (\epsilon \nabla \psi^m, \nabla v) =& -(\mathbf{w}^{m-1} \cdot \nabla \psi^{m-1}, v + \delta\ \mathbf{w}^{m-1} \cdot \nabla v) \\ &+ (S_\alpha(\psi), v + \delta\ \mathbf{w}^{m-1} \cdot \nabla v) \quad \forall v \in W_h \end{aligned} \tag{26}$$

Set $\phi^n = \psi^M$.

The parameter δ is chosen as

$$\delta = \frac{1}{2\sqrt{(\Delta t)^{-2} + |\mathbf{u}|^2 h^{-2}}} = \frac{1}{2\sqrt{(\Delta t)^{-2} + |J^{-1}\mathbf{u}|^2}} \tag{27}$$

where J^{-1} is the inverse Jacobian for the mapping from the reference element to the element in physical coordinates.

For numerical purposes, the sign function $S(\psi)$ has been replaced by a smooth approximation $S_\alpha(\psi)$ given by

$$S_\alpha(\psi) = \frac{\psi}{\sqrt{\psi^2 + \alpha^2}}. \tag{28}$$

The parameter α is on the order of the grid size. Because of this, $\mathbf{w}$, defined as $\mathbf{w} = S_\alpha(\psi_0)\nabla\psi/|\nabla\psi|$ will in practice not be a unit vector inside the smoothed zone of the sign function, where its magnitude will be given by $|S(\psi_0)| < 1$.

Natural boundary conditions are imposed on the boundaries for the advection. Outgoing characteristics of the reinitialization equation keep the inflow boundaries free from disturbances. In the cases where $\mathbf{w}$ does not point outwards at the boundaries, a small modification to $\mathbf{w}$ is added close to the boundary, to ensure that information is propagated out of the domain.

In the reinitialization equation (26), extra numerical diffusion has been added ($\epsilon > 0$). This is needed to stabilize the calculations, since the streamline diffusion modification gives an insufficient diffusion effect close to the zero contour, where $S(\psi_0)$ is small, and where $\mathbf{w}$ therefore is small in magnitude. Such a change, however, negatively affects the conservation of the area fractions of fluid A and B, defined by the positions of the zero level sets.

The most time consuming part of the calculations is the solution of the Navier-Stokes equations. If higher resolution of ϕ is wanted, in order to resolve small scales better in a merging process or to increase the quality of the curvature calculations, it yields much less extra work to only increase the resolution in the advection and reinitialization calculations while retaining the same resolution in the solution for the velocity field, compared to increasing the resolution for the full problem.

In addition to the mesh on which we solve the Navier-Stokes equations, we define a refined mesh, which is obtained by regular subdivision of the first mesh, i.e. by splitting each element into four sub-elements. The advection and reinitialization of the level set function ϕ is performed on this refined mesh. This refinement of the ϕ calculations also yields a more accurate evaluation of the force term $f_\gamma(\mathbf{v})$.

3.1 Discontinuous Density and Viscosity

The density and viscosity fields are easily defined in terms of the level set function ϕ, since ϕ is of different sign in the two fluids. Define

$$\rho(\mathbf{x}) = \rho_B + (\rho_A - \rho_B)\ H(\phi(\mathbf{x})), \tag{29}$$

$$\mu(\mathbf{x}) = \mu_B + (\mu_A - \mu_B)\ H(\phi(\mathbf{x})), \tag{30}$$

where $H(t)$ is the Heaviside function,

$$H(t) = \begin{cases} 0 & \text{for } t < 0, \\ 1/2 & \text{for } t = 0, \\ 1 & \text{for } t > 0, \end{cases} \tag{31}$$

The variational formulation (12) contain integrals with the discontinuous density or viscosity as a factor of some integrands. When evaluating these integrals, we replace the Heaviside function $H(\phi)$ by a smooth approximation $H_w(\phi)$, given by

$$H_w(\phi) = \begin{cases} 1 & \phi > w \\ \nu(\phi/w) & |\phi| \leq w \\ 0 & \phi < -w \end{cases} \tag{32}$$

where $\nu(o/w)$ is a smooth transition function such that $\nu(-1) = 0$ and $\nu(1) = 1$. If nothing else is indicated, the transition function used in the calculations is the fifth order polynomial

$$\nu(\xi) = \frac{1}{2} + \frac{1}{32}\,(45\,\xi - 50\,\xi^3 + 21\xi^5). \tag{33}$$

The reason for this choice is clarified in Section 4, where the error for integration of discontinuous functions is studied.

3.2 Evaluating the Interfacial Force Term $f_\gamma(\mathbf{v})$

For evaluation of the interfacial force term in (12),

$$f_\gamma(\mathbf{v}) = \sigma \int_\gamma \kappa \hat{\mathbf{n}} \cdot \mathbf{v}\; d\gamma, \tag{34}$$

the level-set function ϕ is needed to defined all segments of the interface γ. The curvature κ and normal vectors $\hat{\mathbf{n}}$ can be calculated as

$$\hat{\mathbf{n}} = \frac{\nabla \phi}{|\nabla \phi|}, \qquad \kappa = -\nabla \cdot \hat{\mathbf{n}}. \tag{35}$$

The unit normal vector $\hat{\mathbf{n}}$ always point into fluid A (where $\phi > 0$). The sign of κ determines the direction of the product $\kappa\hat{\mathbf{n}}$, i.e the direction of the surface tension force. When ϕ is an exact distance distance function, it holds that $|\nabla\phi| = 1$.

The line integral (34) is evaluated through a local process. An element will yield a non-zero contribution to this term only if some part of a zero level set of $\phi(\mathbf{x})$ (i.e a segment of γ) is intersecting the element. Each element is splitted into four sub-elements, and the zero level set of ϕ is defined by a linear approximation on this sub-element scale.

As ϕ is advected, the discontinuities of its gradient will create some high frequency dispersion errors of a small magnitude. All oscillating components of the error will be magnified when derivatives are calculated, a fact that complicates the curvature calculations. To avoid this magnification, the high frequencies in ϕ is filtered out, and (35) is applied to that filtered function. We calculate

$$(\tilde{\phi}, v) + (\epsilon\nabla\tilde{\phi}, \nabla v) = (\phi, v) \quad \forall v \in W_h, \tag{36}$$

and from here,

$$\hat{\mathbf{n}} = \frac{\nabla\tilde{\phi}}{|\nabla\tilde{\phi}|}, \qquad \kappa = -\nabla\cdot\hat{\mathbf{n}}. \tag{37}$$

This procedure effectively filters out high frequencies from ϕ to create $\tilde{\phi}$. The function $\tilde{\phi}$ is only used as a step in the curvature calculations, and is not used elsewhere.

The basic tools for using high-order finite elements are available in the library `Albert` [1]. Many of the features used here, for example the streamline diffusion modification and evaluation of the interfacial force term are however not available in this library.

4 Numerical Integration of a Discontinuous Function

The equations governing multiphase flow contain two discontinuous physical variables, the density and the viscosity. When these equations are to be solved numerically, this poses an extra difficulty. For computational reasons, such discontinuities are most commonly smoothed out over a transition zone that cover a few grid cells ([11], [15]). Since we work with a weak form of the equations, it is not the pointwise values of these quantities that are important, but rather how accurate the integrals including these quantities are evaluated.

Let γ be a curve across which discontinuities occur. To define the position of γ, we use a signed distance function ϕ, as described in Section 3, such that the zero level set of ϕ defines γ. Using $\phi(\mathbf{x})$, a general discontinuous function can be defined as

$$\tilde{F}(\mathbf{x}) = \begin{cases} f_1(\mathbf{x}) & \phi(\mathbf{x}) < 0 \\ 1/2\,(f_1(\mathbf{x}) + f_2(\mathbf{x})) & \phi(\mathbf{x}) = 0 \\ f_2(\mathbf{x}) & \phi(\mathbf{x}) > 0 \end{cases} \tag{38}$$

where $f_1(\mathbf{x})$ and $f_2(\mathbf{x})$ are assumed to be smooth. Using the characteristic function $I(\mathbf{x}) = H(\phi(\mathbf{x}))$, this function can be written as $\tilde{F}(\mathbf{x}) = f_1(\mathbf{x}) + H(\phi(\mathbf{x}))(f_2(\mathbf{x}) - f_1(\mathbf{x}))$. The Heaviside function $H(\phi)$ is defined in (31). The first part in the expression of discontinuous function $\tilde{F}(\mathbf{x})$ is smooth. Therefore, what we are interested in analyzing is the integration of the second part. This part can be written as

$$F(\mathbf{x}) = H(\phi(\mathbf{x}))G(\mathbf{x}) \tag{39}$$

where $G(\mathbf{x}) = f_2(\mathbf{x}) - f_1(\mathbf{x})$ is a smooth function.

For the integration of a function $F(\mathbf{x})$ over a triangulated domain, the integral is simply written as a sum of integrals over each triangle. Using the 13-point quadrature rule (cf. Strang and Fix [9], Section 4.3) on each triangle, the quadrature errors for a general smooth function is kept small. Discontinuous functions are however not well approximated by polynomials of any order and the quadrature error will be large if special care is not taken.

There are two main approaches for the evaluation of integrals of discontinuous functions: *i)* Keeping the characteristic function discontinuous, altering the quadrature procedure, and *ii)* to replace the characteristic function by a smooth approximation. In [12], both these approaches were studied. Here, a summary of the results for the second approach is given.

Let quad$(F(\mathbf{x}))$ denote the approximation to $\int_\Omega F(\mathbf{x})\, d\mathbf{x}$ obtained by summation of element integrals evaluated according to the quadrature rule. Replacing $H(\phi)$ by the approximation $H_w(\phi)$, introduced in (32), the total error in the integration of $H(\phi(\mathbf{x}))\, G(\mathbf{x})$ is

$$\begin{aligned}
E_{\mathbf{tot},G} = & \int_\Omega H(\phi(\mathbf{x}))\, G(\mathbf{x})\; d\mathbf{x} - \text{quad}(H_w(\phi(\mathbf{x}))\, G(\mathbf{x})) \\
= & \left(\int_\Omega H(\phi(\mathbf{x}))\, G(\mathbf{x})\; d\mathbf{x} - \int_\Omega H_w(\phi(\mathbf{x}))\, G(\mathbf{x})\; d\mathbf{x}\right) \\
& +\left(\int_\Omega H_w(\phi(\mathbf{x}))\, G(\mathbf{x})\; d\mathbf{x} - \text{quad}(H_w(\phi(\mathbf{x}))\, G(\mathbf{x}))\right) \\
= & \; E_{w,G} + E_{\mathbf{quad},G}.
\end{aligned} \tag{40}$$

where $E_{w,G}$ is the analytical error made by replacing $H(\phi(\mathbf{x}))$ with $H_w(\phi(\mathbf{x}))$, and $E_{\mathbf{quad},G}$ is the numerical error made in the integration of $H_w(\phi(\mathbf{x}))$. Both $E_{w,G}$ and $E_{\mathbf{quad},G}$ depend on the particular choice of the transition function $\nu(\xi)$. In general, the numerical error is large for w small, and decreases with w, while the analytical error increases as w gets larger.

The numerical tests have been performed on a regularly subdivided mesh on $[-1,1]\times[-1,1]$. The quadrature triangles will be isosceles right triangles with the longest side h. The length of the other two sides is $h_\perp$, where $h = \sqrt{2}h_\perp$. The curve γ is initialized as a circle with radius a centered in (x_c, y_c), and the distance function ϕ is defined by

$$\phi(x,y) = a - \sqrt{(x-x_c)^2 + (y-y_c)^2} \tag{41}$$

4.1 Analytical Error

The analytical error $E_{w,G}$ is obtained by integrating the error function $E(\mathbf{x}) = (H(\phi(\mathbf{x})) - H_w(\phi(\mathbf{x})))\, G(\mathbf{x})$ over Ω. We have $E(\mathbf{x}) = 0$ for $\mathbf{x}$ such that $|\phi(\mathbf{x})| > w$. To perform the integration of $E(\mathbf{x})$, we parameterize the region where $|\phi(\mathbf{x})| \le w$. Assume that the zero contour of ϕ can be parameterized by $(x(s), y(s))$, where $s \in [0, 2\pi]$ and $q(s) = \sqrt{x'(s)^2 + y'(s)^2} \neq 0$. The normal to this curve is given by

$$\hat{\mathbf{n}} = \frac{1}{q(s)}(-y'(s),\, x'(s)). \tag{42}$$

Let $n_x(s)$ and $n_y(s)$ denote the x and y components of this normal vector. The domain in which $E(\mathbf{x})$ is non-zero can be parameterized by

$$\Omega_0 = \{\mathbf{x} = (x, y) : x = X(s,t),\ y = Y(s,t),\quad s \in [0, 2\pi],\ t \in [-w, w]\}, \quad (43)$$

where

$$X(s,t) = x_c + x(s) + t n_x(s), \quad (44)$$

$$Y(s,t) = y_c + y(s) + t n_y(s). \quad (45)$$

Under the assumption that $w \max_s |\kappa(s)| < 1$, where

$$\kappa(s) = \frac{x'(s)y''(s) - x''(s)y'(s)}{q(s)^3}, \quad (46)$$

the integral over $E(\mathbf{x}) = (H(\phi(\mathbf{x})) - H_w(\phi(\mathbf{x})))\, G(\mathbf{x})$ can be transformed to an integral in the parameters (s, t).

Denote by $E(t)$, the error function $E(t) = H(t) - H_w(t)$. We introduce what we henceforth call the moments of the error function $E(t)$,

$$M_\alpha(E(t)) = \int_{-w}^{w} E(t)\, t^\alpha \, dt. \quad (47)$$

These moments evaluates as

$$M_\alpha(E(t)) = \int_{-w}^{w} E(t)\, t^\alpha \, dt = w^{\alpha+1}\{\frac{1}{\alpha+1} - \int_{-1}^{1} \nu(\xi)\, \xi^\alpha \, d\xi\}, \quad (48)$$

where $\nu(\xi)$ is the transition function in the definition of $H_w(\phi)$ (32).

Introduce $g(s,t) = G(X(s,t), Y(s,t))$. Since $G(\mathbf{x})$ is assumed to be smooth, and $t \in [-w, w]$, w small, we can expand $g(s,t)$ in a Taylor series for t, centered in $(s, 0)$. The analytical error then evaluates as

$$E_{w,G} = M_0(E(t))\, \{\int_0^{2\pi} q(s)\, g(s,0)\, ds\} + \sum_{\alpha=1}^{\infty} \frac{1}{\alpha!} C_{\alpha,G}\, M_\alpha(E(t)), \quad (49)$$

where

$$C_{\alpha,G} = \int_{s=0}^{2\pi} q(s)\, g_{\alpha t}(s,0)\, ds - \int_{s=0}^{2\pi} q(s)\, \kappa(s)\, g_{(\alpha-1)t}(s,0)\, ds. \quad (50)$$

The sub index α in $g_{\alpha t}$ denotes the number of partial derivatives with respect to t.

Together with the evaluation of the moments (48), this expression yields that the error will be proportional to higher powers of w the more moments of the error function that evaluates as zero. Since w is small, this is a desirable property. The conditions for $\nu(\xi)$ to yield vanishing moments for the error function of the corresponding Heaviside approximation are given by (48). Considering that the Heaviside approximation will be integrated numerically, the number of continuous derivatives of the approximation, given by the number of derivatives of $\nu(\xi)$ that evaluates as zero at $\xi = \pm 1$, will also be important. There are different classes of functions from which $\nu(\xi)$ could be defined. We will study polynomials, and proceed by introducing the definition of a transition polynomial.

Definition 4.1. *Denote by $\nu^{m,k}(\xi)$, the transition polynomial of lowest degree such that*

$$\nu^{m,k}(-1) = 0, \quad \nu^{m,k}(1) = 1, \tag{51}$$

$$(\nu^{m,k})^{(\beta)}(\pm 1) = 0, \quad \beta = 1, \dots, k, \tag{52}$$

and further,

$$\lambda_\alpha^{m,k} = \int_{-1}^{1} \nu^{m,k}(\xi)\ \xi^\alpha\ d\xi - \frac{1}{\alpha + 1} = 0 \quad \alpha = 0, \dots, m. \tag{53}$$

Theorem 4.2. *The transition polynomial $\nu^{m,k}(\xi)$ exists and is uniquely determined by the conditions in Definition 4.1. It is of degree $r = 2\lfloor (m+1)/2 \rfloor + 2k + 1$. Further, $\nu^{m,k}(\xi) = 1/2 + p(\xi)$, where $p(\xi)$ is a polynomial of degree r, containing only odd powers of ξ.*

Proof The proof is given in [12].

□

To each transition polynomial $\nu^{m,k}(\xi)$, we assign a Heaviside approximation $H_w^{m,k}(t)$, and a corresponding error function $E_w^{m,k}(t)$. We have the following definitions:

Definition 4.3. *Denote by $H_w^{m,k}(t)$, the Heaviside approximation defined by*

$$H_w^{m,k}(t) = \begin{cases} 1 & t > w \\ \nu^{m,k}(t/w) & |t| \le w \\ 0 & t < -w \end{cases} \tag{54}$$

where $\nu^{m,k}(\xi)$ was defined in Definition 4.1. Define the corresponding error function $E_w^{m,k}(t)$ by

$$E_w^{m,k}(t) = H(t) - H_w^{m,k}(t). \tag{55}$$

Using the definition of $\nu^{m,k}(\xi)$, we can show the following:

Corollary 4.4. *Let $H_w^{m,k}(t)$ and $E_w^{m,k}(t)$ be as in Definition 4.3. Then it follows that the Heaviside approximation $H_w^{m,k}(t)$ has k continuous derivatives, and*

$$M_\alpha(E_w^{m,k}(t)) = \int_{-w}^{w} E_w^{m,k}(t)\, t^\alpha\, dt = 0, \quad \alpha = 0, \dots, 2\lfloor\frac{m+1}{2}\rfloor. \tag{56}$$

In addition, $M_\alpha(E_w^{m,k}(t)) = 0$ for all α even.

Proof The number of continuous derivatives is simply given by the number of vanishing derivatives of $\nu^{m,k}(\xi)$ at $\xi = \pm 1$, which is k by definition. Further, from (48) and (53),

$$M_\alpha(E_w^{m,k}(t)) = \int_{-w}^{w} E_w^{m,k}(t)\, t^\alpha \, dt = -\lambda_\alpha^{m,k}\, w^{\alpha+1}. \tag{57}$$

From the definition of $\nu^{m,k}(\xi)$, $\lambda_\alpha^{m,k} = 0$ for $0 \leq \alpha \leq m$, and (56) follows for $0 \leq \alpha \leq m$. In addition, $\lambda_\alpha^{m,k} = 0$ for all α even, since $\nu^{m,k}(\xi) = 1/2 + p(\xi)$, where $p(\xi)$ is an odd polynomial. Since $\beta = 2\lfloor (m+1)/2 \rfloor$ is even, $\lambda_\beta^{m,k} = 0$.

□

We now have all the necessary tools to state and show the following theorem considering the analytical error for these polynomial approximations:

Theorem 4.5. *Let $H_w^{m,k}(t)$ and $E_w^{m,k}(t)$ be as in Definition 4.3. Assume that γ, the zero level-set of ϕ, can be parameterized by $\gamma = (x(s), y(s))$, with the curvature $\kappa(s)$ defined by (46). If $(w \max_s |\kappa(s)|) < 1$, then the analytical error for the integration of $H(\phi(\mathbf{x}))\, G(\mathbf{x})$, $G(\mathbf{x}) \in C^\infty(\Omega)$, made when replacing $H(\phi(\mathbf{x}))$ by $H_w^{m,k}(\phi(\mathbf{x}))$,*

$$E_{w,G}^{m,k} = \int_\Omega \{H(\phi(\mathbf{x})) - H_w^{m,k}(\phi(\mathbf{x}))\}\, G(\mathbf{x})\, d\mathbf{x}. \tag{58}$$

is given by

$$\begin{aligned} E_{w,G}^{m,k} &= \sum_{\alpha=\beta/2}^{\infty} \frac{1}{(2\alpha+1)!} C_{2\alpha+1,G}\, M_{2\alpha+1}(E^{m,k}(t)) \\ &= -\frac{1}{(\beta+1)!} C_{\beta+1,G}\, \lambda_{\beta+1}^{m,k}\, w^{\beta+2} + O(w^{\beta+4}), \quad \beta = 2\left\lfloor \frac{m+1}{2} \right\rfloor \end{aligned} \tag{59}$$

with $C_{\beta,G}$ defined by (50) and $\lambda_\beta^{m,k}$ defined by (53).

Proof The general error formula is given in (49). According to Corollary 4.4, $M_\alpha(E_w^{m,k}(t)) = 0$ for $\alpha = 0, \ldots, \beta$ and all α even. The terms retained are therefore those in (59), and the first nonvanishing moment is

$$M_{\beta+1}(E_w^{m,k}(t)) = -\lambda_{\beta+1}^{m,k}\, w^{\beta+2}, \tag{60}$$

with the second non-vanishing moment $M_{\beta+3}(E_w^{m,k}(t)) \sim w^{\beta+4}$.

□

Remark 4.6. If $G(x,y) = G(X(s,t), Y(s,t)) = g(s,t)$ is such that $g(s,t)$ is a polynomial in t of degree n for all fixed s, then $C_{\alpha,G} = 0$ for $\alpha \geq n+2$. For m such that $\beta = 2\lfloor \frac{m+1}{2} \rfloor \geq n+1$, it follows that $E_{w,G}^{m,k} = 0$.

From this follows that if $G(\mathbf{x}) \equiv 1$, any approximation of the Heaviside function with at least two vanishing moments of the error function will introduce no analytical error. The numerical error is in this case the only source of error.

Assuming that w/h is not too small, this error is predicted to show the following dependence:

$$E^{m,k}_{\text{quad}} \sim \frac{h^{k+2}}{w^{k+1}}, \tag{61}$$

given that $k < n$, where n is the order of the quadrature rule used. This can be verified by numerical experiments.

In general, the total error for the integration ($E_{\mathbf{tot},G}$) is the sum of the analytical error ($E_{w,G}$), increasing with the width of the transition zone w, and the numerical error ($E_{\mathbf{quad},G}$), decreasing with w. The numerical error is also decreased as the size of the quadrature triangles decreases.

Figure 2 shows the total error for the approximations $H^{2,k}_w(t)$, $k = 0, 1, 2$, with $G(x, y) = (x - x_c)^2 + (y - y_c)^2$. The approximations are defined by the

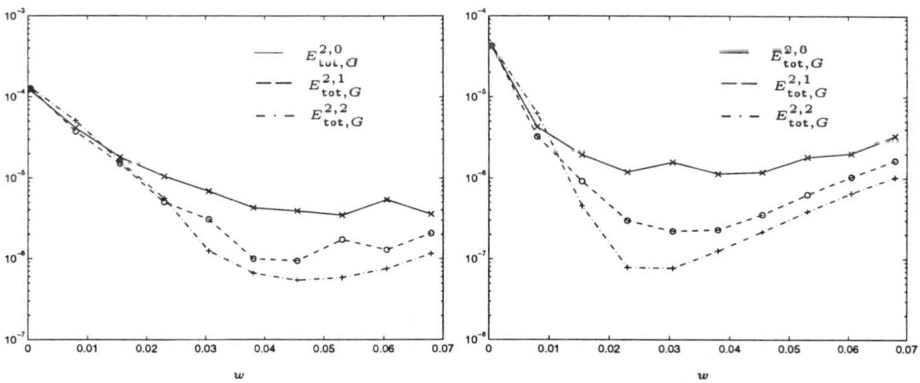

(a) $h_\perp = 0.025$. (b) $h_\perp = 0.0125$.

Figure 2: $E_{\mathbf{tot},G}$ plotted versus w, for the polynomial approximations $H^{2,0}_w(t)$, $H^{2,1}_w(t)$ and $H^{2,2}_w(t)$. $G(x, y) = (x - x_c)^2 + (y - y_c)^2$, $\Delta e = 0.05$. Maximum of error taken over 16 different positions (x_c, y_c).

following transition polynomials:

$$\nu^{2,0}(\xi) = \frac{1}{2} + \frac{1}{8}(9\xi - 5\xi^3) \tag{62}$$

$$\nu^{2,1}(\xi) = \frac{1}{2} + \frac{1}{32}(45\xi - 50\xi^3 + 21\xi^5) \tag{63}$$

$$\nu^{2,2}(\xi) = \frac{1}{2} + \frac{1}{64}(105\xi - 175\xi^3 + 147\xi^5 - 45\xi^7) \tag{64}$$

Two different resolutions have been used, and it is clear that the numerical error decreases as $h_\perp$ is decreased. The analytical error for the approximations

$H_w^{2,k}(t)$ (54) is proportional to w^4. Since this error increases with w, the total error will start to increase with w as soon as the analytical error is dominating. This will happen for a smaller w the smaller the numerical error is.

The transition polynomial $\nu^{2,1}(\xi)$ is of degree 5. Define a different transition polynomial of the same degree,

$$\nu^{4,0}(\xi) = \frac{1}{2} + \frac{1}{128}(225\,\xi - 350\,\xi^3 + 189\,\xi^5). \tag{65}$$

The corresponding error function $E^{4,0}(t)$ has four vanishing moments, and the analytical error $E_{w,G}^{4,0}$ is of order 6 in w, compared to order 4 for $E_{w,G}^{2,1}$. The approximation $H_w^{2,1}(t)$ will however have better numerical properties, since its first derivative is continuous, which is not the case for $H_w^{4,0}(t)$.

In Figure 3, we compare the results for these two polynomial approximations when $G(x,y) = (x-x_c)^2+(y-y_c)^2$. In this case, the analytical error for $H_w^{4,0}(\phi)$ is actually zero due to vanishing derivatives.

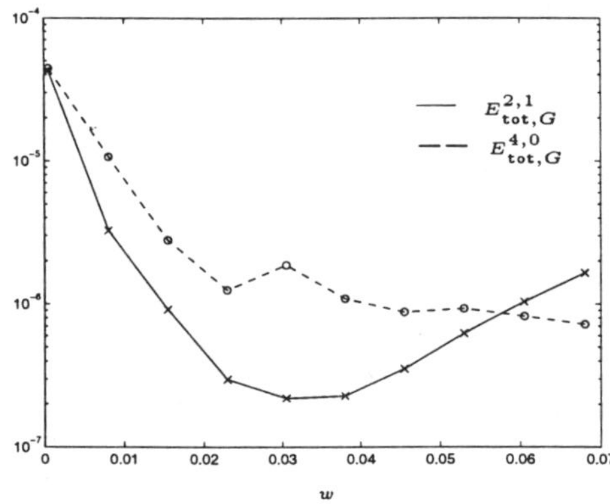

Figure 3: $E_{\mathtt{tot},G}$ plotted versus w for the polynomial approximations $H_w^{2,1}(t)$ and $H_w^{4,0}(t)$. $h_\perp = 0.0125$. $N = 16$.

It is found that $H_w^{2,1}(\phi)$ yields a lower total error for moderate w, even if the analytical error is higher, since the discontinuous first derivative yields a larger numerical error for $H_w^{4,0}(\phi)$. As w increases further however, the total error for $H_w^{2,1}(\phi)$ will eventually grow larger than the total error of $H_w^{4,0}(\phi)$ due to its growing analytical error.

5 Numerical Simulations using the Level-Set Method

We now present a level-set based simulation including topology changes.

The run we present here is a run where we start with two bubbles of fluid A immersed into fluid B, with a layer of fluid A on top of fluid B. The initial configuration is as in Figure 1, with the domain extended up to $y = 6.0$. The fluid is quiescent at $t = 0$. The bubbles will rise in the middle of the domain, and we need the highest resolution in this part of the region, as well as close to the surface. We use the irregular mesh shown in Figure 4 for this simulation.

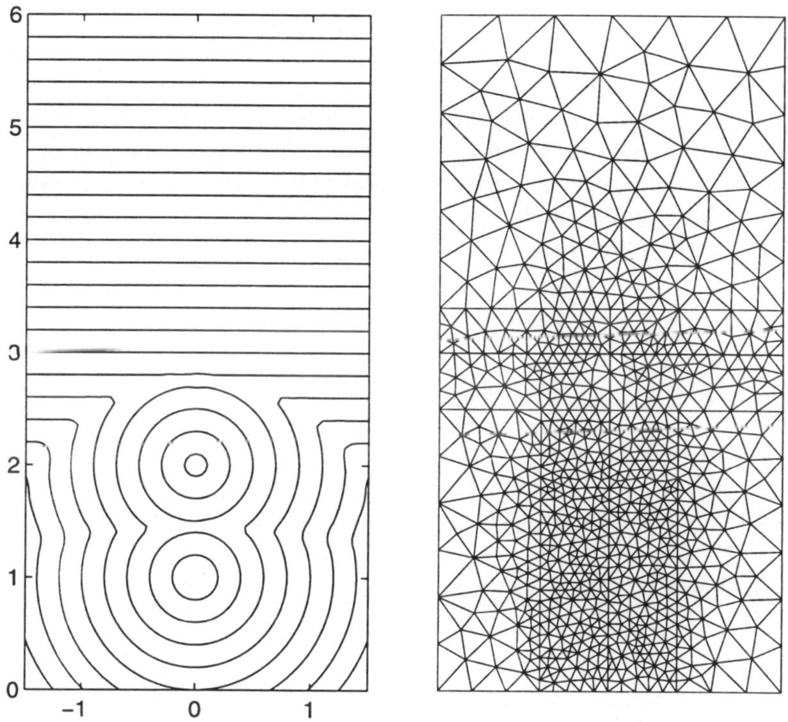

(a) The level set function ϕ. The increment of the contours is 0.2.

(b) The mesh. 1566 elements, with 6 nodes in each element

Figure 4: The level-set function for the initial configuration, and the mesh used in the simulations.

The non-dimensional parameters used to characterize the problem are, in addition to the viscosity and density ratios, μ_B/μ_A and ρ_B/ρ_A, the Morton number M and the Eötvös number Eo. These numbers are defined as

$$M = \frac{g\mu_B^4}{\rho_B\sigma^3}, \quad Eo = \frac{\rho_B g d^2}{\sigma}. \tag{66}$$

Here, μ_B and ρ_B denote the viscosity and density of the outer fluid, respectively, g is the gravitational constant and σ is the surface tension coefficient. $d = 2\sqrt{A/\pi}$, where A is the area of the bubble. The diameter of the largest bubble (the upper one) is normalized to 1. The diameter of the smaller bubble is 0.8. The Morton number and Eötvös numbers given for the flow are based on the largest bubble.

We present the results from a run with $Mo = 0.1$ and $Eo = 10.0$ with density ratio $\rho_B/\rho_A = 100$ and viscosity ratio $\mu_B/\mu_A = 2$. The numerical parameters are as follows: $\Delta t = 5 \cdot 10^{-4}$. In each advection time step one reinitialization step with time step $\Delta\tau = 0.01$ is performed. The diffusion parameter in this procedure is set to $\epsilon = 4 \cdot 10^{-3}$. The smoothing parameter in the Heaviside approximation (32) is set to 0.05, and the smoothing parameter for the sign function in (28), is set to 0.1.

The fluids are quiescent initially. The two bubbles will start to rise due to buoyancy effects, creating a non-zero velocity field. The smaller bubble travels in the wake of the upper, larger bubble, and will rise faster. Eventually, it will catch up with the upper bubble, and the two bubbles will merge, as can be seen in Figure 5. The plots do not cover the top part of the domain. The merged

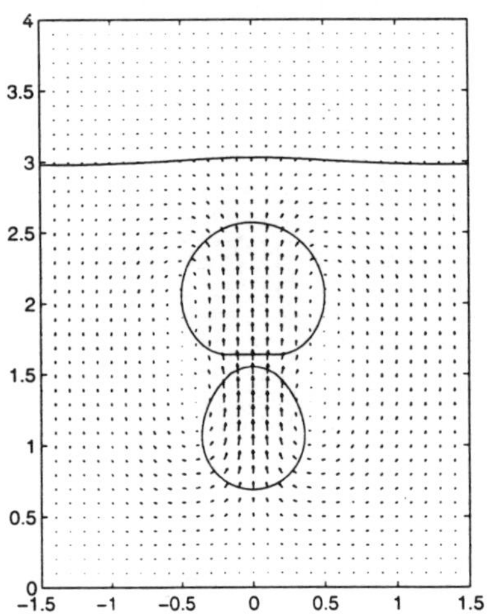

(a) $t = 0.05$. $Re_\infty = 5.52$.

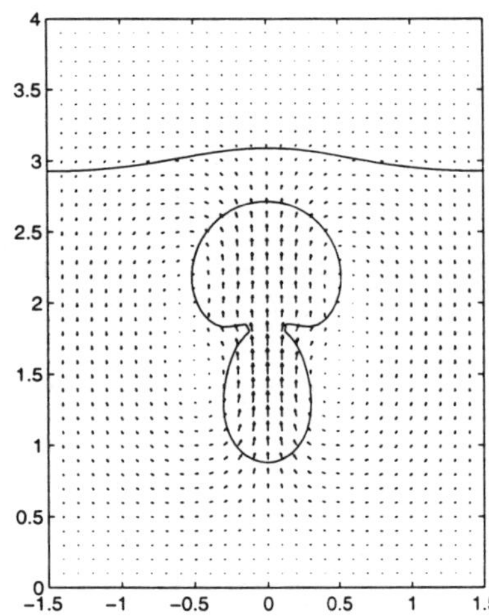

(b) $t = 0.1$. $Re_\infty = 7.65$.

Figure 5: The two bubbles merge. The interfaces plotted together with the corresponding velocity field.

bubble is deforming as it moves closer to the surface. The drainage of fluid B from the region between the two interfaces starts and, finally, the filament between the two interfaces gets so thin that the bubble merge with the surface. This is shown in Figure 6.

Two thin filaments of fluid A are pointing into fluid B after the merge. Local high velocities develop here to smooth the surface out, as can be seen

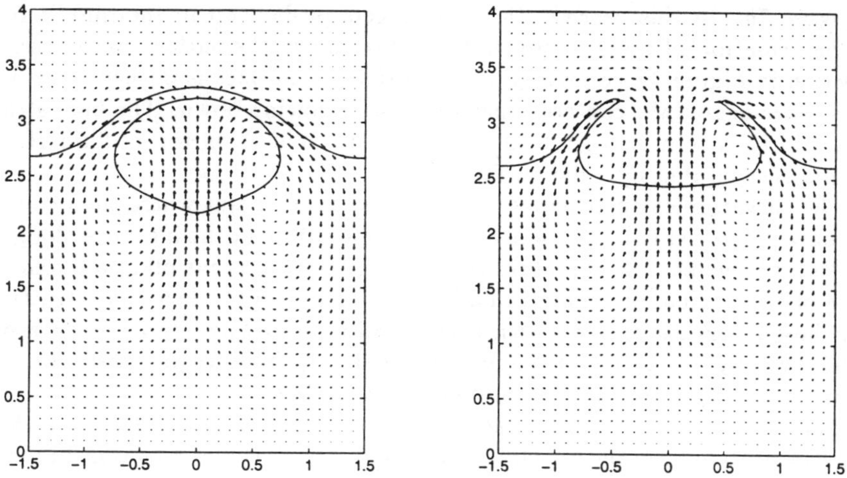

(a) $t = 0.3$. $Re_\infty = 5.07$.

(b) $t = 0.3625$. $Re_\infty = 3.84$.

Figure 6: The interfaces plotted together with the corresponding velocity field.

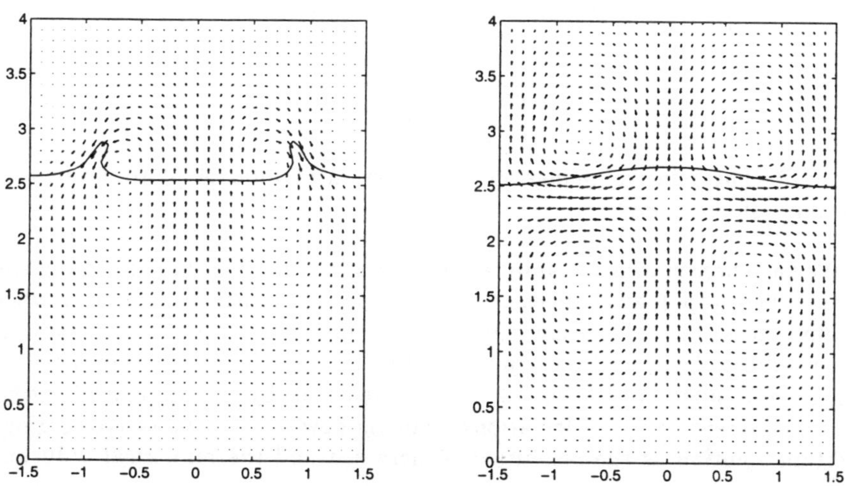

(a) $t = 0.40$. $Re_\infty = 5.064$.

(b) $t = 0.55$. $Re_\infty = 0.86$.

Figure 7: After the last merge. The interface plotted together with the corresponding velocity field.

in Figure 7*a*. In this process, the surface gets pushed up in the middle from the recirculation of fluid. Note that the flow above the surface have changed direction in Figure 7*b*, compared to 7*a*, so that the surface is pushed down to a flat surface again. We can note that the flow is slowing down. The final steady state will be a zero velocity field ($\mathbf{u} = 0$), with fluid A on top of fluid B, the two fluids separated by a flat surface.

The curvature calculations are complicated at the point where the two bubbles have merged (Figure 6*b*), and are not very accurate at the interface along the two thin filaments. The curvature calculated from (37) is cut off at a maximum value of 15.0. This maximum value is motivated from the fact that structures with larger curvatures, i.e. with such small scale details, can not be represented with the resolution of the present mesh. The high frequencies of this cut curvature is thereafter filtered out equivalent to (36).

The area fractions of fluids A and B are not conserved during the simulation. The relative change in the area fraction of fluid A is plotted versus time in Figure 8. The area fraction decreases at first, but increases some at later times. The area fraction of fluid A at time $t = 0.55$, the instant plotted in Figure 7*b*, is 99.3% of the initial area fraction.

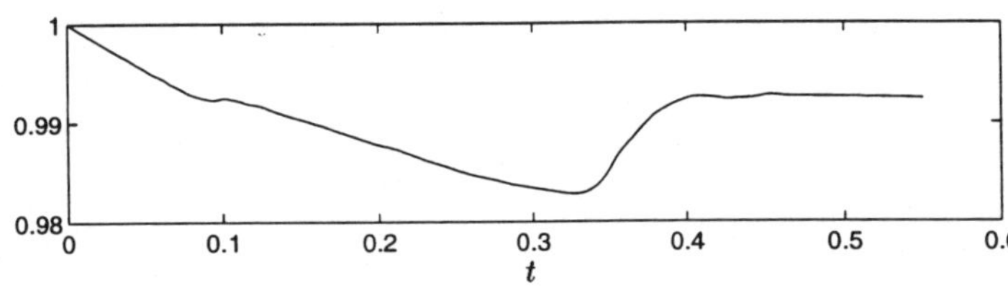

Figure 8: The relative change in area fraction of fluid A plotted versus the non-dimensional time t.

We have here shown the abilities of the level-set method in performing simulations where topology changes occurs. No specific treatment is needed when a merging takes place. The exact time at which merging will occur in a simulation for a fixed set of physical parameters will be however be affected by the resolution, i.e how small scales that can be resolved, and by the amount of artificial diffusion present in the calculations. This is however a process converging to a certain solution as the resolution is increased and the numerical diffusion is decreased.

6 Alternative Methods

The level-set method discussed above is very powerful in simulations involving topology changes. This is much harder with the front-tracking method.

We briefly present the front-tracking method below, and apply it to a single buoyant bubble for which it is both fast and accurate. In Section 6.2, we shall introduce a new method. This segment projection method is also fast and accurate and we apply it here for a single bubble. We shall extend it to include merging in [13].

6.1 The Front-Tracking Method

In the front-tracking method, each interface separating the two fluids (i.e. each segment of γ), is described by a set of discrete points $\{\mathbf{x}^{(l)}\}_{l=1}^{N_I}$, together with a parametric description connecting these points. In this work, a cubic spline-fit has been used, but other descriptions are possible as well.

In order to retain the correct position of the interface, the interface points are advected by the flow, as given by

$$\frac{d\mathbf{x}^{(l)}}{dt} = \mathbf{u}(\mathbf{x}^{(l)}) \quad l = 1, \ldots, N_I. \tag{67}$$

After the points have been advected, a new parametric fit is calculated. This is done for each separate interface. Two different interfaces can never merge automatically. For two parameterizations to merge into one, the discrete points need to be reordered and a parameterization has to be defined from this new, larger set of points.

The second order time-stepping scheme used is based on the implicit Crank-Nicolson scheme, reformulated as an iterative procedure. This time-stepping scheme has been found to provide a good conservation of mass for the two fluids. In this advection procedure, each discrete interface node is individually advected by the local flow. No restrictions are made upon the movement of the points. Depending on the flow, points might cluster at parts of the interface, while other parts might get depleted of points. Points therefore need to be added and deleted as the simulations proceed.

Curvature and normal vectors, needed to define the surface tension forces, can be unambiguously evaluated from the spline parameterization, since the second derivatives are continuous around the curve.

The characteristic function, defining if a point is inside fluid A or fluid B, defines the density and velocity fields. In difference to the level-set method, where $I(\mathbf{x}) = H(\phi(\mathbf{x}))$, the front-tracking method does not provide any such pointwise information. Instead, the parametric description of γ needs to be used in order to determine $I(\mathbf{x})$. This can be done with the notion of orientation of a curve.

Simulations of a singular buoyant bubble show good agreement with results from the level-set method. The front-tracking method conserves the mass (or area, since the density inside the bubble is constant) better than the level-set method. In a typical computation, the decrease compared to the initial area for the level-set method is 1.0%, compared to 0.01% for the front-tracking method, see [12].

6.2 Segment Projection Method

We shall also introduce a new computational technique which can be seen as a compromize between the level-set and front-tracking methods. In the segment projection method a curve γ is given as a union of curve segments γ_j. The segments are chosen such that they can be represented by a function of one coordinate variable. The domain of the independent variables of these functions are projections of the segment onto the coordinate axes.

As an example, the circle γ can be represented by the segments γ_j and the corresponding functions f_j, $j = 1, 2, 3, 4$,

$$\gamma = \bigcup_{j=1}^{4} \gamma_j, \quad \gamma = \{(x, y), \quad x^2 + y^2 = 1\}. \tag{68}$$

The segments are defined by:

$$\gamma_1 = \{(x, y), \quad x = f_1(y) = \sqrt{1 - y^2}, \ |y| \leq \bar{y}\} \tag{69}$$

$$\gamma_2 = \{(x, y), \quad x = f_2(y) = -\sqrt{1 - y^2}, \ |y| \leq \bar{y}\} \tag{70}$$

$$\gamma_3 = \{(x, y), \quad y = f_3(x) = \sqrt{1 - x^2}, \ |x| \leq \bar{x}\} \tag{71}$$

$$\gamma_4 = \{(x, y), \quad y = f_4(x) = -\sqrt{1 - x^2}, \ |x| \leq \bar{x}\}. \tag{72}$$

The functions f_j are described by computational objects containing one dimensional arrays and pointers giving the connectivity of the segments.

As with the standard level-set method quantities like normals and curvature are computed using divided differences. Advection of a time dependent curve by a velocity field $u = (u_1(x, y, t), u_2(x, y, t))$ is given by differential equations. Consider for example the evolution of $f_1 = f_1(y, t)$,

$$\frac{\partial f_1}{\partial t} + u_2 \frac{\partial f_1}{\partial y} = u_1 \tag{73}$$

The segment projection method is similar to front-tracking in that a curve is represented by one dimensional arrays which points on the curve. On the other hand there are similarities with the level-set method in that both methods are based on Eulerian grids and distance functions play important roles.

In all the methods reinitializations are standard in practical computations. The segment projection method requires creation and elimination of segments when the geometry of γ changes with time as, for example, in merging bubbles.

The segment projection method was applied to our multiphase problem, and Figure 9 displays the simulation of a rising bubble.

The method can be extended to $\mathbb{R}^3$ with the segments corresponding to surface patches. Different rules of curve or surface interactions can be approximated. Merging can, for example, be replaced by superposition as in the case of linear wave propagation, see [13].

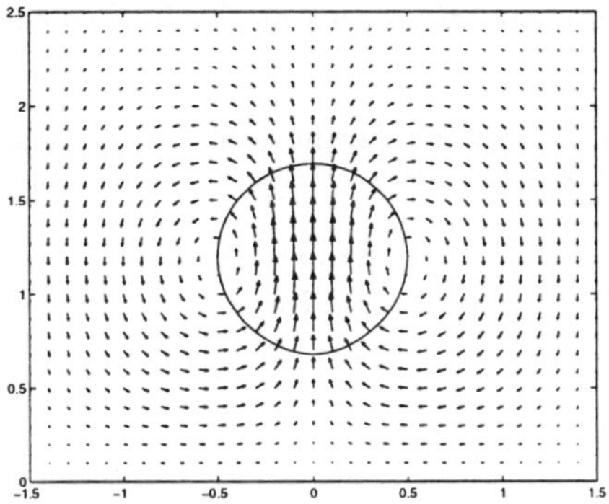

(a) Segment projection, $Re_\infty = 4.48$.

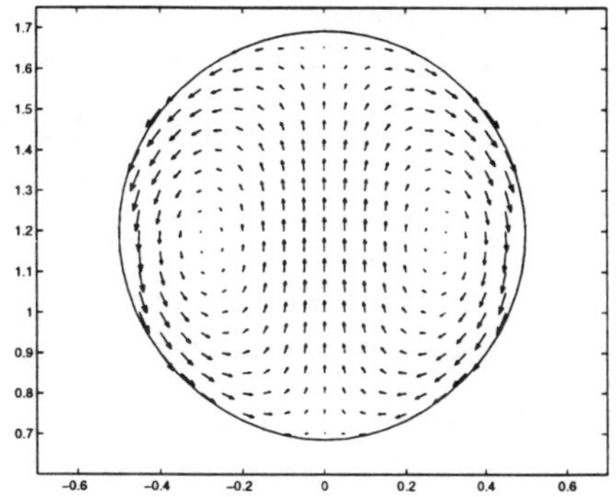

(b) Level-set. $Re = 2.98$.

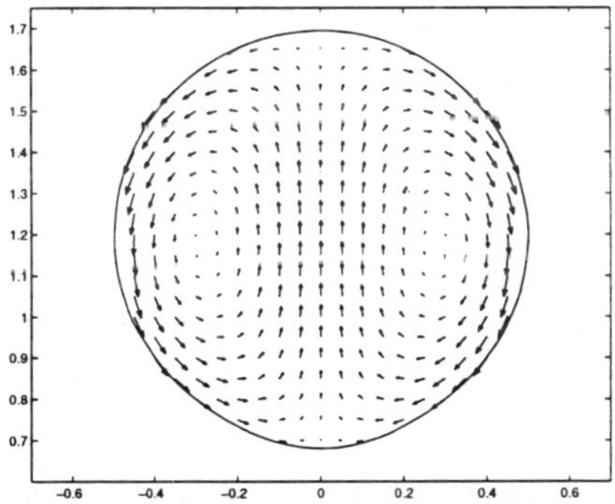

(c) Segment projection. $Re = 2.97$.

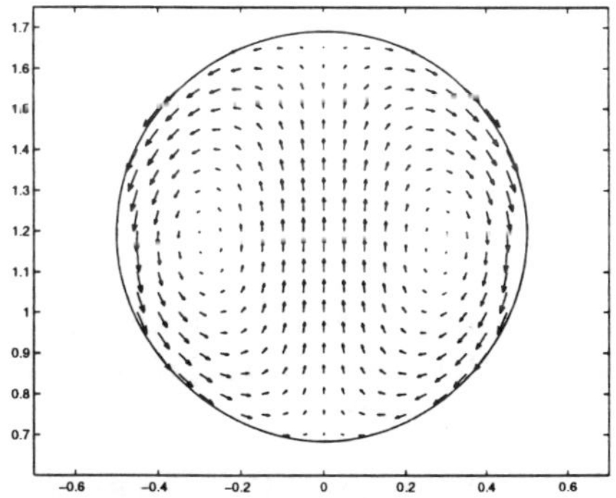

(d) Front-tracking. $Re = 2.97$.

Figure 9: Comparison between the different methods for a single bubble. The velocity field at time $t = 0.1$ is plotted in Figure a). The velocity field relative to the bubble is plotted for the three methods in Figures $b) - d$). Reynolds number Re_∞ is based on the maximum velocity, Re is based on the rise velocity of the bubble.

References

[1] The Albert system is available in the compressed `tar` file `fec.tar.gz` which can be obtained from `http://www.hpc.uh.edu/~babak/fec.tar.gz`.

[2] G. Beylkin. On the Fast Fourier Transform of Functions with Singularities. *Applied and Computational Harmonic Analysis*, 2:363–381, 1995.

[3] S.C. Brenner and L.R. Scott. *The Mathematical Theory of Finite Element Methods.* Springer-Verlag, 1994.

[4] F. H. Harlow and J. E. Welch. Volume-Tracking Methods for Interfacial Flow Calculations. *Physics of Fluids*, 8:2182, 1965.

[5] C. Johnson. *Numerical Solution of Partial Differential Equations by the Finite Element Method.* Cambridge University Press, Lund, Sweden and Cambridge, England, 1987.

[6] S. Osher and J.A. Sethian. Fronts Propagating with Curvature Dependent Speed: Algorithms Based on Hamilton-Jacobi Formulations. *Journal of Computational Physics*, 79:12–49, 1988.

[7] C. S. Peskin. Numerical Analysis of Blood Flow in the Heart. *Journal of Computational Physics*, 25:220–252, 1977.

[8] M. Rudman. Volume-Tracking Methods for Interfacial Flow Calculations. *International Journal for Numerical Methods in Fluids*, 24:671–691, 1997.

[9] G. Strang and G. J. Fix. *An Analysis of the Finite Element Method.* Prentice Hall, Englewood Cliffs, NJ, 1973. now published by Wellesley-Cambridge Press, Wellesley, MA; Cambridge, MA.

[10] M. Sussman, M. Fatemi, P. Smereka, and S. Osher. An Improved Level Method for Incompressible Two-Phase Flows. *Computers & Fluids*, 27:663–680, 1998.

[11] M. Sussman, P. Smereka, and S. Osher. A Level Set Approach for Computing Solutions to Incompressible Two-Phase Flow. *Journal of Computational Physics*, 114:146–159, 1994.

[12] A-K. Tornberg. A Finite Element Based Level Set Method for Multiphase Flow Simulations. Licentiate's thesis, Royal Institute of Technology, Department of Numerical Analysis and Computer Science, September 1998. ISBN 91-7170-317-9, TRITA-NA-9817.

[13] A-K. Tornberg and B. Engquist. The Segment Projection Method for Tracking Interfaces. *To appear.*

[14] A-K. Tornberg, R. W. Metcalfe, R. Scott, and B. Bagheri. A Fluid Particle Motion Simulation Method. In *Computational Science for the 21st Century*, pages 312–321. John Wiley and Sons, New York, 1997.

[15] S.O. Unverdi and G. Tryggvason. A Front-Tracking Method for Viscous, Incompressible, Multi-Fluid Flows. *Journal of Computational Physics*, 100:25–37, 1992.

The Ghost Fluid Method for Viscous Flows[1]

Ronald P. Fedkiw

Computer Science Department
Stanford University, Stanford, California 94305
Email:fedkiw@cs.stanford.edu

Xu-Dong Liu

Department of Mathematics
University of California Santa Barbara, Santa Barbara, California, 93106
Email:xliu@math.ucsb.edu

Keywords: Navier Stokes Equations, Interfaces, Multiphase Flow, Ghost Fluid Method, Viscosity

Abstract: The level set method for compressible flows [13] is simple to implement, especially in the presence of topological changes. However, this method was shown to suffer from large spurious oscillations in [11]. In [4], a new Ghost Fluid Method (GFM) was shown to remove these spurious oscillations by minimizing the numerical smearing in the entropy field with the help of an Isobaric Fix [6] technique. The original GFM was designed for the inviscid Euler equations. In this paper, we extend the formulation of the GFM and apply the extended formulation to the viscous Navier-Stokes equations. The resulting numerical method is robust and easy to implement along the lines of [15].

1 Introduction

In [13], the authors applied the level set method to multiphase compressible flows. The level set function was used as an indicator function and each grid point was designated as one fluid or the other for evaluation of the equation of state. Then numerical fluxes were formed and differenced in the usual manner [15]. The usual virtues of ease of implementation, in particular as regards to topological changes, were apparent. However, in [11], it was shown that this technique produced large spurious oscillations in the pressure and velocity fields. This problem was rectified in [10], [3], and [2] with schemes that involved *explicit* treatment of the appropriate conditions at the interface. As a consequence, these schemes are intricate in one dimension and can only be extended to multiple dimensions with ill-advised dimensional splitting in time. In addition, multilevel time integrators, such as Runge Kutta methods, are difficult to implement for these schemes.

The Ghost Fluid Method [4] avoids these oscillations at multimaterial interfaces without *explicitly* using interface jump conditions. Instead, the GFM

[1]This paper was presented in "Solutions of PDE" Conference in honour of Prof. Roe on the occassion of his 60th birthday, July 1998, Arachaon, France

creates a artificial fluid which *implicitly* induces the proper conditions at the interface. In the flavor of the level set function which gives an *implicit* representation of the interface, the GFM gives an *implicit* representation of the Rankine-Hugoniot jump conditions at the interface. Since the jump conditions are handled *implicitly* by the construction of a ghost fluid, the overall scheme becomes easy to implement in multidimensions without time splitting. In addition, Runge Kutta methods are trivial to apply.

In [4], the GFM was implemented for contact discontinuities in the inviscid Euler equations. In this case, the pressure and normal velocity of the ghost fluid are just copied over from the real fluid in a node by node fashion while the entropy and tangential velocities are defined with a simple partial differential equation for one-sided extrapolation in the normal direction. See [4] for details.

In this paper, we will generalize the GFM and show how this new general technique can be used for material interfaces in the viscous Navier-Stokes equations. The GFM will *implicitly* enforce the jump conditions at the interface by the construction of a ghost fluid. The resulting numerical method is easy to implement in multidimensions (without time splitting) and extends trivially to Runge Kutta methods.

2 Equations

2.1 Navier-Stokes Equations

The basic equations for viscous compressible flow are the Navier-Stokes equations,

$$\vec{U}_t + \vec{F}(\vec{U})_x + \vec{G}(\vec{U})_y + \vec{H}(\vec{U})_z = Vis \tag{1}$$

which can be written in more detail as

$$\begin{pmatrix} \rho \\ \rho u \\ \rho v \\ \rho w \\ E \end{pmatrix}_t + \begin{pmatrix} \rho u \\ \rho u^2 + p \\ \rho uv \\ \rho uw \\ (E+p)u \end{pmatrix}_x + \begin{pmatrix} \rho v \\ \rho uv \\ \rho v^2 + p \\ \rho vw \\ (E+p)v \end{pmatrix}_y + \begin{pmatrix} \rho w \\ \rho uw \\ \rho vw \\ \rho w^2 + p \\ (E+p)w \end{pmatrix}_z = Vis \tag{2}$$

where t is the time, (x, y, z) are the spatial coordinates, ρ is the density, $\vec{V} = < u, v, w >$ is the velocity field, E is the total energy per unit volume, and p is the pressure. The total energy is the sum of the internal energy and the kinetic energy,

$$E = \rho e + \frac{\rho(u^2 + v^2 + w^2)}{2} \tag{3}$$

where e is the internal energy per unit mass. The two-dimensional Navier-Stokes equations are obtained by omitting all terms involving w and z. The one-dimensional Navier-Stokes equations are obtained by omitting all terms involving v, w, y, and z. The inviscid Euler equations are obtained by setting $Vis = 0$.

In general, the pressure can be written as a function of density and internal energy, $p = p(\rho, e)$, or as a function of density and temperature, $p = p(\rho, T)$. In order to complete the model, we need an expression for the internal energy per unit mass. Since $e = e(\rho, T)$ we write

$$de = \left(\frac{\partial e}{\partial \rho}\right)_T d\rho + \left(\frac{\partial e}{\partial T}\right)_\rho dT \tag{4}$$

which can be shown to be equivalent to

$$de = \left(\frac{p - Tp_T}{\rho^2}\right) d\rho + c_v dT \tag{5}$$

where c_v is the specific heat at constant volume. [6]

The sound speeds associated with the equations depend on the partial derivatives of the pressure, either p_ρ and p_e or p_ρ and p_T, where the change of variables from density and internal energy to density and temperature is governed by the following relations

$$p_\rho \rightarrow p_\rho - \left(\frac{p - Tp_T}{c_v\rho^2}\right) p_T \tag{6}$$

$$p_e \rightarrow p_\rho + \left(\frac{1}{c_v}\right) p_T \tag{7}$$

and the sound speed c is given by

$$c = \sqrt{p_\rho + \frac{pp_e}{\rho^2}} \tag{8}$$

for the case where $p = p(\rho, e)$ and

$$c = \sqrt{p_\rho + \frac{T(p_T)^2}{c_v\rho^2}} \tag{9}$$

for the case where $p = p(\rho, T)$.

2.1.1 Viscous Terms

We define the viscous stress tensor as

$$\tau = \begin{pmatrix} \tau_{xx} & \tau_{xy} & \tau_{xz} \\ \tau_{xy} & \tau_{yy} & \tau_{yz} \\ \tau_{xz} & \tau_{yz} & \tau_{zz} \end{pmatrix} \tag{10}$$

where

$$\tau_{xx} = \frac{2}{3}\mu(2u_x - v_y - w_z), \qquad \tau_{xy} = \mu(u_y + v_x) \tag{11}$$

$$\tau_{yy} = \frac{2}{3}\mu(2v_y - u_x - w_z), \qquad \tau_{xz} = \mu(u_z + w_x) \tag{12}$$

$$\tau_{zz} = \frac{2}{3}\mu(2w_z - u_x - v_y), \qquad \tau_{yz} = \mu(v_z + w_y) \tag{13}$$

and μ is the viscosity. In addition, we define

$$\vec{\nabla} \cdot \tau = \begin{pmatrix} \tau_{xx} \\ \tau_{xy} \\ \tau_{xz} \end{pmatrix}_x + \begin{pmatrix} \tau_{xy} \\ \tau_{yy} \\ \tau_{yz} \end{pmatrix}_y + \begin{pmatrix} \tau_{xz} \\ \tau_{yz} \\ \tau_{zz} \end{pmatrix}_z \tag{14}$$

$$\vec{V}\tau = (u\tau_{xx} + v\tau_{xy} + w\tau_{xz}, u\tau_{xy} + v\tau_{yy} + w\tau_{yz}, u\tau_{xz} + v\tau_{yz} + w\tau_{zz}) \tag{15}$$

so that the viscosity terms in the Navier-Stokes equations can be represented as

$$Vis = \begin{pmatrix} 0 \\ \vec{\nabla} \cdot \tau \\ \vec{\nabla} \cdot (\vec{V}\tau) \end{pmatrix} \tag{16}$$

2.1.2 Eigensystem

The Navier-Stokes equations can be thought of as the inviscid Euler equations plus some viscosity terms. We discretize the spatial part of the inviscid Euler equations in the usual way, e.g. ENO [15]. These methods require an eigensystem which we list below. Note that we only list the two dimensional eigensystem, since there are no three dimensional examples in this paper. However, the method works well and it is straightforward to implement in three dimensions as we shall show in a future paper. Once the spatial part of the inviscid Euler equations is discretized, we discretize the viscous terms and use the combined discretization as the right hand side for a time integration method, e.g. we use 3rd order TVD Runge-Kutta [15].

The eigenvalues and eigenvectors for the Jacobian matrix of $\vec{F}(\vec{U})$ are obtained by setting $A = 1$ and $B = 0$ in the following formulas, while those for the Jacobian of $\vec{G}(\vec{U})$ are obtained with $A = 0$ and $B = 1$.

The eigenvalues are

$$\lambda^1 = \hat{u} - c, \quad \lambda^2 = \lambda^3 = \hat{u}, \quad \lambda^4 = \hat{u} + c \tag{17}$$

and the eigenvectors are

$$\vec{L}^1 = \left(\frac{b_2}{2} + \frac{\hat{u}}{2c}, -\frac{b_1 u}{2} - \frac{A}{2c}, -\frac{b_1 v}{2} - \frac{B}{2c}, \frac{b_1}{2}\right) \tag{18}$$

$$\vec{L}^2 = (1 - b_2, b_1 u, b_1 v, -b_1) \tag{19}$$

$$\vec{L}^3 = (\hat{v}, B, -A, 0) \tag{20}$$

$$\vec{L}^4 = \left(\frac{b_2}{2} - \frac{\hat{u}}{2c}, -\frac{b_1 u}{2} + \frac{A}{2c}, -\frac{b_1 v}{2} + \frac{B}{2c}, \frac{b_1}{2}\right) \tag{21}$$

$$\vec{R}^1 = \begin{pmatrix} 1 \\ u - Ac \\ v - Bc \\ H - \hat{u}c \end{pmatrix}, \quad \vec{R}^2 = \begin{pmatrix} 1 \\ u \\ v \\ H - \frac{1}{b_1} \end{pmatrix} \tag{22}$$

$$\vec{R}^3 = \begin{pmatrix} 0 \\ B \\ -A \\ -\hat{v} \end{pmatrix}, \quad \vec{R}^4 = \begin{pmatrix} 1 \\ u + Ac \\ v + Bc \\ H + \hat{u}c \end{pmatrix} \tag{23}$$

where

$$q^2 = u^2 + v^2, \quad \hat{u} = Au + Bv, \quad \hat{v} = Av - Bu \tag{24}$$

$$\Gamma = \frac{p_e}{\rho}, \quad c = \sqrt{p_\rho + \frac{\Gamma p}{\rho}}, \quad H = \frac{E + p}{\rho} \tag{25}$$

$$b_1 = \frac{\Gamma}{c^2}, \qquad b_2 = 1 + b_1 q^2 - b_1 H \tag{26}$$

The eigensystem for the one-dimensional equations is obtained by setting $v = 0$.

2.2 Level Set Equation

We use the level set equation

$$\phi_t + \vec{V} \cdot \vec{\nabla}\phi = 0 \tag{27}$$

to keep track of the interface location as the zero level of ϕ. In this equation, the level set velocity, $\vec{V}$, is chosen to be the local fluid velocity which is a natural choice when the interface is a simple contact discontinuity or a non-reacting material interface. In general ϕ starts out as the signed distance function, is advected by solving equation 27 using the methods in [9], and then is reinitialized using

$$\phi_t + S(\phi_o)\left(|\vec{\nabla}\phi| - 1\right) = 0 \tag{28}$$

to keep ϕ approximately equal to the distance function, $|\vec{\nabla}\phi| = 1$, near the interface where we need additional information. We note that our method allows us to solve equation 27 independently of the Navier-Stokes equations. That is, equation 27 can be solved directly using the method in [9], and the eigensystem listed above does not depend on ϕ, since we will be solving only one phase problems with any given eigensystem (see the later sections). For more details on the level set function see [4, 13, 14, 16].

2.3 Equation of State

For an ideal gas $p = \rho RT$ where $R = \frac{R_u}{M}$ is the specific gas constant, with $R_u \approx 8.31451\frac{J}{molK}$ the universal gas constant and M the molecular weight of the gas. Also valid for an ideal gas is $c_p - c_v = R$ where c_p is the specific heat at constant pressure. Additionally, gamma as the ratio of specific heats $\gamma = \frac{c_p}{c_v}$. [8]

For an ideal gas, equation 5 becomes

$$de = c_v dT \tag{29}$$

and assuming that c_v does not depend on temperature (calorically perfect gas), we integrate to obtain

$$e = c_v T \tag{30}$$

where we have set e to be zero at $0K$. Note that e is not uniquely determined, and we could choose any value for e at $0K$ (although one needs to use caution when dealing with more than one material to be sure that integration constants are consistent with the heat release in any chemical reactions that occur).

Note that we may write

$$p = \rho RT = \frac{R}{c_v}\rho e = (\gamma - 1)\rho e \tag{31}$$

for use in the eigensystem.

3 The GFM for Inviscid Flow

We use the level set function to keep track of the interface. The zero level marks the location of the interface, while the positive values correspond to one fluid and the negative values correspond to the other fluid. Each fluid satisfies the inviscid Euler equations ($Vis = 0$) described in the last section with different equations of state for each fluid. Based on the work in [9], the discretization of the level set function can be done independent of the two sets of Euler equations. Besides discretizing equation 27 we need to discretize two sets of Euler equations. This will be done with the help of ghost cells.

Any level set function defines two separate domains for the two separate fluids, i.e. each point corresponds to one fluid or the other. Our goal is to define a ghost cell at every point in the computational domain. In this way, each grid point will contain the mass, momentum, and energy for the real fluid that exists at that point (according to the sign of the level set function) and a ghost mass, momentum, and energy for the other fluid that does not really exist at that point (the fluid on the other side of the interface). Once the ghost cells are defined, we can use standard methods, e.g. see [15], to update the Euler equations at every grid point for both fluids. Then we advance the level set function to the next time step and use the sign of the level set function to determine which of the two updated fluid values should be used at each grid point.

Consider a general time integrator for the Euler equations. In general, we construct right hand sides of the ordinary differential equation for both fluids based on the methods in [15], then we advance the level set to the next time level and pick one of the two right hand sides to use for the Euler equations based on the sign of the level set function. This can be done for every step and every combination of steps in a multistep method. Since both fluids are solved for at every grid point, we just choose the appropriate fluid based on the sign of the level set function.

To summarize, use ghost cells to define each fluid at every point in the computational domain. Update each fluid separately in multidimensional space for one time step or one substep of a multistep time integrator with standard methods. Then update the level set function independently using the real fluid velocities and the sign of the level set function to decide which of the two answers is the valid answer at each grid point. Keep the valid answer and discard the other so that only one fluid is defined at each grid point.

Lastly, we note that only a band of 3 to 5 ghost cells on each side of the interface is actually needed by the computational method depending on the stencil and movement of the interface. One can optimize the code accordingly.

3.1 Defining Values at the Ghost Cells

In [4], the GFM was implemented for a contact discontinuity in the inviscid Euler equations. In that case, it was apparent that the pressure and normal velocity were continuous, while tangential velocity was continuous in the case of a no-slip interface but discontinuous for shear waves. It was also apparent that the entropy was discontinuous.

For variables that are continuous across the interface, we define the values of the ghost fluid to be equal to the values of the real fluid at each grid point. Since these variables are continuous, this node by node population will implicitly capture the interface values of the continuous variables.

Note that the discontinuous variables are governed by a linearly degenerate eigenvalue. Thus, they move with the speed of the interface and information in these variables should not cross the interface. In order to avoid numerical smearing of these variables, we use one sided extrapolation to populate the values in the ghost fluid. Note that the work in [6] shows that one does not have to deal directly with the entropy. There are a few options for the choice of the variable used in extrapolation, ranging from density to temperature.

The extrapolation of the discontinuous variables is carried out in the following fashion. Using the level set function, we can define the unit normal at every grid point as

$$\vec{N} = \frac{\vec{\nabla}\phi}{|\vec{\nabla}\phi|} = < n_1, n_2, n_3 > \tag{32}$$

and then we can solve the advection equation

$$I_t \pm \vec{N} \cdot \vec{\nabla} I = 0 \tag{33}$$

for each variable I that we wish to extrapolate. Note that the "±" sign is chosen to be of one sign to propagate one fluid's values into one ghost region and the opposite sign to propagate the other fluid's values into the other ghost region. Of course this partial differential equation is solved with a Dirichlet type boundary condition so that the real fluid values do not change. This equation only needs to be solved for a few time steps to populate a thin band of ghost cells needed for the numerical method.

In addition, we use the Isobaric Fix [6] to minimize "overheating" when necessary. In order to do this, we adjust our Dirichlet type boundary condition in the procedure above so that a band of real fluid values can change their entropy.

When the need arises to extrapolate the tangential velocity, we achieve this by first extrapolating the entire velocity field, $\vec{V} = < u, v, w >$. Then, at every cell in the ghost region we have two separate velocity fields, one from the real fluid and one from the extrapolated fluid. For each velocity field, the normal component of velocity, $V_N = \vec{V} \cdot \vec{N}$, is put into a vector of length three, $V_N \vec{N}$, and then we use a complementary projection idea [7] to define the two dimensional velocity field in the tangent plane by another vector of length three, $\vec{V} - V_N \vec{N}$. Finally, we take the normal component of velocity, $V_N \vec{N}$, from the real fluid and the tangential component of velocity, $\vec{V} - V_N \vec{N}$, from the extrapolated fluid and add them together to get our new velocity to occupy the ghost cell. This new velocity is our ghost fluid velocity.

Once the ghost fluid values are defined as outlined above, they can be use to assemble the conserved variables for the ghost fluid.

4 Generalization of the Ghost Fluid Method

For a simple contact discontinuity in the inviscid Euler equations, we were able to separate the variables into two sets based on their continuity at the interface. The continuous variables were copied into the ghost fluid in a node by node fashion in order to capture the correct interface values. The discontinuous variables were extrapolated in a one sided fashion to avoid numerical dissipation. We wish to apply this idea to a general interface, moving at speed D in the normal direction, separating two general materials. That is, we need to know which variables are continuous for this general interface.

Conservation of mass, momentum, and energy can be applied to the interface in order to abstract some continuous variables. One can place a flux on the interface oriented tangent to the interface so that material that passes through this flux passes through the interface. Then this flux will move with speed D (the interface speed in the normal direction), and the mass, momentum, and energy which flows into this flux from one side of the interface must flow back out the other side of the interface. That is, the mass, momentum, and energy flux in this moving reference frame are continuous variables. Otherwise, there would be a mass, momentum, or energy sink at this interface and conservation would be violated. We will denote mass, momentum, and energy flux in this moving reference frame as F_ρ, $\vec{F}_{\rho\vec{V}}$, and F_E respectively. The statement that these variables are continuous is essentially the Rankine-Hugoniot jump conditions for an interface moving with speed D in the normal direction. Instead of applying the Rankine-Hugoniot jump conditions explicitly to the interface, we will use the fact that F_ρ, $\vec{F}_{\rho\vec{V}}$, and F_E are continuous to define a ghost fluid that captures the interface values of these variables.

Remark: Note that numerically F_ρ, $\vec{F}_{\rho\vec{V}}$, and F_E may not be continuous. This could occur from initial data or wave interactions. However, since we treat F_ρ, $\vec{F}_{\rho\vec{V}}$, and F_E as though they were continuous in the numerical method, numerical dissipation will smooth them out. In fact, this numerical dissipation will help to guarantee the correct numerical solution.

Remark: The level set function is only designed to represent interfaces where the interface crosses the material at most once [14]. Simple material interfaces that move with the local material velocity never cross over material. If one material is being converted into another then the interface may include a regression rate for this conversion. If the regression rate for this conversion of one material into another is based on some sort of chemical reaction, then the interface can pass over a material exactly once changing it into another material. The same chemical reaction cannot occur to a material more than once. The GFM for deflagration and detonation discontinuities is presented in [5].

Remark: Shocks may be seen as a conversion of an uncompressed material to a compressed material. In this case, D would be the shock speed. This method could be used to follow a lead shock, but since shocks can pass over a material more than once, all subsequent shocks must be captured. In fact, this turns out to be useful and we will examine the use of this method for shocks in a future paper [1].

Remark: In the general case, $\vec{F}_{\rho\vec{V}}$ and F_E will include general mechanical stress terms on the interface besides viscosity and pressure, e.g. surface tension, and material models. We will consider these general mechanical stress terms in future papers. In this paper, pressure and viscosity will be the only mechanical stress terms on the interface.

Remark: In the general case, F_E will include general thermal stress on the interface, e.g. thermal conductivity. We will consider thermal stress in a future paper.

To define F_ρ, $\vec{F}_{\rho\vec{V}}$, and F_E, one takes the equations and writes them in conservation form for mass, momentum, and energy. The fluxes for these variables are then rewritten in the reference frame of a flux which is tangent to the interface by simply taking the dot product with the normal direction

$$\left[\left\langle \vec{F}(\vec{U}), \vec{G}(\vec{U}), \vec{H}(\vec{U})\right\rangle - \begin{pmatrix} \vec{0} \\ \tau \\ \vec{V}\tau \end{pmatrix}\right] \cdot \vec{N} \tag{34}$$

where $\vec{0} =< 0,0,0 >$. Note that the dot product with $\vec{N}$ is equivalent to multiplication by the transpose of $\vec{N}$. We use the superscript T to designate the transpose and rewrite the above expression as

$$\begin{pmatrix} \rho \\ \rho\vec{V}^T \\ E \end{pmatrix} V_N + \begin{pmatrix} 0 \\ (pI-\tau)\vec{N}^T \\ \vec{V}(pI-\tau)\vec{N}^T \end{pmatrix} \tag{35}$$

where $V_N = \vec{V}\cdot\vec{N}$ is the local fluid velocity normal to the interface and I is the three by three identity matrix. Then the measurements are taken in the moving reference frame (speed D) to get

$$\begin{pmatrix} \rho \\ \rho\left(\vec{V}^T - D\vec{N}^T\right) \\ \rho e + \frac{\rho|\vec{V}-D\vec{N}|^2}{2} \end{pmatrix} (V_N - D) + \begin{pmatrix} 0 \\ (pI-\tau)\vec{N}^T \\ (\vec{V}-D\vec{N})(pI-\tau)\vec{N}^T \end{pmatrix} \tag{36}$$

from which we define

$$F_\rho = \rho(V_N - D) \tag{37}$$

$$\vec{F}_{\rho\vec{V}} = \rho\left(\vec{V}^T - D\vec{N}^T\right)(V_N - D) + (pI-\tau)\vec{N}^T \tag{38}$$

$$F_E = \left(\rho e + \frac{\rho|\vec{V}-D\vec{N}|^2}{2}\right)(V_N - D) + (\vec{V}-D\vec{N})(pI-\tau)\vec{N}^T \tag{39}$$

as continuous variables for use in the GFM. That is, we will define the ghost fluid in a node by node fashion by solving the system of equations

$$F_\rho^G = F_\rho^R \tag{40}$$

$$\vec{F}_{\rho\vec{V}}^G = \vec{F}_{\rho\vec{V}}^R \tag{41}$$

$$F_E^G = F_E^R \tag{42}$$

at each grid point. Note that the superscript "R" stands for a real fluid value at a grid point, while the superscript "G" stands for a ghost fluid value at a grid point. Since F_ρ^R, $\vec{F}_{\rho\vec{V}}^R$, F_E^R, $\vec{N}$, and D are known at each grid point, these can be substituted into equations 40, 41, and 42, leaving ρ^G, $\vec{V}^G$, p^G, e^G, and τ^G undetermined. The appropriate equation of state can be used to eliminate one of these variables.

4.1 Non-reacting Material Interfaces

In this paper, we are interested in simple material interfaces where the interface moves with the fluid velocity only. That is, we use the local fluid velocity, $\vec{V}$, in the level set equation as the level set velocity.

The level set velocity, $\vec{V}$, is defined everywhere by a continuous function where the value of $\vec{V}$ at the interface is equal to the value that moves the interface at the correct interface velocity. This global definition of $\vec{V}$ is the one we use to find D for use in solving equations 40, 41, and 42. In a node by node fashion, we define $D = \vec{V} \cdot \vec{N}$ as the velocity of the interface in the normal direction. In this way, we capture the correct value of D at the interface.

Using $D = \vec{V}^R \cdot \vec{N} = V_N^R$ in equation 40 yields

$$\rho^G (V_N^G - V_N^R) = 0 \tag{43}$$

implying that $V_N^G = V_N^R$. That is, the normal component of the ghost fluid velocity should be equal to the normal component of the real fluid velocity at each point. This allows us to simplify equations 38 and 39 to

$$\vec{F}_{\rho\vec{V}} = (pI - \tau)\vec{N}^T \tag{44}$$

$$F_E = (\vec{V} - V_N\vec{N})(pI - \tau)\vec{N}^T = -(\vec{V} - V_N\vec{N})\tau\vec{N}^T \tag{45}$$

If the flow were inviscid, then $\tau = 0$ and equation 41 becomes $p^G = p^R$ implying that the pressure of the ghost fluid should be equal to the pressure of the real fluid at each point. Equation 42 is trivially satisfied and we are left with some freedom. As shown earlier, the entropy should be extrapolated in the normal direction along with an Isobaric Fix [6] to minimize "overheating". The tangential velocities may be extrapolated for a shear wave or copied over node by node to enforce continuity of the tangential velocities for a "no-slip" boundary condition.

5 The Viscous Stress Tensor

Before we define τ^G, we compute τ at each grid point by first computing the spatial derivatives of the velocities at each grid point. For a "no-slip" boundary condition, we use standard second order central differencing in order to compute the derivatives of the velocities. Note that this will consist of differencing across the interface, and discontinuous velocity profiles will become continuous due to numerical dissipation. For a "slip" boundary condition, we avoid differencing across the interface in order to allow jumps in the velocity field. If a grid point and both of its neighbors have the same sign of ϕ (meaning that they are all on the same side of the interface), then we use a standard second order central difference. If a grid point only has one neighbor with the same sign of ϕ, then we only use that neighbor and a first order one-sided difference. If a grid point has a different sign of ϕ than both of its neighbors, we are in an underresolved region and cannot compute the derivative. Our only choice is to set the derivative to zero. Once all the spatial derivatives are computed, we combine them with the appropriate value of μ depending on the sign of ϕ.

After computing τ at all the grid points, we use use equation 33 to extrapolate each term in τ across the interface. Then, at every grid point in the ghost region we have two separate values for τ, one from the real fluid and one from the extrapolated fluid. Equations 41 and 42 tell us that part of the viscous stress tensor is continuous and part is discontinuous. We write

$$\tau = \tau_C + \tau_D \tag{46}$$

where τ_C is the continuous part of τ, and τ_D is the discontinuous part of τ. Then we take the real fluid value of τ_C and the extrapolated fluid value of τ_D and add them together to get τ^G as our ghost fluid value at each grid point.

Consider a specific grid point where the unit normal is $\vec{N}$. We could choose a basis for the the two dimensional tangent plane with two orthogonal unit tangent vectors $\vec{T_1}$ and $\vec{T_2}$. Then the viscous stress tensor can be written in this new coordinate system with a simple transformation

$$\begin{pmatrix} \tau_{nn} & \tau_{nt_1} & \tau_{nt_2} \\ \tau_{nt_1} & \tau_{t_1t_1} & \tau_{t_1t_2} \\ \tau_{nt_2} & \tau_{t_1t_2} & \tau_{t_2t_2} \end{pmatrix} = \begin{pmatrix} \vec{N} \\ \vec{T_1} \\ \vec{T_2} \end{pmatrix} \begin{pmatrix} \tau_{xx} & \tau_{xy} & \tau_{xz} \\ \tau_{xy} & \tau_{yy} & \tau_{yz} \\ \tau_{xz} & \tau_{yz} & \tau_{zz} \end{pmatrix} \begin{pmatrix} \vec{N} \\ \vec{T_1} \\ \vec{T_2} \end{pmatrix}^T \tag{47}$$

where we have defined our transformation matrix Q by

$$Q = \begin{pmatrix} \vec{N} \\ \vec{T_1} \\ \vec{T_2} \end{pmatrix} \tag{48}$$

with $Q^TQ = QQ^T = I$.

5.1 Momentum Terms

We use the transformation Q to put equation 44 into a better coordinate system

$$Q\vec{F}_{\rho\vec{V}} = \begin{pmatrix} p - \tau_{nn} \\ -\tau_{nt_1} \\ -\tau_{nt_2} \end{pmatrix} \tag{49}$$

illustrating that $p - \tau_{nn}$, τ_{nt_1} and τ_{nt_2} are continuous variables across the interface. Thus τ_C should contain τ_{nt_1} and τ_{nt_2} and the jump in τ_{nn} can be used to find the pressure of the ghost fluid.

We define the continuous part of τ as

$$\tau_C = Q^T \begin{pmatrix} 0 & \tau_{nt_1} & \tau_{nt_2} \\ \tau_{nt_1} & 0 & 0 \\ \tau_{nt_2} & 0 & 0 \end{pmatrix} Q \tag{50}$$

and compute τ_C through the equation

$$\tau_C = \hat{\tau}_C + \hat{\tau}_C^T \tag{51}$$

where

$$\hat{\tau}_C = Q^T \begin{pmatrix} 0 & 0 & 0 \\ \tau_{nt_1} & 0 & 0 \\ \tau_{nt_2} & 0 & 0 \end{pmatrix} Q = Q^T \begin{pmatrix} 0 \\ \tau_{nt_1} \\ \tau_{nt_2} \end{pmatrix} \vec{N} \tag{52}$$

is easy to find. Note that equation 47 shows that

$$\begin{pmatrix} \tau_{nn} \\ \tau_{nt_1} \\ \tau_{nt_2} \end{pmatrix} = Q\tau\vec{N}^T \tag{53}$$

and using this in equation 52 gives

$$\hat{\tau}_C = Q^T \left[Q\tau\vec{N}^T - \begin{pmatrix} \tau_{nn} \\ 0 \\ 0 \end{pmatrix} \right] \vec{N} = \tau\vec{N}^T\vec{N} - \vec{N}^T\tau_{nn}\vec{N} \tag{54}$$

which can be further simplified to

$$\hat{\tau}_C = (\tau - \tau_{nn}I)\vec{N}^T\vec{N} \tag{55}$$

finally allowing us to define

$$\tau_C = (\tau - \tau_{nn}I)\vec{N}^T\vec{N} + \vec{N}^T\vec{N}(\tau - \tau_{nn}I) \tag{56}$$

From equation 47 we have $\tau_{nn} = \vec{N}\tau\vec{N}^T$ and we use this in equation 56 to compute τ_C. We define τ^G by adding the real fluid value of τ_C to the extrapolated fluid value of $\tau_D = \tau - \tau_C$.

We can find the pressure by solving the equation

$$p^G - \tau_{nn}^G = p^R - \tau_{nn}^R \tag{57}$$

for p^G using the extrapolated value of τ to define τ_{nn}^G.

We stress that equation 56 and the entire numerical algorithm for the viscous stress tensor only depends on the normal, $\vec{N}$, of Q. We have used a basis free projection technique to eliminate the dependence on $\vec{T_1}$ and $\vec{T_2}$ similar to the work in [7].

5.2 Energy Terms

We use equation 47 to rewrite equation 45 as

$$F_E = -(\vec{V} - V_N\vec{N})Q^T \begin{pmatrix} \tau_{nn} \\ \tau_{nt_1} \\ \tau_{nt_2} \end{pmatrix} = -(0, V_{t_1}, V_{t_2}) \begin{pmatrix} \tau_{nn} \\ \tau_{nt_1} \\ \tau_{nt_2} \end{pmatrix} \tag{58}$$

where V_{t_1} and V_{t_2} are the velocities in the tangent directions T_1 and T_2 respectively. Thus, we need

$$F_E = -(V_{t_1}\tau_{nt_1} + V_{t_2}\tau_{nt_2}) \tag{59}$$

to be continuous across the interface.

When the viscosity is zero or the flow is uniform, we have $\tau_{nt_1} = \tau_{nt_2} = 0$ and equation 59 is trivially satisfied. In this case the tangential velocities may or may not be continuous. That is, we may have a shear wave in which case we use a "slip" boundary condition.

When the viscosity is nonzero and the flow is not uniform, $\tau_{nt_1} \neq 0 \neq \tau_{nt_2}$ and we need to define V_{t_1} and V_{t_2} so that equation 59 is satisfied. This can be done with the "no-slip" boundary condition which requires V_{t_1} and V_{t_2} to be continuous across the interface. Since V_N is necessarily continuous, the "no-slip" boundary condition requires the entire velocity field, $\vec{V}$, to be continuous.

5.3 Constructing Fluxes

We will discuss the the numerical discretization in the x-direction, after which the y-direction and z-direction can be inferred by symmetry. We will need a spatial discretization for

$$\begin{pmatrix} 0 \\ \tau_{xx} \\ \tau_{xy} \\ \tau_{xz} \\ u\tau_{xx} + v\tau_{xy} + w\tau_{xz} \end{pmatrix}_x \tag{60}$$

which will be achieved by differencing fluxes. In order to construct the appropriate flux function (which is just equation 60 without the x derivative) we need values of τ_{xx}, τ_{xy}, τ_{xz} and $\vec{V}$ at the cell walls which are centered between the grid points. We define these functions at the cell walls by the average of the values at the neighboring grid points.

Using the arithmetic average to define τ_{xx}, τ_{xy}, τ_{xz} and $\vec{V}$ gives the standard definition of μ, v_y, w_z, u_y, and u_z, but gives a rather wide stencil for u_x, v_x, and w_x. We wish to avoid this wide stencil in regions of the real fluid away from the interface. If a flux in the real fluid has both of its neighboring grid points in the real fluid, then we replace arithmetic average of u_x, v_x, and w_x with the centered difference computed using these two grid points. This gives the standard compact stencil which avoids "odd-even" decoupling.

6 Examples

In our numerical examples we use third order ENO-LLF with 3rd order TVD Runge Kutta methods [15].

6.1 Example 1

We use a $1m$ square domain with 20 grid points in each direction with linear extrapolation of all variables as a first order boundary condition. We use $\gamma = 1.4$, $M = .029\frac{kg}{mol}$, $\rho = 1\frac{kg}{m^3}$, $u = 0\frac{m}{s}$, and $p = 1 \times 10^5 Pa$ for both gases. The interface is defined by $\phi = x - .5$ which makes the interface a straight line defined by $x = .5$, i.e. $\phi = 0$. We initialize the velocity tangent to the interface by $v = 300\sin(2\pi\phi)$ as shown in figure 1. A cross-sectional view of the tangential velocity is shown in 2.

We set the viscosity in both fluids to $100\frac{kg}{ms}$ and show the solution at $t = .001s$ in figure 3. Since both viscosities are equal, the spatial profile of the tangential velocity is linear. We used the "no-slip" boundary condition.

In figure 4 we change the viscosity on the left to $50\frac{kg}{ms}$ and show the solution at $t = .0025$. Once again the spatial profile of the tangential velocity is linear, however there is a kink in the linear function at the interface due to the jump in viscosity. We used the "no-slip" boundary condition.

In figure 5 we change the viscosity on the left to zero, which in turn implies the use of a "slip" boundary condition. At $t = .001s$, the viscous fluid on the right is seeking a flat profile (uniform flow) as shown in figure 5. Note that the discontinuous tangential velocity is indicative of a shear wave.

6.2 Example 2

In this example we use $\phi = -y + 2x - .5$ which defines the $\phi = 0$ interface by the straight line $y = 2x - .5$. This interface is rotated 22.5 degrees clockwise from the interface in the last example. In addition, we change the initial velcoity field to $u = 300\sin(\frac{\pi\phi}{.9d00})\sin(67.5^o)\frac{m}{s}$ and $v = 300\sin(\frac{\pi\phi}{.9d00})\cos(67.5^o)\frac{m}{s}$ as shown in figure 6. We choose a line perpendicular to the interface and plot a cross-section of the tangential velocities in 7. We treat the boundary of the problem by linear extrapolation in the direction defined by the line $y = 2x$. This leaves four boundary values in the upper left hand corner and the lower right hand corner undefined, and we define them by constant extrapolation from the interior corner points, i.e. points (2,19) and (19,2) respectively.

As in the last example, we show the solution at $t = .001s$ in figure 8 for the case where both fluids have a viscosity of $100\frac{kg}{ms}$. In figure 9, we show the solution at $t = .0025$ where the fluid on the left had its viscosity changed to $50\frac{kg}{ms}$. In figure 10, we show the solution at $t = .001s$ where the fluid on the left has zero viscosity. Once again, the first two cases had a "no-slip" boundary condition while the last case has a "slip" boundary condition. We get the expected results in each case.

6.3 Example 3

In this example we return to the interface defined by $\phi = x - .5$ in Example 1. All initial data is the same as in Example 1, except that we define a uniform normal velocity of $u = 50\frac{m}{s}$ so that the interface will move to the right, crossing grid nodes. The initial velcoity field and a cross-section of the tangential velocity are shown in figure 11 and 12 respectively.

We show the solution at $t = .00175s$ for the case where both fluids have viscosities of $100\frac{kg}{ms}$ in figure 13. $x = .6m$. We change the viscosity of the fluid on the left to $50\frac{kg}{ms}$ and show the results at $t = .0025$ in figure 14. We change the viscosity of the fluid on the left to zero and show the results at $t = .00175s$ in figure 15. The first two cases had a "no-slip" boundary condition while the last case has a "slip" boundary condition. In the first two cases, the interface has crossed two grid nodes and is located near $x = .6$, while in the last case the interface has only crossed one grid node.

7 Acknowledgements

Research supported in part by ONR N00014-97-1-0027 and ONR N00014-97-1-0968.

References

[1] Aslam,T.D., *A Level Set Algorithm for Tracking Discontinuities in Hyperbolic Conservation Laws II: Systems of Equations*, (in preparation).

[2] Cocchi, J.-P., Saurel, S., *A Riemann Problem Based Method for the Resolution of Compressible Multimaterial Flows*, Journal of Computational Physics, vol. 137, (1997) pp. 265-298.

[3] Davis, S., *An interface tracking method for hyperbolic systems of conservation laws*, Applied Numerical Mathematics, 10 (1992) 447-472.

[4] Fedkiw, R., Aslam, T., Merriman, B., and Osher, S., *A Non-Oscillatory Eulerian Approach to Interfaces in Multimaterial Flows (The Ghost Fluid Method)*, J. Computational Physics, vol. 152, n. 2, 457-492 (1999).

[5] Fedkiw, R., Aslam, T., and Xu, S., *The Ghost Fluid Method for deflagration and detonation discontinuities*, J. Computational Physics 154, n. 2, 393-427 (1999).

[6] Fedkiw, R., Marquina, A., and Merriman, B., *An Isobaric Fix for the Overheating Problem in Multimaterial Compressible Flows*, J. Computational Physics 148, 545-578 (1999).

[7] Fedkiw, R., Merriman, B., and Osher, S., *Efficient characteristic projection in upwind difference schemes for hyperbolic systems (The Complementary Projection Method)*, J. Computational Physics, vol. 141, 22-36 (1998).

[8] Fedkiw, R., Merriman, B., and Osher, S., *High accuracy numerical methods for thermally perfect gas flows with chemistry*, J. Computational Physics 132, 175-190 (1997).

[9] Fedkiw, R., Merriman, B., and Osher, S., *Simplified Discretization of Systems of Hyperbolic Conservation Laws Containing Advection Equations*, J. Computational Physics, 157, 302-326 (2000).

[10] Fedkiw, R., Merriman, B., and Osher, S., *Numerical methods for a one-dimensional interface separating compressible and incompressible flows*, Barriers and Challenges in Computational Fluid Dynamics, pp 155-194, edited by V, Venkatakrishnan, M. Salas, and S. Chakravarthy, Kluwer Academic Publishers (Norwell,MA), 1998.

[11] Karni, S., *Multicomponent Flow Calculations by a Consistent Primitive Algorithm*, Journal of Computational Physics, v. 112, 31-43 (1994).

[12] Liu, X-D., and S. Osher, *Convex ENO High Order Schemes Without Field-by-Field Decomposition or Staggered Grids*, J. Comput Phys, v142, pp 304-330, (1998).

[13] Mulder, W., Osher, S., and Sethian, J.A., *Computing Interface Motion in Compressible Gas Dynamics*, J. Comput. Phys., v. 100, 209-228 (1992).

[14] Osher, S. and Sethian, J.A., *Fronts Propagating with Curvature Dependent Speed: Algorithms Based on Hamilton-Jacobi Formulations*, Journal of Comput. Phys., vol. 79, n. 1, pp. 12-49, (1988).

[15] Shu, C.W. and Osher, S., *Efficient Implementation of Essentially Non-Oscillatory Shock Capturing Schemes II (two)*, Journal of Computational Physics; Volume 83, (1989), pp 32-78.

[16] Sussman, M., Smereka, P. and Osher, S., *A level set approach for computing solutions to incompressible two-phase flow*, J. Comput. Phys., v. 114, (1994), pp. 146-154.

[17] Wardlaw, A., "Underwater Explosion Test Cases", IHTR 2069, 1998

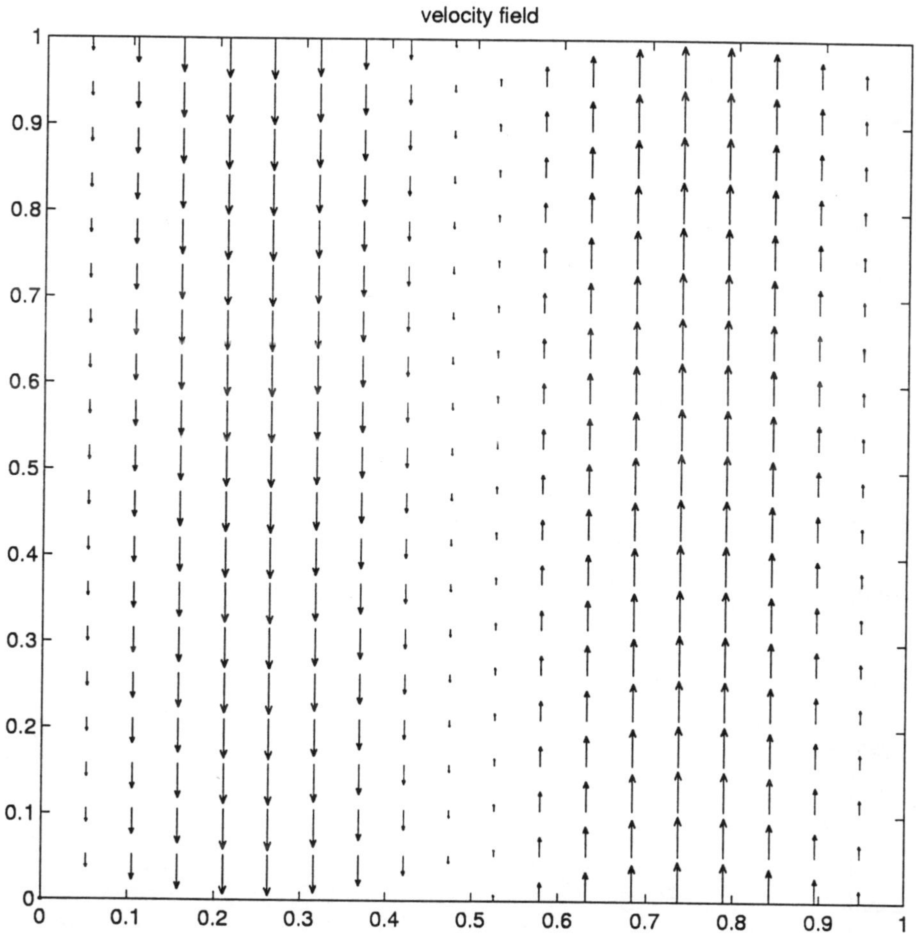

Figure 1: Initial velocity field

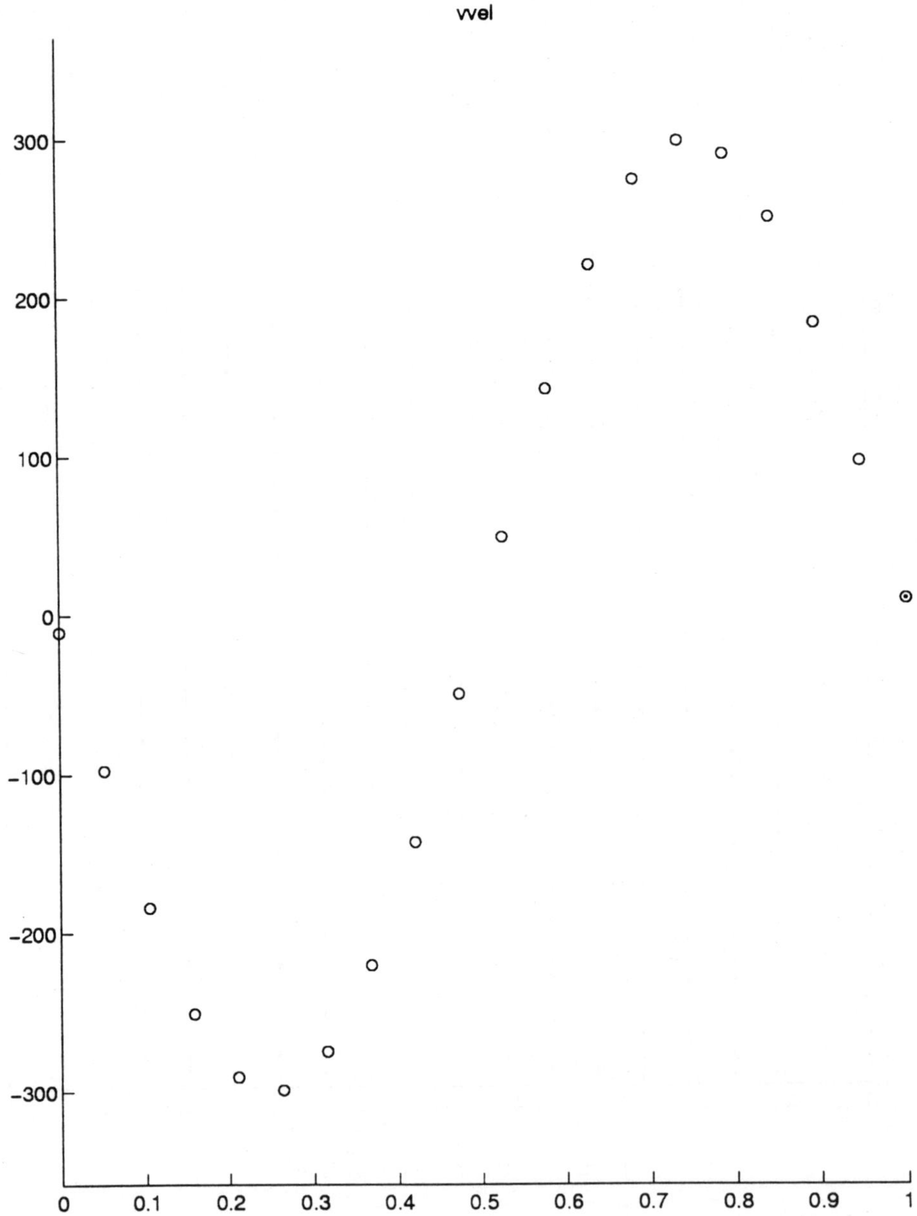

Figure 2: Initial tangetial velocity profile

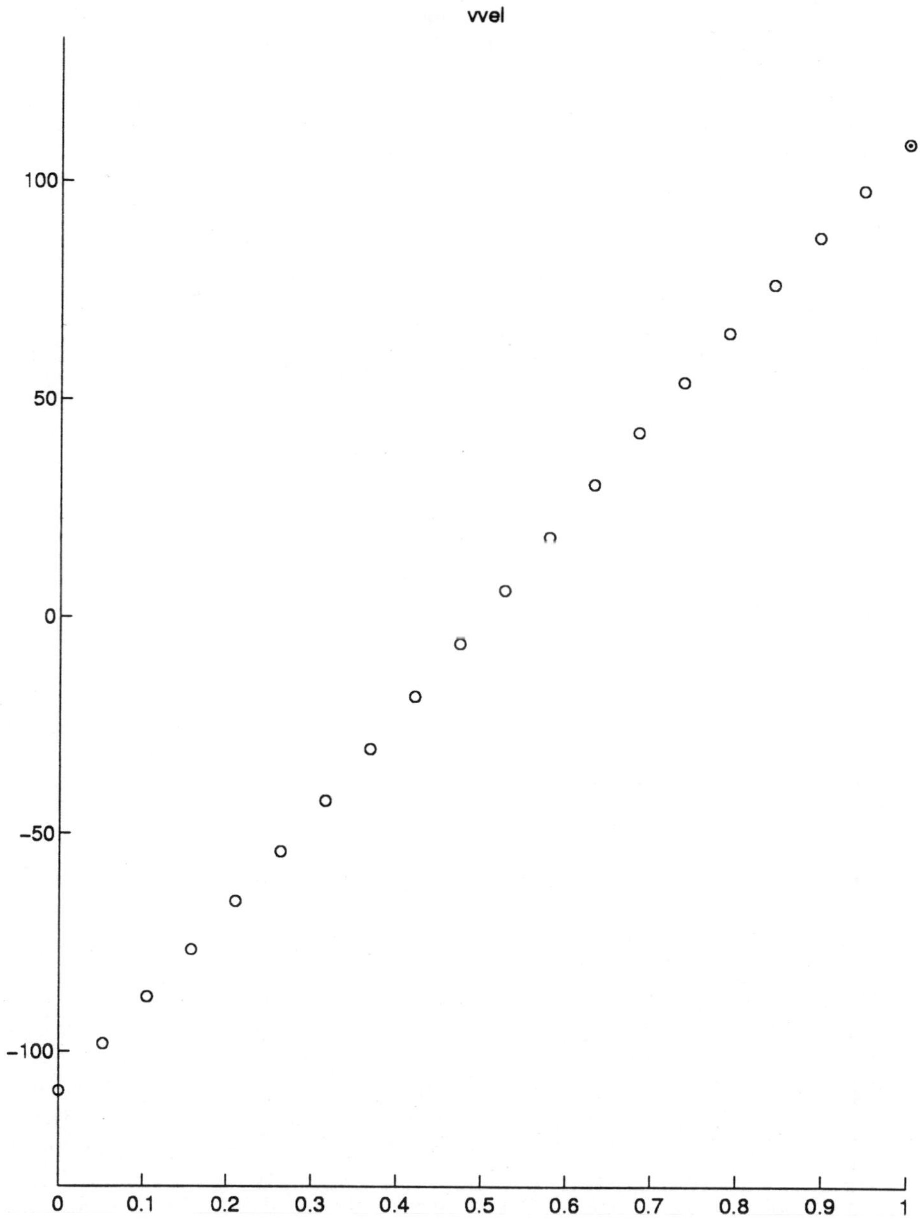

Figure 3: $\mu_{left} = \mu_{right} = 100$, "no-slip"

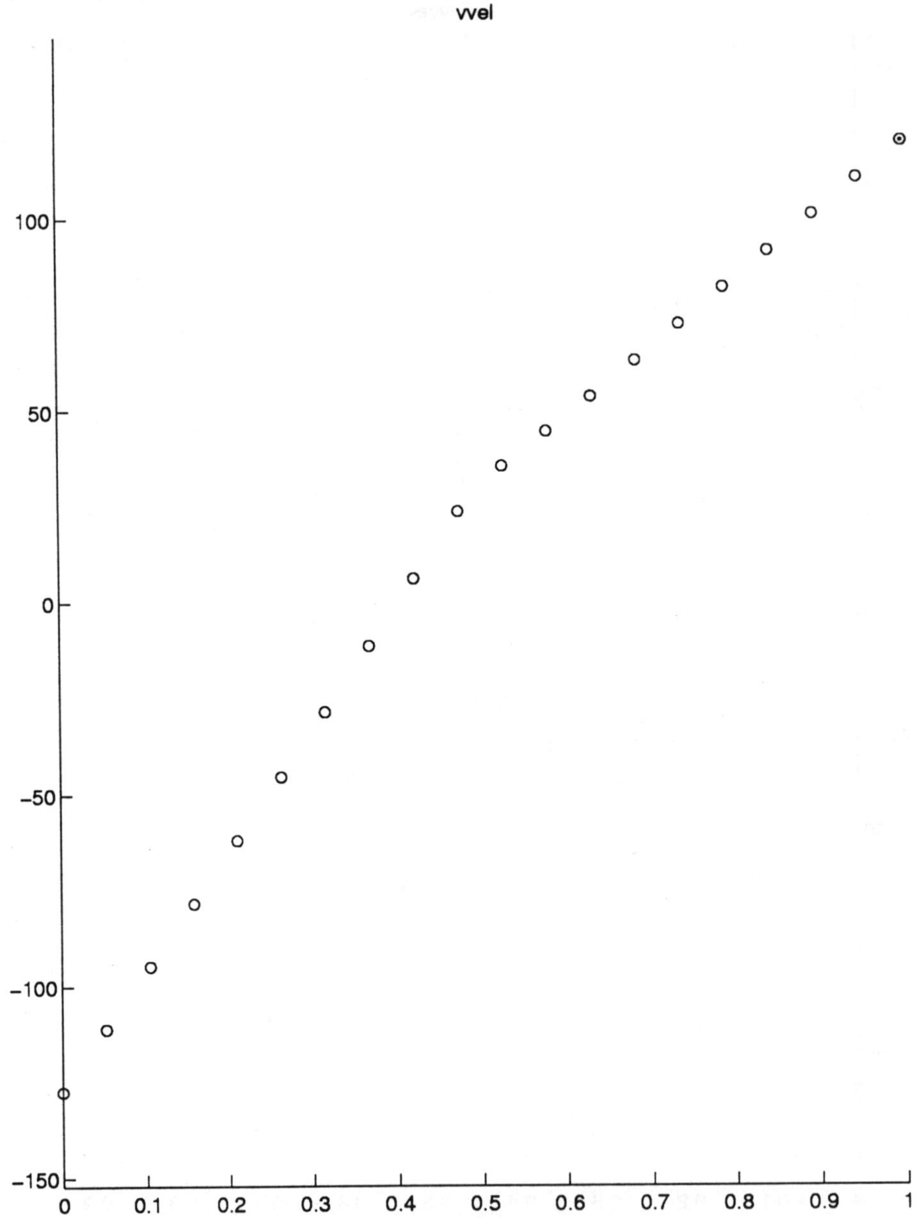

Figure 4: $\mu_{left} = 50$, $\mu_{right} = 100$, "no-slip"

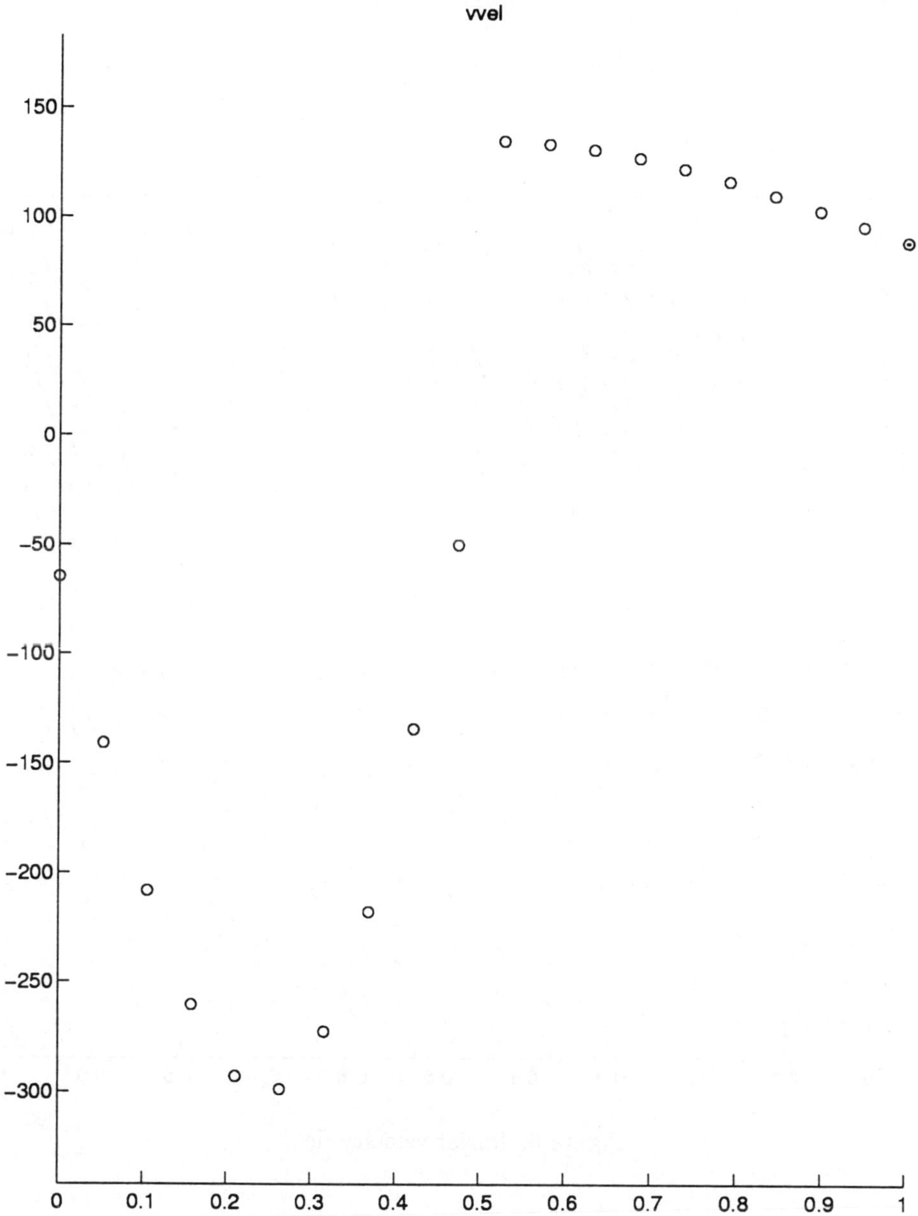

Figure 5: $\mu_{left} = 0$, $\mu_{right} = 100$, "slip"

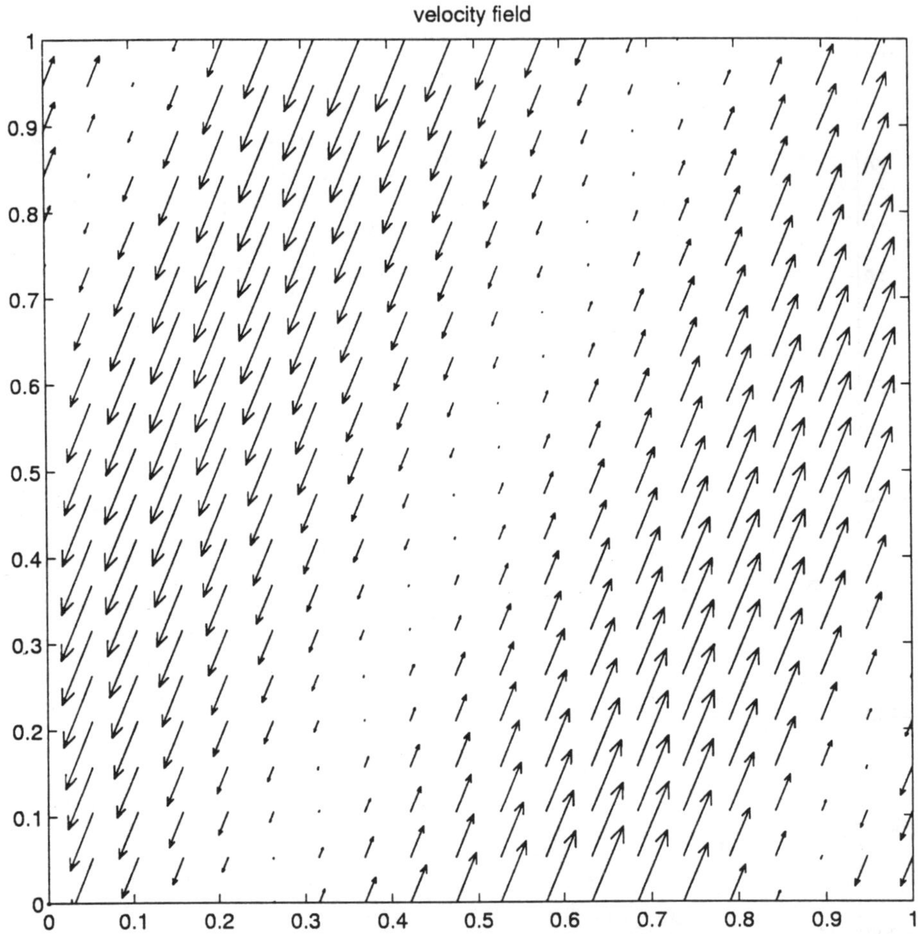

Figure 6: Initial velocity field

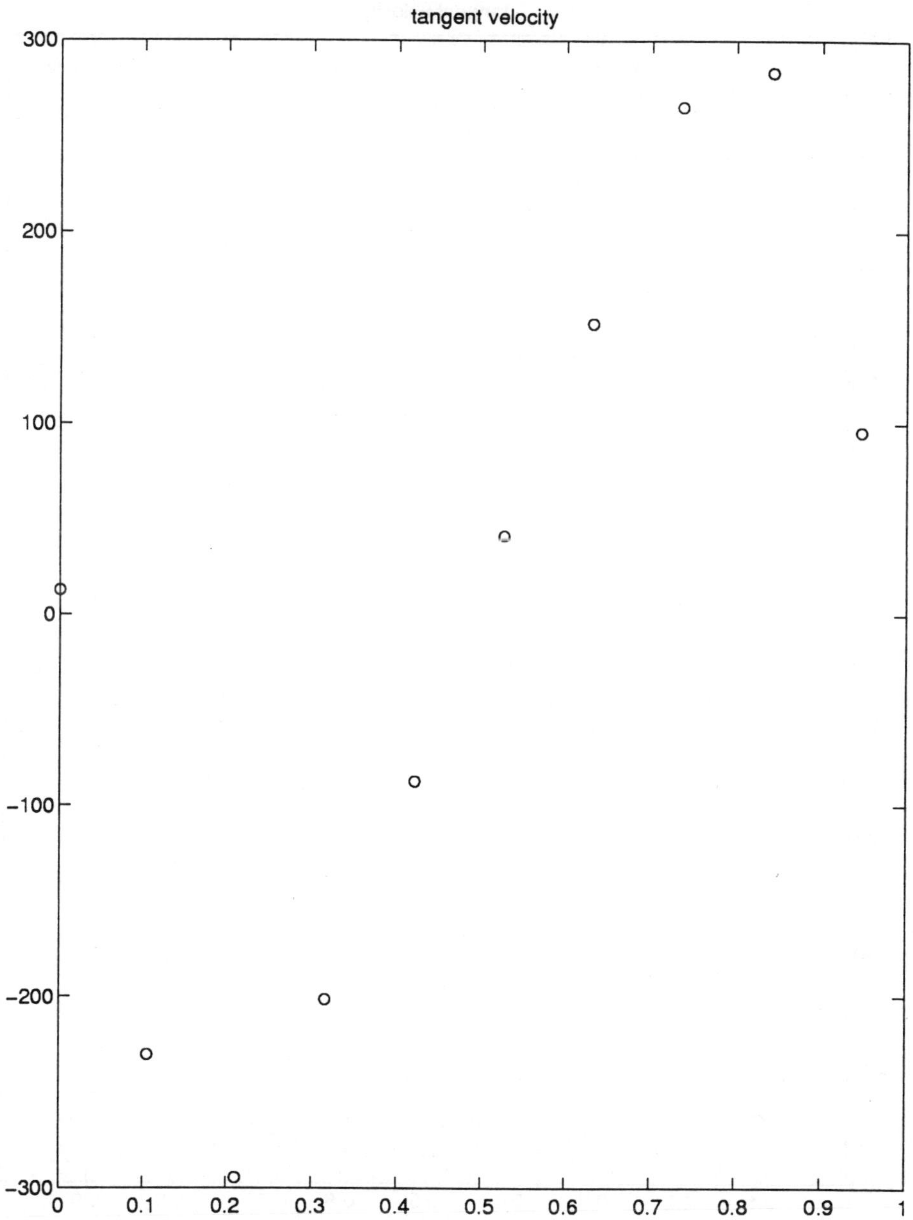

Figure 7: Initial tangetial velocity profile

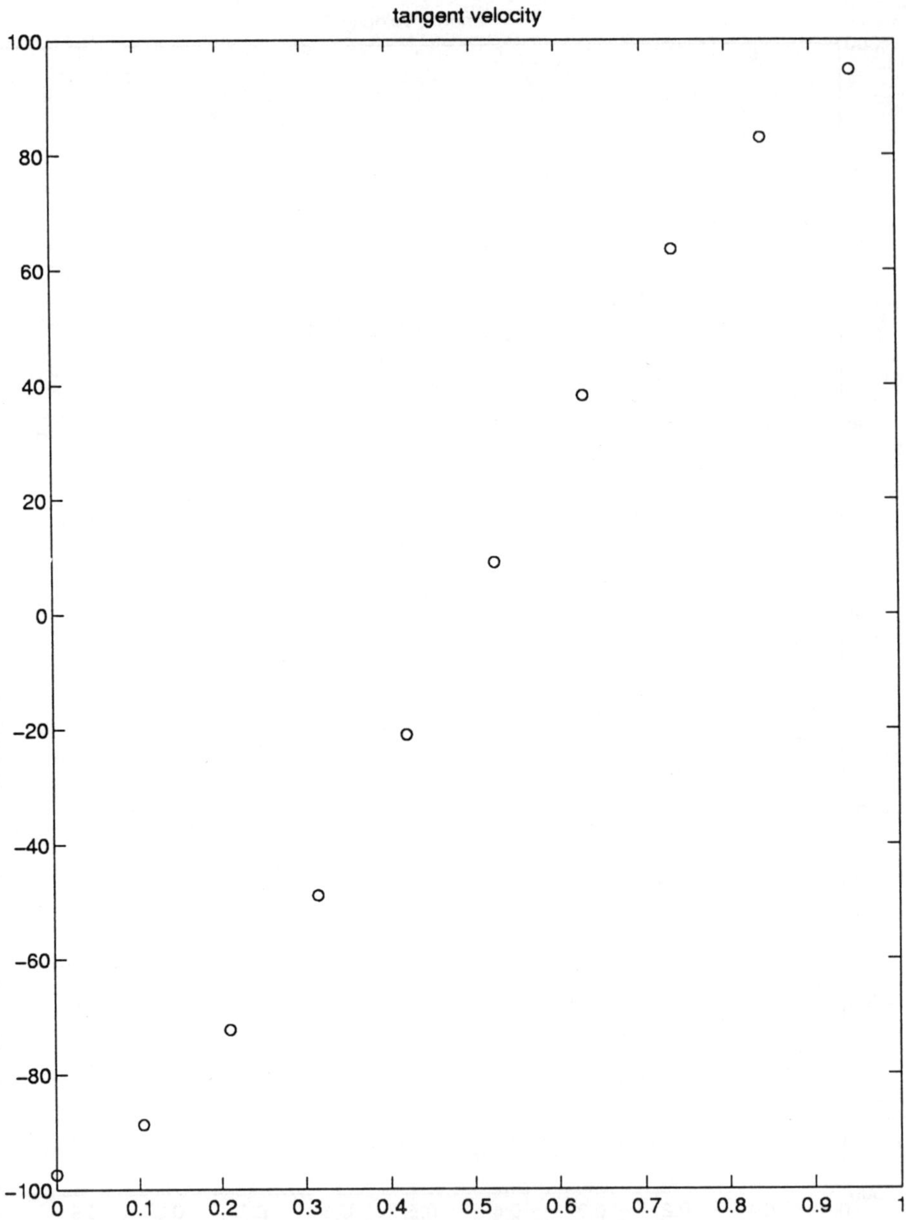

Figure 8: $\mu_{left} = \mu_{right} = 100$, "no-slip"

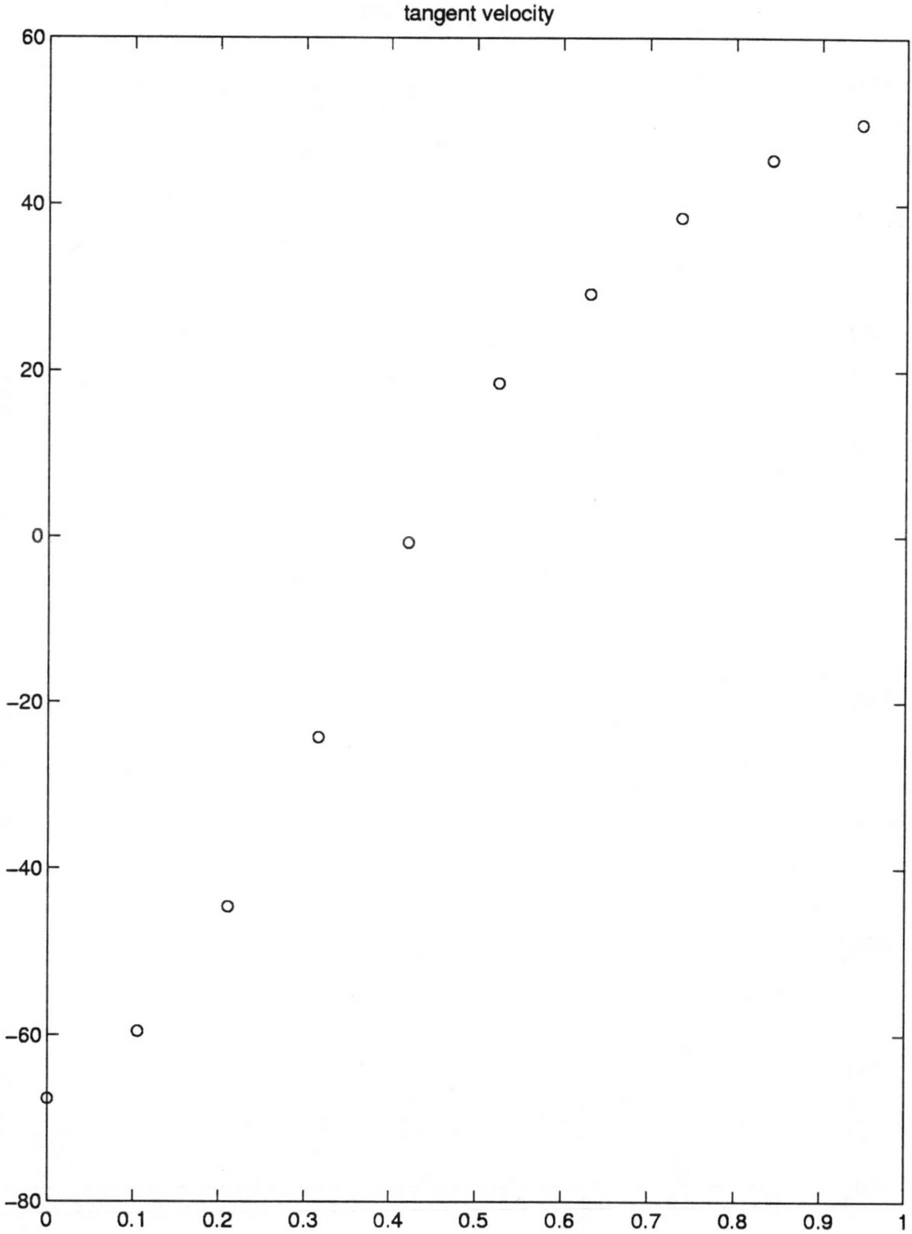

Figure 9: $\mu_{left} = 50$, $\mu_{right} = 100$, "no-slip"

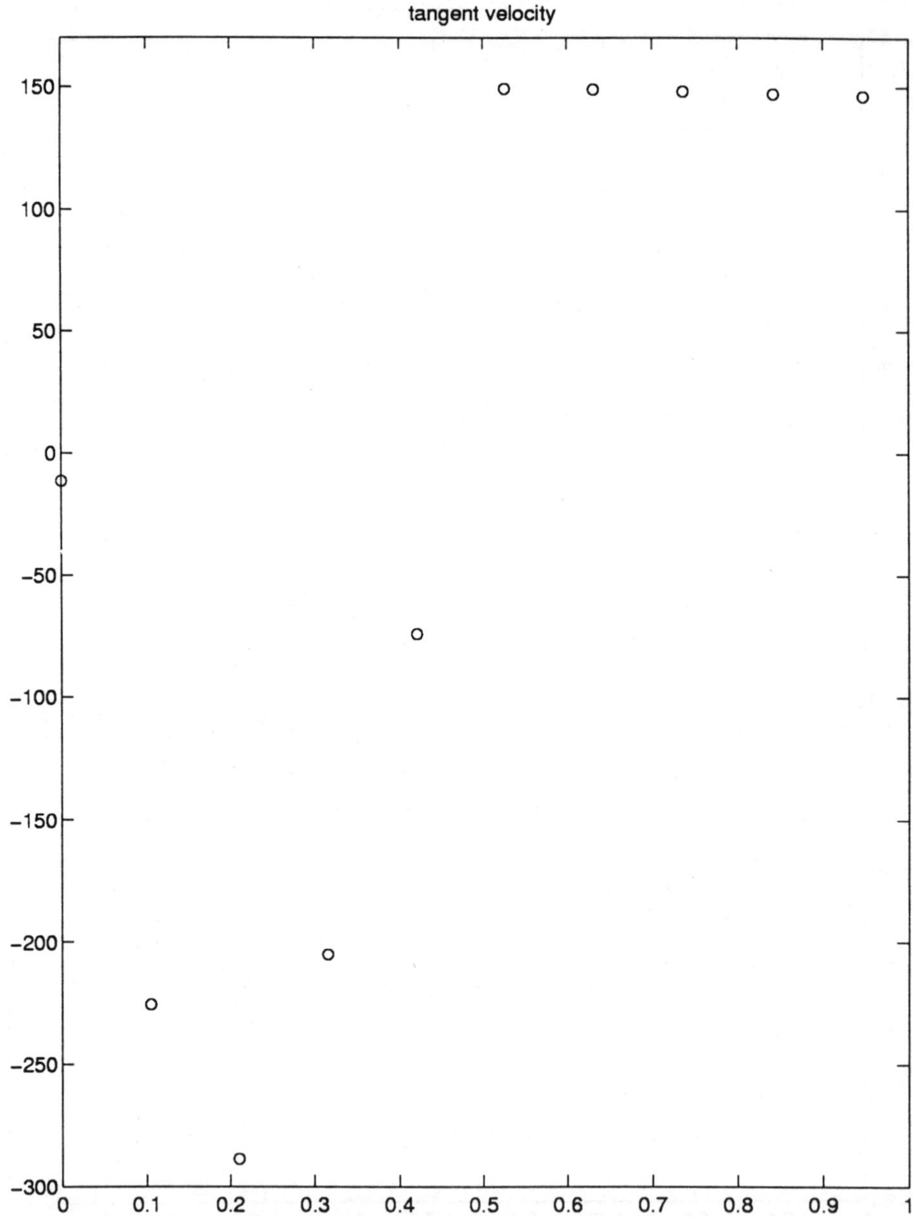

Figure 10: $\mu_{left} = 0$, $\mu_{right} = 100$, "slip"

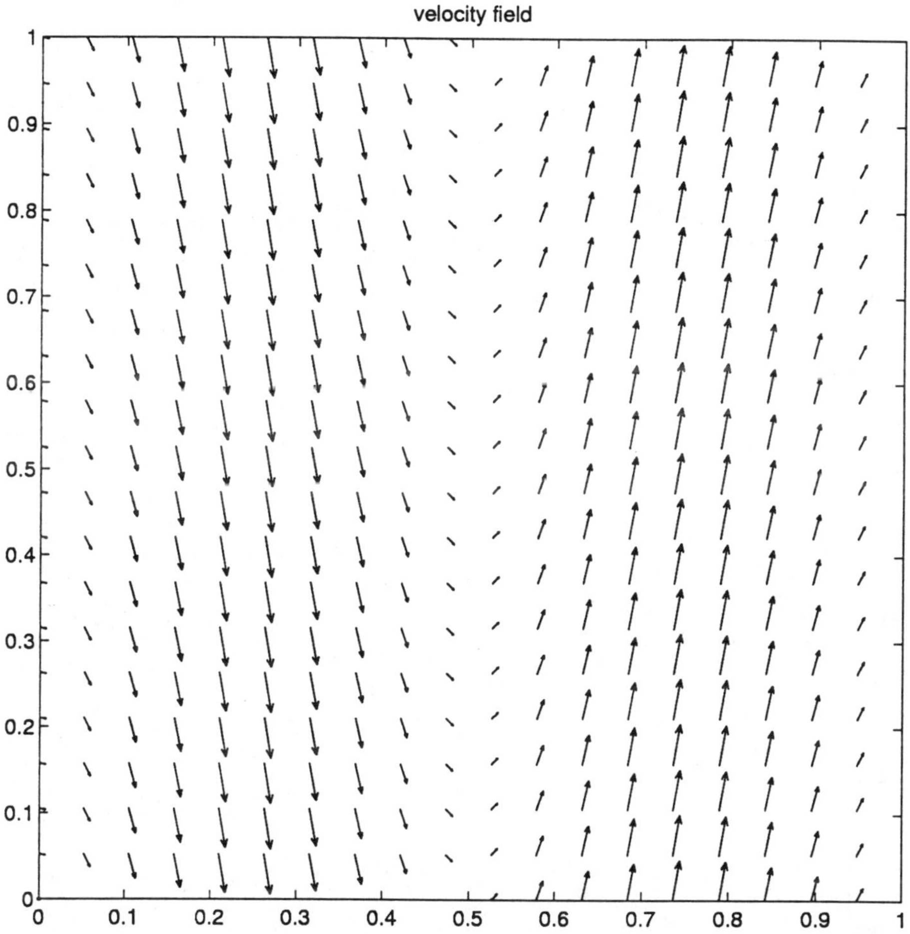

Figure 11: Initial velocity field

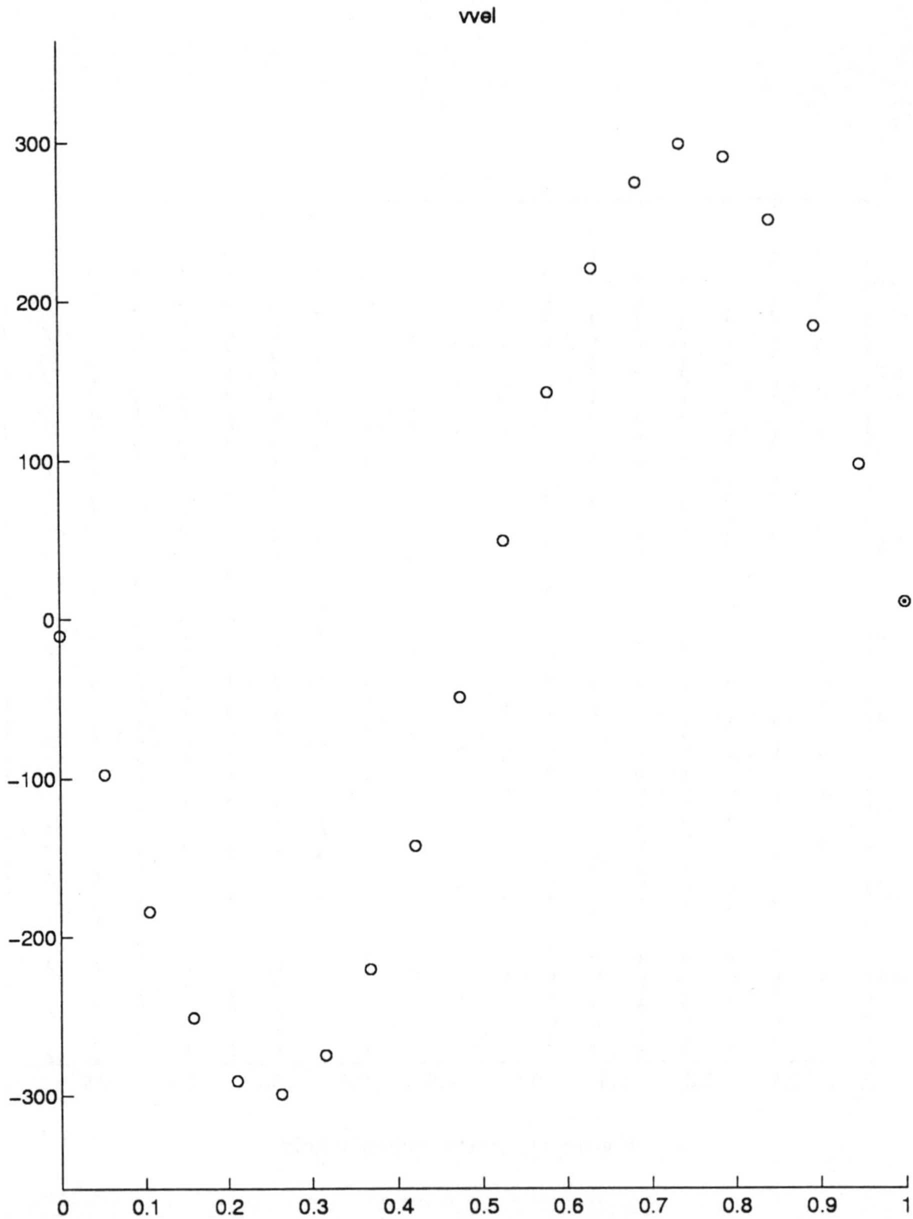

Figure 12: Initial tangetial velocity profile

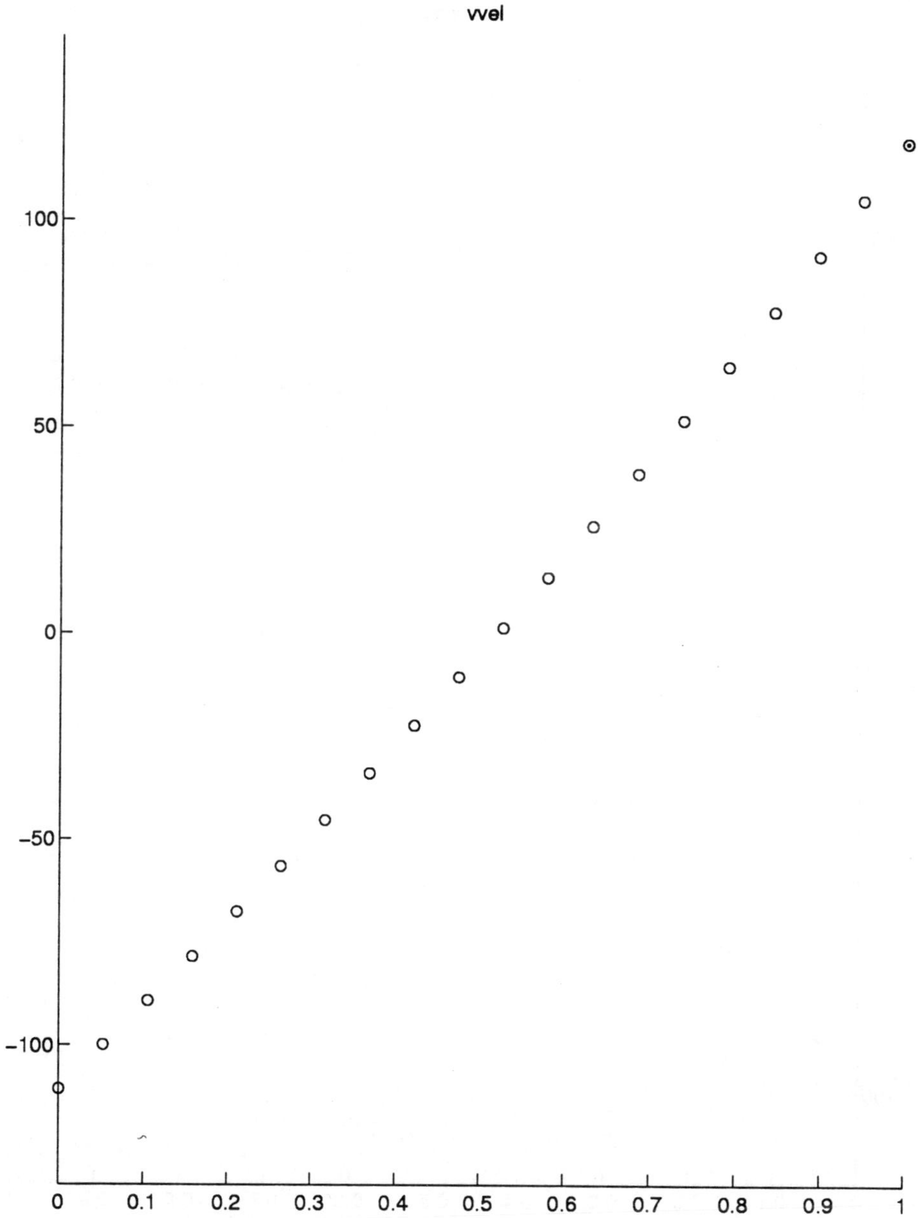

Figure 13: $\mu_{left} = \mu_{right} = 100$, "no-slip"

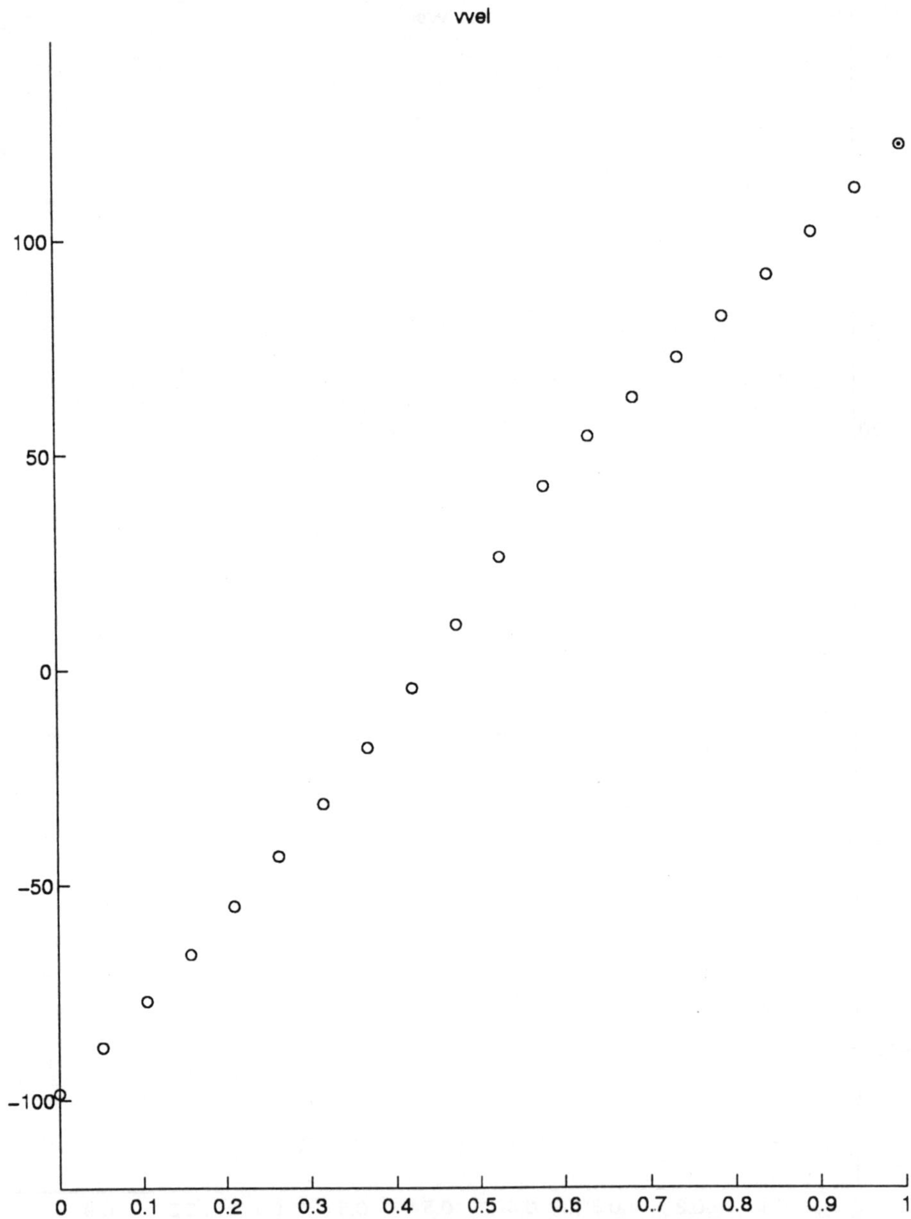

Figure 14: $\mu_{left} = 50$, $\mu_{right} = 100$, "no-slip"

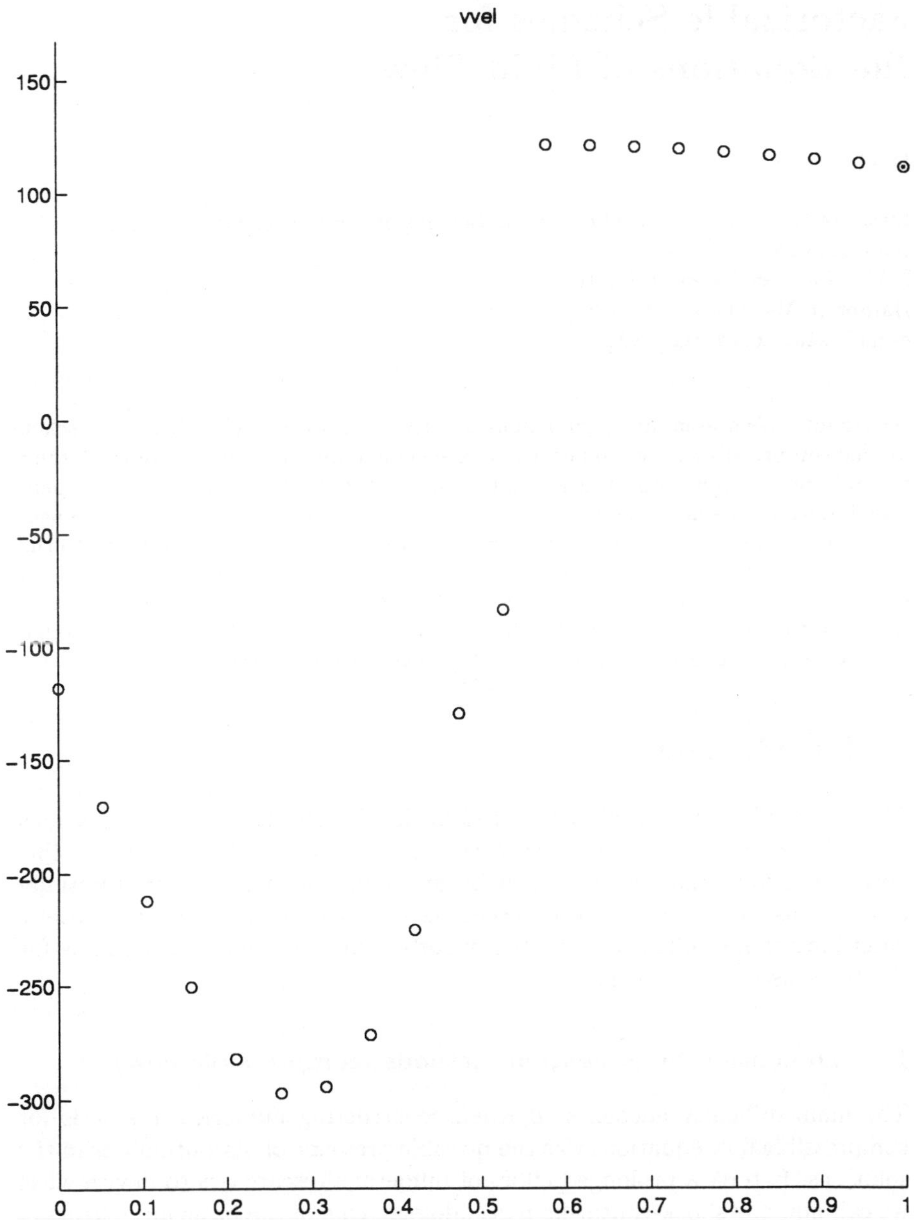

Figure 15: $\mu_{left} = 0$, $\mu_{right} = 100$, "slip"

Factorizable Schemes for the Equations of Fluid Flow

David Sidilkover[1]

Institute for Computer Applications in Science and Engineering (ICASE),
Mail Stop 132C,
NASA Langley Research Center,
Hampton, VA 23681-2199, USA
e-mail: sidilkov@icase.edu.

Abstract. We present an upwind high-resolution factorizable (UHF) discrete scheme for the compressible Euler equations. The scheme approximates equations in their general conservative form and is related to the family of genuinely multidimensional upwind schemes developed previously and demonstrated to have good shock-capturing capabilities. A unique property of this scheme is that in addition to the aforementioned features it is also *factorizable*, i.e. it allows to distinguish betwen full-potential and advection factors at the *discrete* level. The latter property facilitates the construction of optimally efficient multigrid solvers. This is done through a relaxation procedure that utilizes the factorizability property.

1 Introduction

The standard numerical methods used for incompressible flow computations do not have much in common with the standard methods for compressible flow. The explanation of this fact is because difficulties encountered initially in each case were of a very different nature. We shall begin with briefly describing these difficulties and the evolution of the numerical methods for the two classes of problems.

1.1 Dimension-by-dimension methods (compressible flow)

The main difficulty encountered when constructing numerical methods for compressible flow equations was the possible presence of discontinuities in the solutions. It took a prolonged effort of numerous researchers to devise what we call now the shock-capturing methodology. One of the basic ingredients of the shock-capturing schemes is the so-called (approximate) Riemann solvers (or their alternative – flux-splitting techniques) which are used to construct a first order accurate scheme. Another important ingredient is the so-called high-resolution mechanism, that allows one to combine higher order accuracy with shock-capturing capabilities, i.e., to circumvent Godunov's theorem. It appears in the form of interpolation (or extrapolation) based on a certain smoothness monitor, that is usually implemented in the form of a flux-limiter.

Many of the relevant issues could be studied on one-dimensional (unsteady) model problems.

The methods developed for one-dimensional problems were extended later to multidimensions in the most straightforward way – on a dimension-by-dimension basis.

One practical need was to compute steady flow, both external and internal, in multidimensions. Steady problems are generally solved through (pseudo-) time evolution. In other words, the problems are treated in a hyperbolic (with respect to time) sense. Multigrid methods became widely used as a mean to accelerate convergence to the steady-state. The key ingredient of a multigrid algorithm is the smoother, i.e., a relaxation procedure that efficiently reduces the high-frequency error content. The difficulty, however, is that the high-frequency error content may be nearly invisible in the residuals (poor measure of h-ellipticity) of high-resolution schemes constructed on a dimension-by-dimension basis. This makes it inherently impossible to construct an efficient smoother.

Another major difficulty is that, in general, no distinction can be made between the different co-factors at the discrete level, i.e., the standard high-resolution schemes are not *factorizable.* This deficiency of the standard discretizations also contributes to the poor performance of standard multigrid solvers. It also leads to the loss of accuracy and deteriorating computational efficiency in the case of low speed compressible flow.

Discrete schemes that are only partially based on the dimension-by-dimension approach were presented in [5], [6], [9] and [13]. In these schemes, some of the second order corrections appear in the form of terms approximating mixed derivatives. The remaining corrections are based on dimension-by-dimension high-resolution mechanism. Schemes of this type when tuned for steady-state computations would have a better h-ellipticity property than the standard ones. This direction, however, has not been explored.

1.2 Projection-type methods (incompressible flow)

Incompressible flow in one dimension is trivial. Therefore, numerical analysts had to address the multidimensional problems right away. Incompressible flow problems (both steady and unsteady) contain an elliptic part, which rules out the use of an explicit scheme.

One of the earlier approaches was to substitute the continuity equations by the pressure Poisson equation, derived using the momentum and the continuity equations (see a summary in [10]). By using such a representation, decoupling of the elliptic factor (in the form of the pressure Poisson equation) from the rest of the system can be achieved. This formulation has been used in conjunction with multigrid as well. In [20], it was demonstrated that by deriving and treating the boundary conditions for the pressure properly one can obtain the ideal multigrid efficiency both for inviscid and viscous cases. Some different versions of the algorithm are discussed in [19] and in

[14], where some steps in deriving the pressure Poisson equation are made at the discrete level. Two variants of the algorithm are being pursued: one for structured body-fitted grids, another for unstructured grids (see [14]). Ideal multigrid efficiency was demonstrated for both on several test cases. A possible extension of this approach to the compressible subsonic case is discussed in [19].

Several methods exist based on discretizing the equations of incompressible flow in their usual (primitive) form. It is interesting to note that one of the earlier methods (MAC) (see [8]) relies on a staggered grid discretization and uses a pressure Poisson equation derived at the discrete level to update the pressure.

A known property of a vector field is that it can be represented as a sum of its irrotational and solenoidal components. The attractive feature of the staggered grid discretization is that this property can be imitated on discrete level (second-order accurate approximation to the Cauchy-Riemann equations) without getting an odd-even instability. It is problematic, though, to achieve the same on non-staggered grids.

When solving the equations of incompressible flow in their primitive form, the important part of the process is to satisfy the continuity equation, i.e., to project the velocity field onto the divergence-free manifold or, in other words, to discard the irrotational component of the velocity field. Such a process is the main part of the *projection* method [4], whose original version was based on non-staggered grids. It was reformulated later using staggered grids. A convenient way to perform the projection step is to introduce an auxiliary variable potential and to solve the resulting Poisson equation. This was first done in [7].

A commonly used algorithm for solving the incompressible flow equations is the SIMPLE algorithm by Patankar & Spalding [12]. It is interesting to note that this algorithm was extended to the compressible (subsonic) case (see [11],[1]).

Applying multigrid methods for solving the fluid flow problems is one of the main subjects of the landmark work [2]. Although a few various possibilities are mentioned, the approach, systematically studied and advocated, is based on the staggered-grid discretization and an auxiliary potential variable. It was shown that the proper treatment (Distributive Gauss-Seidel relaxation - shortly DGS) results in the decoupling of different co-factors of the system. It was demonstrated later in [3] that an optimal MG efficiency can be obtained for the incompressible Euler equations. Some ways of generalizing these ideas to the compressible flow were sketched out in [2] as well.

The Vorticity-Potential (or vorticity and streamfunction in two dimensions) formulation of the flow equations seems very attractive, since the different co-factors of the equations decouple. The well known difficulty associated with this formulation, though, is the derivation and treatment of the boundary conditions for vorticity. Another difficulty associated with this approach is that the components of the vorticity vector in three dimensions need to satisfy a certain *compatibility* condition. Also, it is highly problematic to obtain a numerical scheme with shock-capturing capabilities using the Vorticity-Potential formulation.

The ideal multigrid efficiency for the compressible Euler equations in the subsonic case was demonstrated first in [23] using the *canonical variables* formulation of the equations [22]. However, this formulation cannot be generalized to viscous and unsteady cases.

1.3 About this paper

In order to achieve optimal multigrid efficiency, we have to use projection-type methods to solve the discrete equations, i.e., methods that distinguish between the different co-factors of the equations.

The attempts ([1],[23]) to apply projection-type methods to solve the compressible flow equations were limited to the subsonic case, since the discretizations used have no shock-capturing capabilities. On the other hand the projection-type methods cannot be applied in conjunction with the standard upwind dimension-by-dimension discretizations, since the latter are not *factorizable* (see [19]), i.e., no distinction between the different factors can be made at the discrete level.

There is a need for a *factorizable* discretization scheme with good shock-capturing capabilities. The scheme should rely on a high resolution mechanism and it should be sensitive to the high-frequency error content (h-elliptic) at the same time. The flow equations should also be approximated in their usual conservative form, since using special formulations leads to the loss of generality. The separation between the treatment of co-factors should be achieved through the relaxation procedure, which may rely on some special auxiliary variables and/or special forms of equations. It may look as if too many (seemingly contradictory) requirements need to be satisfied by a single discretization. The purpose of this paper is to describe a construction of such a discrete scheme.

A search for a genuinely multidimensional upwind scheme was motivated initially by the necessity to devise a discretization that has the high-resolution and h-ellipticity property at the same time. Such a scheme for scalar advection problems was proposed in [15],[21]. This approach was extended to the Euler equations in [17,16,18] in the context of unstructured triangular grids. It was demonstrated that a simple pointwise Gauss-Seidel relaxation is stable when applied directly to the high-resolution discrete equation (contrary to the case of the standard schemes) and has good smoothing properties. The latter is a manifestation of the h-elipticity property of the discretization.

It was pointed out in [19] that the genuinely multidimensional approach towards discretizing the compressible flow equations leads to schemes that are, in addition to other desirable properties, *factorizable*. In this paper we describe such a scheme, i.e. a scheme which is Upwind, High-resolution and

Factorizable (UHF). Also, we develop a relaxation procedure that utilizes the factorizability property of the scheme.

2 Preliminaries

In this section we briefly review the genuinely two-dimensional advection scheme, introduce some basic properties of the Euler equations and discuss some standard discretization schemes for the fluid flow equations.

2.1 Scalar advection schemes

Consider a steady-state conservation law in two dimensions

$$-\varepsilon \Delta u + f(u)_x + g(u)_y = 0 \tag{1}$$

where $\varepsilon > 0$ is an infinitesimally small number. A general conservative dis-

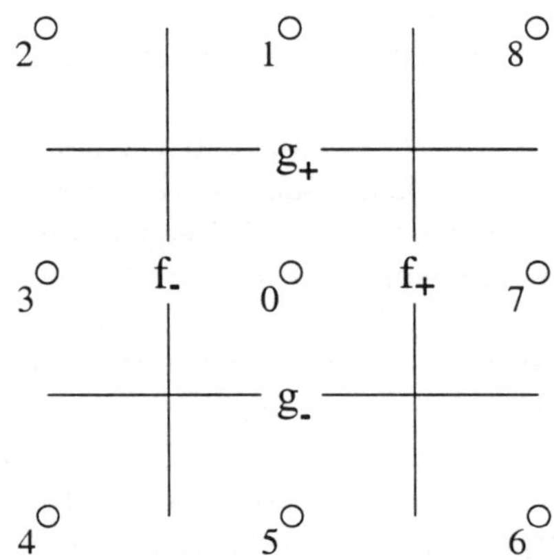

Fig. 1. Computational grid segment and a control volume.

cretization of (1) is given by

$$h[(f_+ - f_-) + (g_+ - g_-)] = 0 \tag{2}$$

where $f_\pm, g_\pm$ are the numerical fluxes. A particular discretization scheme for (1) can be given by defining these numerical fluxes.

A numerical scheme can be written in the form

$$u_0 = \sum_i C_i u_i. \tag{3}$$

Definition 1. The scheme is said to be of the positive type if $C_i \geq 0$.

We shall illustrate the construction of some discretization scheme on the linear constant coefficient version of (1)

$$-\varepsilon \Delta u + a u_x + b u_y = 0. \tag{4}$$

A first order upwind scheme can be given by the following numerical fluxes

$$\begin{aligned} f_-^u &= \tfrac{1}{2} a(u_0 + u_3) - \tfrac{1}{2}|a|(u_0 - u_3) \\ g_-^u &= \tfrac{1}{2} b(u_0 + u_5) - \tfrac{1}{2}|b|(u_0 - u_5). \end{aligned} \tag{5}$$

Assume for simplicity that

$$b \geq a \geq 0.$$

A second order scheme with a genuinely multidimensional flavor is given by

$$\begin{aligned} f_- &= f_-^u - \tfrac{1}{2} b(u_3 - u_4) \\ g_- &= g_-^u - \tfrac{1}{2} a(u_5 - u_4). \end{aligned} \tag{6}$$

This scheme, however, is not of the positive type.

In order to combine the positivity property with the second order accuracy we need to incorporate the second order correction in a nonlinear fashion. A "straightforward" way of doing this is to modify the fluxes in the following way

$$\begin{aligned} f_- &= f_-^u - \tfrac{1}{2} b(u_3 - u_4)\psi(Q_-) \\ g_- &= g_-^u - \tfrac{1}{2} a(u_5 - u_4)\psi(S_-), \end{aligned} \tag{7}$$

which gives a positive second order scheme if the limiter function ψ satisfies the inequality

$$0 \leq \psi(Q) \leq 1; \quad 0 \leq \frac{\psi(Q)}{Q} \leq 1, \tag{8}$$

$$\psi(1) = 1 \tag{9}$$

and the arguments of the limiter function are defined by

$$Q_- = -\frac{a(u_0 - u_3)}{b(u_3 - u_4)}$$

$$S_- = -\frac{b(u_0 - u_5)}{a(u_5 - u_4)}.$$

Using the following identities

$$\begin{aligned} b(u_3 - u_4)\psi(Q_-) &\equiv -a(u_0 - u_3)\psi(Q_-)/Q_- \\ a(u_5 - u_4)\psi(S_-) &\equiv -b(u_0 - u_5)\psi(S_-)/R_-, \end{aligned} \tag{10}$$

it is easy to see that this scheme is no better than the standard high-resolution schemes, since under certain circumstances it becomes identical to a central scheme (no h-ellipticity property). The fact that it is *positive* for a non-compressive limiter only (i.e., no artificial compression can be added) is a disadvantage as well.

Note, that a part of the second order correction can be added with no limiter while still resulting in a positive (first order) scheme - the so-called N-scheme (see [15,21])

$$\begin{aligned} f_-^N &= f_-^u - \tfrac{1}{2}a(u_3 - u_4) \\ g_-^N &= g_-^u - \tfrac{1}{2}a(u_5 - u_4) \end{aligned} \tag{11}$$

A second order high-resolution scheme can be obtained by adding the remaining part of the correction in a nonlinear fashion

$$\begin{aligned} f_- &= f_-^N - \tfrac{1}{2}(b-a)(u_3 - u_4)\psi(R_-) \\ g_- &= g_-^N \end{aligned} \tag{12}$$

where

$$R_- = -\frac{a(u_5 - u_4)}{b(u_3 - u_4)}.$$

It is easy to see ([21]) that (12) defines a positive scheme if

$$0 \leq \psi(R) \leq 2; \quad 0 \leq \frac{\psi(R)}{R} \leq 2. \tag{13}$$

2.2 Euler equations

The quasilinear form of the steady Euler equations in two dimensions can be written in the following way

$$L\mathbf{u} = 0 \tag{14}$$

where $\mathbf{u} = (s, u, v, p)^T$ is the vector of unknown functions (namely, entropy, horizontal velocity component, vertical velocity component and pressure, respectively),

$$L = \begin{pmatrix} q & 0 & 0 & 0 \\ 0 & \rho q & 0 & \partial_x \\ 0 & 0 & \rho q & \partial_x \\ 0 & \rho c^2 \partial_x & \rho c^2 \partial_y & q \end{pmatrix}, \tag{15}$$

ρ stands for density, c for the speed of sound and q denotes the advection operator (with velocity (u, v)). We shall consider the case of subsonic flow ($u^2 + v^2 \leq c^2$) throughout the paper. It is sufficient for the purpose of the analysis in this paper to consider the linear constant coefficient case.

Determining the type of a system of partial differential equations can be done formally by computing the determinant of matrix L and examining its principal part

$$\det L = (\rho q)^2 \cdot (q^2 - c^2 \Delta). \tag{16}$$

The first multiplier in (16) is the advection operator (times the density) in the power equal to the dimension of the problem. The second multiplier is the full-potential operator, which, in subsonic case, is of elliptic type. Thus, the steady Euler system of equations is of the mixed hyperbolic-elliptic type.

One-dimensional case Consider a first order upwind scheme for the one-dimensional Euler equations. Without loss of generality we consider the primitive variable formulation

$$L\mathbf{u} = 0 \tag{17}$$

where $\mathbf{u} = (s, u, p)^T$ and

$$L = \begin{pmatrix} u\partial_x & 0 & 0 \\ 0 & \rho u \partial_x & \partial_x \\ 0 & \rho c^2 \partial_x & u\partial_x \end{pmatrix}. \tag{18}$$

A first order upwind scheme approximating (17) is given by

$$L^h \mathbf{u}^h = 0 \tag{19}$$

where the discrete variables

$$\mathbf{u}^h = (s^h, u^h, p^h)^T \tag{20}$$

and

$$L^h = \begin{pmatrix} -\frac{h}{2}|u|\partial^h_{xx} + Q^{2h} & 0 & 0 \\ 0 & \rho(-\frac{h}{2}c\partial^h_{xx} + Q^{2h}) & \partial^{2h}_x - \frac{h}{2}\frac{u}{c}\partial^h_{xx} \\ 0 & \rho c^2(\partial^{2h}_x - \frac{h}{2}\frac{u}{c}\partial^h_{xx}) & -\frac{h}{2}c\partial_{xx} + Q^{2h} \end{pmatrix}, \tag{21}$$

where h is a mesh size, ∂^h_{xx} is a central approximation of the second derivative, ∂^{2h}_x is a central approximation of the first derivative and $Q^{2h} = u\partial^{2h}_x$ is the advection operator.

Factorization The determinant of L^h:

$$\det(L^h) = \rho(-\frac{h}{2}|u|\partial^h_{xx} + Q^{2h})[(c^2 - u^2)\partial^h_{xx}] \tag{22}$$

The first factor is the upwind scheme approximating the advection operator corresponding to the entropy equation. The Full-Potential factor is approximated by a "short" central difference. The issue of factorization appears to be trivial in this case, since the momentum and the pressure equations correspond solely to the elliptic factor.

Two-dimensional case Since the equation for entropy decouples in the form of the advection equation, whose numerical treatment is straightforward, we can consider without loss of generality the case of isentropic flow. We can also assume for simplicity that $\rho = 1$. Two-dimensional steady isentropic linearized Euler equations are now given by the following

$$L\mathbf{u} = 0, \tag{23}$$

where $\mathbf{u} = (u, v, p)^T$ and

$$L = \begin{pmatrix} q & 0 & \partial_x \\ 0 & q & \partial_y \\ c^2\partial_x & c^2\partial_y & q \end{pmatrix}. \tag{24}$$

The FDA (First Differential Approximation or Modified equations) corresponding to the first order upwind scheme, constructed on the dimension-by-dimension basis, is given by

$$L_1^{FDA} = \begin{pmatrix} (\mathbf{q} - \frac{h}{2}(c - |u|)\partial_{xx}) & 0 & \partial_x - \frac{h}{2}\frac{u}{c}\partial_{xx} \\ 0 & (\mathbf{q} - \frac{h}{2}(c - |v|)\partial_{yy}) & \partial_y - \frac{h}{2}\frac{v}{c}\partial_{yy} \\ c^2(\partial_x - \frac{h}{2}\frac{u}{c}\partial_{xx}) & c^2(\partial_y - \frac{h}{2}\frac{v}{c}\partial_{yy}) & Q - \frac{h}{2}c\Delta \end{pmatrix} \tag{25}$$

where $\mathbf{q}$ is an FDA of the first order upwind advection scheme. It is easy to verify that L_1^{FDA} does not preserve the factorizability property of L, i.e., in the $\det(L_1^{FDA})$ we cannot distinguish between the FDAs corresponding to advection and full-potential factors. The same applies, of course, to the corresponding discrete scheme.

The dimension-by-dimension scheme (whose FDA is given by (25)) can be upgraded to higher order accuracy by introducing the standard nonlinear high-resolution correction. However, by doing so, we obtain a scheme that is not only not factorizable but also can be insensitive to some high-frequency error components (has a poor measure of h-ellipticity). All this makes it practically impossible to construct an efficient solver for the resulting discrete equations.

3 Factorizable scheme

In order to combine the high-resolution and h-ellipticity properties the extension of the ideas leading to the genuinely multidimensional advection scheme need to be generalized to systems of equations. Such a generalization on triangular grids was proposed in [17,16,18]. As well as in the scalar case, the second-order corrections are given by the terms approximating mixed derivatives. These corrections may rely on two-dimensional limiters. It was demonstrated that the Gauss-Seidel relaxation is stable when applied directly to the high-resolution discrete equations (unlike in the case of standard dimension-by-dimension schemes). This indicated that these new schemes have stability properties superior to the standard ones.

It was pointed out in [16] that some of these corrections can be added without limiters, in a linear fashion. Indeed, similar to the scalar case (see [21] and also §2.1), in order to obtain a scheme with better linear stability properties, it is desirable to include as much of the second order correction as possible without limiters. This is especially important in the context of structured grids ("9-point box" stencil); otherwise one can obtain a scheme that is no better than the standard ones (see §2.1).

However, unlike the scalar case, it is unclear what is a suitable criterion for determining which parts of the corrections do not have to rely on the limiters. The difficulty here is that the notions of maximum principle and positivity do not extend to systems of equations in a suitable way.

We recall here the observation that was made in [19]. Consider the FDA corresponding to the second order genuinely multidimensional upwind scheme with all the second order corrections added in a linear fashion

$$L_2^{FDA} = \begin{pmatrix} \mathbf{q} - \frac{h}{2}(c-|u|)\partial_{xx} & -\frac{h}{2}(c-|v|)\partial_{xy} & \partial_x - \frac{h}{2}\frac{1}{c}Q_x \\ -\frac{h}{2}(c-|u|)\partial_{xy} & \mathbf{q} - \frac{h}{2}(c-|v|)\partial_{yy} & \partial_y - \frac{h}{2}\frac{1}{c}Q_y \\ c^2\partial_x - \frac{h}{2}cQ_x & c^2\partial_y - \frac{h}{2}cQ_y & Q - \frac{h}{2}c\Delta \end{pmatrix}. \tag{26}$$

It can be verified that (26) is factorizable.

The implication of this is that the multidimensional corrections not only lead to a second-order scheme, but some of them are also responsible for the *factorizability*. Since it is a desirable property, the latter corrections should be included without limiters.

There exist many discretizations that correspond to (26). Note that not all of them are factorizable. We would now like to construct an actual discrete cheme which is factorizable as well.

The technical difficulty here is that the following differential equalities

$$\begin{aligned} \partial_{xx}\partial_{yy} &= \partial_{xy}\partial_{xy} \\ \partial_{xx}\partial_y &= \partial_{xy}\partial_x \\ \partial_{yy}\partial_x &= \partial_{xy}\partial_y \end{aligned} \tag{27}$$

need to have discrete analogs. Approximating the derivatives $\partial_x, \partial_{xx}, \partial_y, \partial_{yy}$ by standard central finite differences and discretizing the mixed derivatives in some reasonable way, we can see that the discrete analog of (27) does not hold.

Let us introduce some non-standard finite differences

$$\bar{\partial}_x = \frac{1}{8}\frac{1}{h}\begin{pmatrix} -1 & 0 & 1 \\ -2 & 0 & 2 \\ -1 & 0 & 1 \end{pmatrix}; \quad \bar{\partial}_y = \frac{1}{8}\frac{1}{h}\begin{pmatrix} 1 & 2 & 1 \\ 0 & 0 & 0 \\ -1 & -2 & -1 \end{pmatrix} \tag{28}$$

$$\bar{\partial}_{xx} = \frac{1}{4}\frac{1}{h^2}\begin{pmatrix} 1 & -2 & 1 \\ 2 & -4 & 2 \\ 1 & -2 & 1 \end{pmatrix}; \quad \bar{\partial}_{yy} = \frac{1}{4}\frac{1}{h^2}\begin{pmatrix} 1 & 2 & 1 \\ -2 & -4 & -2 \\ 1 & 2 & 1 \end{pmatrix} \tag{29}$$

$$\bar{\partial}_{xy} = \frac{1}{4}\frac{1}{h^2}\begin{pmatrix} -1 & 0 & 1 \\ 0 & 0 & 0 \\ 1 & 0 & -1 \end{pmatrix}. \tag{30}$$

Let us denote also

$$\bar{\Delta} = \frac{1}{4}\frac{1}{h^2}\begin{pmatrix} 1 & 2 & 1 \\ 2 & -12 & 2 \\ 1 & 2 & 1 \end{pmatrix} \tag{31}$$

and

$$\begin{aligned} \bar{Q}^h_x &= u\bar{\partial}_{xx} + v\bar{\partial}_{xy} \\ \bar{Q}^h_x &= u\bar{\partial}_{xy} + v\bar{\partial}_{yy}. \end{aligned} \tag{32}$$

It is easy to see that the discrete analog of (27) holds if the derivatives are approximated by the finite differences as defined by (28), (29) and (30).

Consider now the following discrete scheme that corresponds to the FDA given by (26)

$$L^h = \begin{pmatrix} \mathbf{q}^h - \frac{h}{2}(c-|u|)\bar{\partial}^h_{xx} & -\frac{h}{2}(c-|v|)\bar{\partial}^h_{xy} & \bar{\partial}^h_x - \frac{h}{2}\frac{1}{c}\bar{Q}^h_x \\ -\frac{h}{2}(c-|u|)\bar{\partial}^h_{xy} & \mathbf{q}^h - \frac{h}{2}(c-|v|)\bar{\partial}^h_{yy} & \bar{\partial}^h_y - \frac{h}{2}\frac{1}{c}\bar{Q}^h_y \\ c^2\partial^h_x - \frac{h}{2}cQ_x & c^2\partial^h_y - \frac{h}{2}cQ_y & Q^h - \frac{h}{2}c\bar{\Delta}^h \end{pmatrix}. \tag{33}$$

It is factorizable and its determinant is given by

$$\begin{aligned} \det(L^h) = \mathbf{q}^h[(Q^h - \frac{h}{2}c\bar{\Delta}^h)(\mathbf{q}^h - \frac{h}{2}(c-|u|)\bar{\partial}^h_{xx} - \frac{h}{2}(c-|v|)\bar{\partial}^h_{yy}) \\ -(\bar{\partial}^h_x - \frac{h}{2}\frac{1}{c}\bar{Q}^h_x)(c^2\partial^h_x - \frac{h}{2}cQ_x) \\ -(\bar{\partial}^h_y - \frac{h}{2}\frac{1}{c}\bar{Q}^h_y)(c^2\partial^h_y - \frac{h}{2}cQ_y)] \end{aligned}$$

Note that the discrete approximation of the elliptic factor becomes the so-called "skewed" discrete Laplacian (not an h-elliptic operator) when the flow speed (Mach number) goes to zero. This difficulty can be dealt with in various ways. The simplest option is to slightly modify the scheme so that the problem mentioned above disappears.

Denote the total velocity

$$U = |\sqrt{u^2 + v^2}|,$$

the Mach number

$$M = \frac{U}{c}$$

and the "augmented" Mach number

$$\bar{M} = \begin{cases} M, & \text{if } M > Ch \\ Ch & \text{otherwise} \end{cases} \tag{34}$$

where $C > 0$ is a constant.

We can "rescale" the artificial dissipation terms, obtaining the following discrete scheme, which we shall call UHF (Upwind, High-resolution, Factorizable)

$$L = \begin{pmatrix} \bar{\mathbf{q}}^h - \frac{h}{2}(U - |u|)\bar{\partial}^h_{xx} & -\frac{h}{2}(U - |v|)\bar{\partial}^h_{xy} & \bar{\partial}^h_x - \frac{h}{2}\frac{M}{c}\bar{Q}^h_x \\ -\frac{h}{2}(U - |u|)\bar{\partial}^h_{xy} & \bar{\mathbf{q}}^h - \frac{h}{2}(U - |v|)\bar{\partial}^h_{yy} & \bar{\partial}^h_y - \frac{h}{2}\frac{M}{c}\bar{Q}^h_y \\ c^2(\partial^h_x - \frac{h}{2}\frac{1}{cM}Q_x) & c^2(\partial^h_y - \frac{h}{2}\frac{1}{cM}Q_y) & Q^h - \frac{h}{2}\frac{c}{M}\bar{\Delta}^h \end{pmatrix} \tag{35}$$

where

$$\bar{\mathbf{q}}^h = \mathbf{q}^h - \frac{1}{2}h^2 cC\Delta^h, \tag{36}$$

is the discrete advection operator augmented by a second-order small dissipative term and Δ^h is the usual "5-point-star" discrete Laplacian. This "regularization," obviously, prevents the advection scheme from degenerating at a stagnation point. It plays an additional important role, though, which will be explained later in this section.

Before discussing some properties of the UHF scheme and computing its determinant, we shall introduce the simplified notation

$$\begin{aligned} \boldsymbol{\partial}^h_x &= \partial^h_x - \tfrac{h}{2}\tfrac{1}{cM}Q^h_x \\ \boldsymbol{\partial}^h_y &= \partial^h_y - \tfrac{h}{2}\tfrac{1}{cM}Q^h_y \end{aligned} \tag{37}$$

$$\begin{aligned} \bar{\boldsymbol{\partial}}^h_x &= \bar{\partial}^h_x - \tfrac{h}{2}\tfrac{M}{c}\bar{Q}^h_x \\ \bar{\boldsymbol{\partial}}^h_y &= \bar{\partial}^h_y - \tfrac{h}{2}\tfrac{M}{c}\bar{Q}^h_y \end{aligned} \tag{38}$$

and

$$\tilde{\mathbf{q}}^h = Q^h - \frac{h}{2}\frac{c}{M}\bar{\Delta}^h, \tag{39}$$

$$\hat{\mathbf{q}}^h = \bar{\mathbf{q}}^h - \frac{h}{2}((U - |u|)\bar{\partial}^h_{xx} + (U - |v|)\bar{\partial}^h_{yy}). \tag{40}$$

Then the UHF scheme (35) can be written as

$$L = \begin{pmatrix} \bar{\mathbf{q}}^h - \frac{h}{2}(U - |u|)\bar{\partial}^h_{xx} & -\frac{h}{2}(U - |v|)\bar{\partial}^h_{xy} & \bar{\boldsymbol{\partial}}^h_x \\ -\frac{h}{2}(U - |u|)\bar{\partial}^h_{xy} & \bar{\mathbf{q}}^h - \frac{h}{2}(U - |v|)\bar{\partial}^h_{yy} & \bar{\boldsymbol{\partial}}^h_y \\ c^2\boldsymbol{\partial}^h_x & c^2\boldsymbol{\partial}^h_y & \tilde{\mathbf{q}}^h \end{pmatrix} \tag{41}$$

and its determinant

$$\det(L^h) = \bar{\mathbf{q}}^h \cdot [c^2(\boldsymbol{\partial}^h_x\bar{\boldsymbol{\partial}}^h_x + \boldsymbol{\partial}^h_y\bar{\boldsymbol{\partial}}^h_y) - \tilde{\mathbf{q}}^h \cdot \hat{\mathbf{q}}^h]. \tag{42}$$

Let us take a closer look at what happens at a stagnation point ($U = u = v = 0$)

$$\bar{\mathbf{q}}^h = -\frac{1}{2}h^2 cC\Delta^h \tag{43}$$

$$\hat{\mathbf{q}}^h = -\frac{1}{2}h^2 cC\Delta^h \tag{44}$$

$$\tilde{\mathbf{q}}^h = -\frac{c}{2C}\bar{\Delta}^h. \tag{45}$$

It becomes plain that

$$\tilde{\mathbf{q}}^h \cdot \hat{\mathbf{q}}^h = \frac{1}{4}h^2 c^2 \bar{\Delta}^h \Delta^h, \tag{46}$$

which is an h-elliptic operator. We can conclude now that for the UHF scheme (41) (or (35)) the approximation of the Full-Potential factor remains h-elliptic in the case of vanishing velocity. Moreover, it can be verified that the Full-Potential factor is approximated in this case by a "regular" five-point star discrete Laplacian Δ^h.

4 The relaxation procedure

Having constructed a *factorizable* discretization scheme for the Euler equations, an important question becomes how to use this property in order to obtain an efficient algorithm for solving the discrete equations.

Introduce the auxiliary variables, namely, the discrete stream-function and potential by the following

$$\begin{pmatrix} u^h \\ v^h \\ p^h \end{pmatrix} = \mathcal{M}^h \cdot \begin{pmatrix} \Psi^h \\ \Phi^h \end{pmatrix} \tag{47}$$

where

$$\mathcal{M}^h = \begin{pmatrix} \partial_y & \bar{\partial}^h_x \\ -\partial^h_x & \bar{\partial}^h_y \\ 0 & \hat{\mathbf{q}}^h \end{pmatrix} \tag{48}$$

We can also define the discrete vorticity

$$\omega^h = \mathrm{curl}^h \cdot (u^h, v^h)^T = (\bar{\partial}^h_y, -\bar{\partial}^h_x) \cdot (u^h, v^h)^T \tag{49}$$

or

$$\omega^h = (\bar{\partial}^h_x \partial^h_x + \bar{\partial}^h_y \partial^h_y)\Psi^h. \tag{50}$$

Introduce also the following matrix operator

$$\mathcal{P}^h = \begin{pmatrix} \bar{\partial}_y^h & -\bar{\partial}_x^h & 0 \\ 0 & 0 & 1 \end{pmatrix}. \tag{51}$$

It is easy to see that applying the discrete curl^h operator to the momentum equations (pre-multiplying L^h by $\mathcal{P}^h$) and performing the substitution of variables according to (47) (post-multiplying L^h by $\mathcal{M}^h$), we obtain

$$\mathcal{P}^h \cdot L^h \cdot \mathcal{M}^h = \begin{pmatrix} \bar{\mathbf{q}}^h \cdot (\partial_x^h \bar{\partial}_x^h + \partial_y^h \bar{\partial}_y^h) & 0 \\ 0 & c^2(\partial_x^h \bar{\partial}_x^h + \partial_y^h \bar{\partial}_y^h) - \tilde{\mathbf{q}}^h \cdot \hat{\mathbf{q}}^h \end{pmatrix}. \tag{52}$$

In other words we end up with "solving" the system

$$\begin{pmatrix} \bar{\mathbf{q}}^h \cdot (\partial_x^h \bar{\partial}_x^h + \partial_y^h \bar{\partial}_y^h) & 0 \\ 0 & c^2(\partial_x^h \bar{\partial}_x^h + \partial_y^h \bar{\partial}_y^h) - \tilde{\mathbf{q}}^h \cdot \hat{\mathbf{q}}^h \end{pmatrix} \cdot \begin{pmatrix} \Psi^h \\ \Phi^h \end{pmatrix} = 0 \tag{53}$$

or

$$\begin{pmatrix} \bar{\mathbf{q}}^h & 0 \\ 0 & c^2(\partial_x^h \bar{\partial}_x^h + \partial_y^h \bar{\partial}_y^h) - \tilde{\mathbf{q}}^h \cdot \hat{\mathbf{q}}^h \end{pmatrix} \cdot \begin{pmatrix} \omega^h \\ \Phi^h \end{pmatrix} = 0. \tag{54}$$

The *Distributive* relaxation procedure at a point amounts to computing updates to the discrete potential and vorticity (or streamfunction) according to (54) and to translating them into corresponding updates of the velocity components and the pressure at the point of interest and its neighbors according to matrix (48) and (47).

Note that the operator $(\partial_x^h \bar{\partial}_x^h + \partial_y^h \bar{\partial}_y^h)$ is not h-elliptic. Therefore, the relaxation procedure described above will not smooth certain high-frequency error components. This lack of h-ellipticity seems unavoidable for obtaining the desired factorization. However, a simple remedy exits for this trouble. It is to augment each sweep of the Distributive relaxation by a sweep of a point Collective Gauss-Seidel relaxation. The latter will smooth the problematic error components.

It might also be useful to perform the point collective relaxation, instead of Distributive, on and near the boundary, thus avoiding the difficulty of imposing boundary conditions for vorticity.

5 Discussion and conclusions

When discretizing a system of partial differential equations, being concerned just with obtaining an approximation of a certain order of accuracy may not necessarily lead to a satisfactory result. It may be very useful to make sure that the discretization also imitates some fundamental properties of the PDEs. We have outlined a construction of a discretization scheme that not only approximates the compressible Euler equations, but imitates their factorizability property. This paves the way towards construction of optimally efficient multigrid solvers and also should alleviate the problem associated with computation of low-speed flow using the standard shock-capturing schemes.

This paper addressed the subsonic case only. The constructed UHF scheme, though, is related to the family of the genuinely multidimensional upwind schemes constructed previously (see [17,16,18]). Schemes of this type were demonstrated to have excellent shock capturing capabilities in transonic-supersonic cases together with maintaining the h-ellipticity property. Therefore, we do not anticipate any difficulties in extending the factorizable scheme to the transonic-supersonic regimes. This is a subject of future work.

We also developed a relaxation procedure that uses the factorizability property of the scheme for the purpose of obtaining ideal multigrid efficiency. This procedure relies on auxiliary potential and vorticity (or streamfunction) variables. Note that using only an auxiliary potential variable (as in the incompressible case) is not sufficient yet to decouple the different factors (or obtaining upper- or lower-triangular matrix of difference operators). This is due to the bulk viscosity-like terms, which are a part of the artificial dissipation in the constructed scheme and are essential for its factorizability and second order accuracy. Another auxiliary variable is needed - vorticity (or streamfunction).

The approach we introduce in this paper seems very general. The future work will be devoted generalizing it to three dimensions, to Navier-Stokes equations and to unsteady problems.

The proposed approach also essentially unifies the numerical treatment of the steady incompressible and compressible flow, since, on one hand, it belongs to the class of projection-type methods, on the other hand its extension to the supersonic case has shock-capturing capabilities.

Acknowledgements. The author would like to thank Tom Roberts, Jim Thomas, Dimitri Mavriplis and Manny Salas for reading the manuscript and making numerous helpful suggestions and comments.

References

1. B. J. Braams. Modelling of a transport problem in plasma physics. In J. G. Verwer, editor, *Topics in Applied Numerical Analysis.* Centrum voor Wiskunde en Informatica, 1984.
2. A. Brandt. Multigrid techniques: 1984 guide with applications to fluid dynamics. The Weizmann Institute of Science, Rehovot, Israel, 1984.
3. A. Brandt and I. Yavneh. Accelerated multigrid convergence and high reynolds recirculating flows. *SIAM J. Sci. Statist. Comput.*, 14:607–626, 1993.
4. A. J. Chorin. A numerical method for solving incompressible viscous problems. *J. Comp. Phys.*, 2:12–26, 1967.
5. P. Colella. Multidimensional upwind methods for hyperbolic conservation laws. Technical Report LBL-17023, Lawrence Berkeley Report, 1984.

6. P. Colella. Multidimensional upwind methods for hyperbolic conservation laws. *J. Comp. Phys.*, 87:171, 1990.
7. F. H. Harlow and A. A. Amsden. A simplified MAC technique for incompressible fluid flow calculations. *J. Comp. Phys.*, 6:322–325, 1970.
8. F. H. Harlow and J. E. Welch. –. *Phys. Fluids*, 8:2182–2189, 1965.
9. R. J. LeVeque. High resolution finite volume methods on arbitrary grids via wave propagation. *J. Comp. Phys.*, 78, 1988.
10. S. A. Orszag, M. Israeli, and M. O. Deville. Numerical simulations of viscous incompressible flows. *Ann. Reviews in Fluid Mech.*, 6:281–318, 1974.
11. S. V. Patankar. *Numerical Heat Transfer and Fluid Flow.* Hemisphere, 1980.
12. S. V. Patankar and D. B. Spalding. A calculation procedure for heat, mass and momentum transfer in three-dimensional parabolic flow. *Int. J. Heat and Mass Transfer*, 15:1787–1806, 1972.
13. Yu. B. Radvogin. Quasi-monotonous multidimensional difference schemes with second order accuracy. Soviet Academy of Science Preprint, 1991.
14. T. W. Roberts, R. C. Swanson, and D. Sidilkover. An algorithm for ideal multigrid convergence for the the steady Euler equations, 1999.
15. D. Sidilkover. *Numerical solution to steady-state problems with discontinuities.* PhD thesis, The Weizmann Institute of Science, Rehovot, Israel, 1989.
16. D. Sidilkover. A genuinely multidimensional upwind scheme and efficient multigrid solver for the compressible euler equations. Report No. 94-84, ICASE, 1994.
17. D. Sidilkover. A genuinely multidimensional upwind scheme for the compressible Euler equations. In J. Glimm, M. J. Graham, J. W. Grove, and B. J. Plohr, editors, *Proceedings of the Fifth International Conference on Hyperbolic Problems: Theory, Numerics, Applications.* World Scientific, June 1994.
18. D. Sidilkover. Multidimensional upwinding and multigrid. AIAA 95-1759, June 19-22, 1995. 12th AIAA CFD meeting, San Diego.
19. D. Sidilkover. Some approaches towards constructing optimally efficient multigrid solvers for the inviscid flow equations. *Computers & Fluids*, 28:551–571, 1999.
20. D. Sidilkover and U. Ascher. A multigrid solver for the steady-state Navier-Stokes equations using the pressure-Poisson formulation. *Matematica Aplicada e Computational*, 14:21–35, 1995.
21. D. Sidilkover and A. Brandt. Multigrid solution to steady-state 2d conservation laws. *SIAM J. Numer. Anal.*, 30:249–274, 1993.
22. S. Ta'asan. Canonical forms of multidimensional steady inviscid flows. Report No. 93-94, ICASE, 1993.
23. S. Ta'asan. Canonical-variables multigrid method for steady-state euler equations. Report No. 94-14, ICASE, 1994.

Evolution Galerkin Methods as Finite Difference Schemes

K. W. Morton

Universities of Bath and Oxford, UK

Key Words: CHARACTERISTIC GALERKIN, EVOLUTION GALERKIN, LAGRANGE GALERKIN, RECOVERY PROCEDURE, SUPERCONVERGENCE, SUPRACONVERGENCE

Abstract. Using a Galerkin projection onto a piecewise constant approximation space is exactly equivalent to using cell averages in a finite volume scheme. This is the starting point to relating evolution Galerkin methods to finite difference schemes. First order schemes are often identical, but the evolution Galerkin formation is more general — arbitrary time steps, nonuniform meshes, multidimensions. For higher order, recovery procedures are often equivalent to reconstruction methods used with finite differences.

1 Introduction

As far as I can tell from my old diaries, I first met Phil Roe twenty years ago when he visited me at the University of Reading on 29th August, 1978. I clearly remember that the visit had its origin in a heated discussion at a sherry party: I had recently joined the university and was keen to forge links between university mathematicians (who were at that time reluctant to get involved outside academia) and those in industry with practical problems to solve (who were equally reluctant to concede that university expertise might be relevant); the university had a loose link with the Royal Aircraft Establishment at Farnborough, which was only a twenty minute drive away, and Ken Winter (head of the Aerodynamics Division) was at this Vice-Chancellor's get-together party.

The upshot of the conversation was that Ken said "I have this guy Phil Roe in my division who has been working on some ideas which he seems to think it would be helpful to discuss with a numerical analyst"; and I replied that I had a member of my group who wanted to take a sabbatical so that he could get his research moving again — perhaps they could work together. That person was Mike Baines and his collaboration with Phil Roe has continued to this day, as you will hear in the next talk.

Mike recently sent me a copy of two letters from Phil to me, which follow up ideas from previous meetings and correspondence during that summer and autumn twenty years ago. He was clearly bubbling with excitement at the prospect of "wiggle-free" solutions obtained from nonlinear algorithms even for linear problems. Equally clearly, with my many other commitments as Head of Department etc. etc., I was finding it difficult to keep up with him (e.g. he said " I do realise ... I am probably throwing information at you faster than you can respond")!

I also remember very clearly when Phil Roe first met Stan Osher. That was in the coffee room of the Maths Dept. at Reading in March 1982. Stan was visiting us for a few weeks and giving a series of lectures on conservation laws; as you can imagine, we all learned much from the resulting lively discussion of the relative merits of their two approaches to devising appropriate Riemann solvers.

It was these two past events and memories that prompted my choice of topic for today's talk. By one simple change of viewpoint, right from the beginning of our association Phil had transformed our concept of algorithms to approximate hyperbolic problems. We had thought that to model $u_t + f_x(u) = 0$ we wrote

$$U_i^{n+1} = U_i^n - \frac{\Delta t}{\Delta x}[h(U_i^n, U_{i+1}^n) - h(U_{i-1}^n, U_i^n)] \tag{1}$$

with $h(U, U) \equiv f(U)$, and programmed a "node loop" over i that updated each U_i^n in turn according to the definition of $h(\cdot,\cdot)$. However, he thought in terms of a "cell loop" over $i + \frac{1}{2}$ which allocated the "fluctuation" $f(U_{i+1}^n) - f(U_i^n)$ in each cell to the nodes i on the left or $i + 1$ on the right according to some solution-dependent prescription.

Stan, on the other hand, introduced us to the idea of splitting $f(\cdot)$ into $f^+(\cdot) + f^-(\cdot)$, the monotone increasing and decreasing components of the flux function and using an upwind scheme on each. I was working on finite element methods, particularly Petrov-Galerkin methods, and wondering whether they could be of value in hyperbolic problems. It seemed that both Stan's schemes and Phil's ideas could be related to what we were then beginning to call characteristic Galerkin methods, and more generally evolution Galerkin methods. Moreover, as a result of Phil recently refereeing a paper of mine on this subject, we have taken up the lively collaboration which in my partial retirement I can fully respond to in a way that I could not twenty years earlier.

2 Scalar advection in one dimension

In the solution of elliptic problems, finite element methods are distinguished by being based on variational principles rather than local satisfaction of the differential equation. Thus for the scalar advection equation

$$u_t + au_x = 0, \qquad a \equiv a(u, x, t), \tag{2}$$

it is natural for a finite element approximation to start from the more global Lagrangian formulation

$$\left(\frac{\partial}{\partial t} + a\frac{\partial}{\partial x}\right) u = 0; \tag{3}$$

this implies that there is a family of trajectories $dX/dt = a$ in the (x,t)-plane along each of which u is constant. Suppose that between time levels t_n and t_{n+1} we approximate these by straight line trajectories, $X(t_n;x) = x, X(t_{n+1};x) = y$, so that we introduce the convenient notation, with $\Delta t = t_{n+1} - t_n$,

$$y = x + a(u(x,t_n), x, t_n)\Delta t. \tag{4}$$

Of course, in the nonlinear case in which a depends on u, the true evolution operator $E(\Delta t)$ which takes $u(\cdot,t)$ to $u(\cdot,t+\Delta t)$ has to incorporate the formation and evolution of shocks. However, imposing constancy of the solution along approximate characteristic trajections given by (4) defines an *approximation evolution operator* which we denote by E_Δ,

$$(E_\Delta u)(y) = u(x); \tag{5}$$

we will later consider the interpretation of this definition when characteristics cross.

2.1 Lagrange Galerkin methods

Now suppose we combine this global approximation to the differential equation with the finite element approximation at the time level t_n,

$$U^n(x) = \sum_{(j)} U_j^n \phi_j(x), \tag{6}$$

where $\{\phi_j(x)\}$ are local basis functions and $\{U_j^n\}$ are expansion coefficients. Typical choices of basis function are (i) *piecewise constant*, in which $\phi_j(x) \equiv 1$ in the *cell* $(x_{j-\frac{1}{2}}, x_{j+\frac{1}{2}})$ centred at x_j, and zero elsewhere; and (ii) *continuous piecewise linear*, in which $\phi_j(x_j) = 1$ and $\phi_j(x_i) = 0$ for all other *nodes* $x_i \neq x_j$. Suppose that the approximation $U^{n+1}(x)$ of the form (6), at time level t_{n+1}, is given by two steps: approximate evolution of $U^n(x)$ by the E_Δ of (5); and *Galerkin projection* back onto the approximation space spanned by the basis functions $\{\phi_j\}$. Then we obtain the *Lagrange Galerkin method*, which can be written symbolically as

$$U^{n+1} = P_h E_\Delta U^n, \tag{7}$$

where P_h denotes the Galerkin projection, or more fully in terms of the inner product $(U^{n+1}, \phi_i) \equiv \int U^{n+1}(y)\phi_i(y)dy$ as

$$(U^{n+1}, \phi_i) = \int U^n(x)\phi_i(y)dy. \tag{8}$$

Note that the integral on the right is nontrivial, because of the relationship (4) between x and y with $u(x,t_n)$ replaced by $U^n(x)$, and will normally be approximated by an appropriate quadrature formula — see [18] for the resulting difficulties.

Generalisations of (4) and (8) to multidimensions are straightforward, and this method has had its greatest use in the solution of the incompressible Navier-Stokes equations — see [1] and [21] — and in porous media flow [7].

2.2 Incremental form; characteristic Galerkin methods

However, to explore the relationship to finite difference methods it is better to use an incremental formulation. Let us consider the scalar conservation law

$$u_t + f_x = 0, \quad f \equiv f(u), \tag{9}$$

so that $a(u) \equiv f'(u)$. Then we can manipulate the difference between the Galerkin projections at two time levels as follows: using the relationship between x and y given by (4) ,

$$\begin{aligned}(U^{n+1} - U^n, \phi_i) &= \int U^n(x)[\phi_i(y)dy - \phi_i(x)dx] \\ &= \int U^n(x)d \int_x^y \phi(s)ds \\ &= -\int \left[\int_x^y \phi_i(s)ds\right] dU^n \\ &= -\Delta t \int \Phi_i^n \left[a(U^n)dU^n = df(U^n)\right],\end{aligned} \tag{10}$$

where we have introduced the averaged basis function

$$\Phi_i^n(x) = \frac{1}{a(U^n(x))\Delta t} \int_x^y \phi_i(s)ds. \tag{11}$$

We can write the result as

$$(U^{n+1} - U^n, \phi_i) + \Delta t(df(U^n), \Phi_i^n) = 0. \tag{12}$$

Here the integrals, especially where $U^n(x)$ is discontinuous, should be interpreted as Riemann-Stieltjes integrals along a continuous parameterisation of the (U^n, x) - graph. Such a method is called a *characteristic Galerkin method* and has been studied in detail in [12], [13].

The simplest example of (12) is obtained with piecewise constant ϕ_i, so that U_i^n represents a *cell average*. Then if h_i is the cell size and the CFL conditions $a(U_i^n)\Delta t \leq h_{i+1}, a(U_i^n)\Delta t \geq -h_{i-1}$ are satisfied for all cells, characteristics which start in one cell cannot reach beyond its two neighbours and the scheme reduces to the Engquist-Osher difference scheme [8]

$$h_i(U_i^{n+1} - U_i^n) + \Delta t[\Delta_{-x} f^+(U_i^n) + \Delta_{+x} f^-(U_i^n)] = 0, \tag{13}$$

where $f^{\pm}(u) = \int^u a^{\pm}(v)dv$ and Δ_{+x}, Δ_{-x} are the forward and backward difference operators. Also, the $df(U^n)$ in (12) can be indentified with Phil Roe's *fluctuations*, and the inner product with Φ_i^n determines the allocation to update (U^n, ϕ_i).

2.3 Finite volume formulation

Specialisation to piecewise constants leads naturally to a finite volume (or Godunov) formulation. This can be obtained from the intermediate equation (10) by rewriting the basis function as the difference between two Heaviside functions

$$\phi_i(x) = H_{i+\frac{1}{2}}(x) - H_{i-\frac{1}{2}}(x). \tag{14}$$

Then that part of the integral from $H_{i+\frac{1}{2}}$, and for which $a(U^n) > 0$, can be written as

$$\begin{aligned}
&- \int dU^n \int_x^{x+a\Delta t} H_{i+\frac{1}{2}}(s)ds = - \int dU^n \max\{0, \min[a\Delta t, x_{i+\frac{1}{2}} - x]\} \\
&= - \int a(U^n)dU^n \max\{0, \min[\Delta t, (x_{i+\frac{1}{2}} - x)/a \equiv \tau]\} \\
&= \int_0^{\Delta t} f(U^n(x = x_{i+\frac{1}{2}} - a\tau))d\tau, \qquad (15)
\end{aligned}$$

where we have replaced adU by df and integrated by parts by introducing $\tau = (x_{i+\frac{1}{2}} - x)/a(U^n(x))$. Combining (15) with the corresponding integral over that part of dU^n where $a < 0$, and dividing by Δt, defines an average flux $F^{n+\frac{1}{2}}_{i+\frac{1}{2}}$ through the cell edge $x_{i+\frac{1}{2}}$ between t_n and t_{n+1}; hence (12) can be written in the equivalent finite volume form

$$h_i(U_i^{n+i} - U_i^n) + \Delta t[F^{n+\frac{1}{2}}_{i+\frac{1}{2}} - F^{n+\frac{1}{2}}_{i-\frac{1}{2}}] = 0, \tag{16}$$

in which what were spatial integrals in (12) have become time integrals and

$$F^{n+\frac{1}{2}}_{i+\frac{1}{2}} \approx \frac{1}{\Delta t} \int_0^{\Delta t} f(u(x_{i+\frac{1}{2}}, t_n + \tau))d\tau. \tag{17}$$

This formulation turns out to be the most useful for generalisations to more complicated equations, and also for a sharp error analysis — see below. It also provides the best means for comparison with alternative schemes: for example, expansion of the integrand of (17) in a Taylor series, with time derivatives replaced by spatial derivatives through the differential equation and replacement of these by divided differences, leads directly to various two-step Lax-Wendroff difference schemes.

However, before moving on to these developments, especially the use of recovery techniques to achieve higher than first order accuracy, we should note that the construction in (12) is exactly equivalent to the use of *Brenier's transport collapse operator* [2] in approximating the evolution operator for the conservation law (9). Thus the formation and evolution of shocks are approximated by allowing the (u, x)-graph of the solution to become multi-valued (so-called overturned manifolds) and then averaging over the result; so shocks are smeared. In fact, characteristics are allowed to cross as many cells as necessary and the schemes are ***unconditionally stable***; time step restrictions arise only from considerations of accuracy. The most important practical consequence is that local mesh refinement can be introduced while using the same global time step. The theoretical consequences of the link to Brenier's work is the establishment of a rigorous convergence theory based on L_1 contraction and TVD properties — see [3, 12, 13].

2.4 Linear, constant coefficient, advection

To investigate, and improve, the accuracy of these methods, which we collect under the generic title of evolution Galerkin schemes, it is helpful to start with Fourier analysis and the assumption that the characteristic speed a is constant and the mesh uniform. Then, if we decompose the CFL number $\nu = a\Delta t/h$ into its integer and noninteger parts, $\nu = m + \widehat{\nu}$ with $\widehat{\nu} \in [0,1)$, it is easy to see that the scheme (7), (8), (12) or (16) becomes

$$\begin{aligned} U_i^{n+1} &= U_{i-m}^n - \widehat{\nu}\Delta_- U_{i-m}^n \\ &\equiv (1-\widehat{\nu})U_{i-m}^n + \widehat{\nu}U_{i-m-1}^n, \end{aligned} \tag{18}$$

which reduces to the familiar first order upwind scheme when $m = 0$.

The use of continuous piecewise linear basis functions leads to a scheme first derived by Lesaint [9] and since rediscovered (and renamed) by many authors, namely

$$(1+\tfrac{1}{6}\delta^2)\tilde{U}_i^{n+1} = [(1+\tfrac{1}{6}\delta^2) - \widehat{\nu}\Delta_0 + \tfrac{1}{2}\widehat{\nu}^2\delta^2 - \tfrac{1}{6}\widehat{\nu}^3\Delta_-\delta^2]\tilde{U}_{i-m}^n. \tag{19}$$

Note that here we have used $\tilde{U}_i^n$ for the expansion coefficients in (6), so that we can continue to make use of U_i^n as the cell averages in the developments below.

In both (18) and (19), the values of U^n used in the update come from points which are symmetrically arranged either side of the interval (x_{i-m-1}, x_{i-m}), in which lies the foot of the characteristic drawn back from the point (x_i, t_{n+1}) to time level t_n: one point is either side in (18), and two either side in (19). Piecewise constants can be regarded as B-splines of order 1 and piecewise linears as B-splines of order 2; and the sequence of schemes can be extended indefinitely by using higher order splines. They are all unconditionally stable and splines of order s give an accuracy of order $2s-1$, a phenomenon of ***superconvergence*** first noted by Thomée and Wendroff [24].

So why, and in what sense, do we obtain third order accuracy from using just linear basis functions? The most useful and convincing answer lies in observations by Childs and Morton [3] based on introducing a recovery stage into the overall procedure. Suppose that, as with a finite volume formulation, we start at each time level with a piecewise constant approximation U^n with cell average values $\{U_j^n\}$. Then we can construct an expansion $\tilde{U}^n$ in quadratic splines (i.e. $s = 3$), with expansion coefficients $\tilde{U}_j^n$, such that the cell average values are preserved: with R_h denoting this ***recovery stage***,

$$\tilde{U}^n = R_h U^n \text{ such that } P_h \tilde{U}^n = U^n. \tag{20}$$

Note that quadratic splines, which have a continuous first derivative as well as function value at the cell boundaries $\{x_{j+\frac{1}{2}}\}$, have the same number of free parameters as piecewise constant splines; and the inner product of a piecewise quadratic spline with a piecewise constant, that is implied by the projection P_h in (20), is the same as that between two piecewise linear splines — this results from a convolution property of splines that holds only on a uniform mesh. Hence $P_h \tilde{U}^n = U^n$ can be written as

$$(1 + \tfrac{1}{6}\delta^2)\tilde{U}_i^n = U_i^n, \tag{21}$$

with the three-point operator on the left corresponding exactly to the tridiagonal mass matrix resulting from the use of a continuous piecewise linear finite element approximation. Moreover, as Childs and Morton observed this is also true of all the operators on the right of (19): thus solution of the implicit system (21) can be followed by the explicit update

$$U_i^{n+1} = U_{i-m}^n - [\hat{\nu}\Delta_0 - \tfrac{1}{2}\hat{\nu}^2\delta^2 + \tfrac{1}{6}\hat{\nu}^3\Delta_-\delta^2]\tilde{U}_{i-m}^n, \tag{22}$$

with the $\{\tilde{U}_j^n\}$ being consistently interpreted as the coefficients in the quadratic spline recovery function. Consistently with (20) we can write

$$U^{n+1} = P_h E_\Delta R_h U^n \text{ or } \tilde{U}^{n+1} = R_h P_h E_\Delta \tilde{U}^n \tag{23}$$

as alternative symbolic representations of the two-stage process.

Many important consequences follow from this reformulation, both for this simple model problem but more importantly for more general problems:

- The superconvergent third order accuracy of (19) or (22) is now fully explained — it is the quadratic spline that gives an $O(h^3)$ accurate approximation to the solution u.
- Moreover, this accuracy can now be achieved on nonuniform meshes; the scheme obtained from piecewise linears is not the same on a nonuniform mesh and is not third order accurate.
- As mentioned earlier, replacement of the exact integration of products of piecewise linears, that leads to the scheme (19), by a quadrature formula introduces small intervals of instability (see [18]); but these can be completely overcome by the application of the same formulae to the integration of a product of piecewise constants with quadratic splines as in the alternative form (22).
- Approximate solution of the implicit equations (21) gives greater efficiency at an easily assessed cost in the accuracy of the recovery.
- A whole hierarchy of alternative finite difference schemes can be generated through the use of alternative recovery stages: the closest is the use of continuous piecewise quadratics on each cell as in the PPM scheme of Colella and Woodward [5]; simpler discontinuous piecewise quadratics yields the explicit scheme of Warming, Kutler and Lomax

[26]; and of course discontinuous piecewise linears can be used as in the MUSCL scheme of van Leer [25], with a reduction of accuracy to second order.

- The importance of these alternative recovery procedures is that they contain more free parameters which introduces the possibility of the recovery being ***solution-adaptive***, and the preservation of important solution properties such as positivity, monotonicity, TVD etc.; thus there is a direct link to the wide variety of schemes based on slope limiters, flux limiters etc.
- The explicit introduction of the cell averages $\{U_i^n\}$ means that one can always use the more flexible finite volume formulation, with (22) for $m = 0$ being an example of (16) with

$$F_{i+\frac{1}{2}}^{n+\frac{1}{2}} = a[\mu - \tfrac{1}{2}\hat{\nu}\delta + \tfrac{1}{6}\hat{\nu}^2\Delta_-\delta]\tilde{U}_{i+\frac{1}{2}}^n \tag{24}$$

where μ is the averaging operatore $\mu U_{i+\frac{1}{2}} = \frac{1}{2}(U_i + U_{i+1})$; thus the recovery procedure can be regarded as applied to the fluxes.

3 More dimensions, sytems of equations

The result of the last section is that we take as the generic form of an evolution Galerkin scheme for the scalar conservation law $u_t + f_x = 0$ a finite volume formulation similar to (16), namely

$$h_i(U_i^{n+1} - U_i^n) + \Delta t[\tilde{F}_{i+\frac{1}{2}}^{n+\frac{1}{2}} - \tilde{F}_{i-\frac{1}{2}}^{n+\frac{1}{2}}] = 0, \tag{25}$$

where the recovered fluxes will generally take the form

$$\tilde{F}_{i+\frac{1}{2}}^{n+\frac{1}{2}} = \frac{1}{\Delta t}\int_0^{\Delta t} f(E_\tau R_h U^n(x_{i+\frac{1}{2}}))d\tau, \tag{26}$$

where R_h is the recovery operator and E_τ is an approximate evolution step of length τ.

Extending this scheme to the two-dimensional scalar equation

$$u_t + f_x + g_y = 0, \quad f' = a(u), \quad g' = b(u), \tag{27}$$

on a rectangular mesh is quite straightforward: the fluxes through the cell sides are again give by (26), but with an additional averaging in the y-direction, and those through the top and bottom of a similar form with $g(\cdot)$ replacing $f(\cdot)$. Indeed, a triangular, quadrilateral or any other mesh can be handled in the same way. The difficulties lie in devising appropriate recovery and evolution procedures; some possibilities can be found in [6].

For the purpose of the present paper, however, we consider only a rectangular mesh (with cell sides $\Delta x_i, \Delta y_j$) and begin with the unrecovered, first order case, corresponding to the Engquist-Osher scheme (13). For simplicity we assume both $a(u) \geq 0$ and $b(u) \geq 0$, so that in the evolution step the (i, j) cell is influenced by those to its left and below; and, again for simplicity, we assume the time steps sufficiently small for only near neighbours to interact. Then the generalisation of (13) is

$$\begin{aligned} \Delta x_i \Delta y_j (U_{ij}^{n+1} - U_{ij}^n) + \Delta t[\Delta y_j \Delta_{-x} f(U_{ij}^n) + \Delta x_i \Delta_{-y} g(U_{ij}^n)] \\ - \Delta t^2 \Delta_{-x} \Delta_{-y} h(U_{ij}^n) = 0, \end{aligned} \tag{28}$$

where the cross-differenced corner flux $h(\cdot)$ is given by

$$h(u) = \int^u a(v)b(v)dv. \tag{29}$$

The corner terms in (28) are crucial to extending the good properties of the scheme from one dimension to two — in particular, the unconditional stability of the general scheme and the TVD property. The scheme was first formulated in [17] and its properties have been studied in [13].

The particular form (29) for the corner flux $h(\cdot)$ arises directly from approximating the evolution operator by constancy of the solution along the charcteristics. However, the importance of having a cross difference corner term such as this has been widely recognised by many authors — e.g., Radvogin and Zaitsev [22], Colella [4] and LeVeque [10]. The formulation given above has the advantage that it generalises directly to nonmonotone fluxes, triangular meshes, arbitrary time steps, and higher order accuracy through the use of recovery procedures. Moreover, when f and g are nonmonotone we can introduce $a^\pm$ and $b^\pm$, as in (13), to define four corner fluxes $h^{\pm\pm}$ as in (29) to give the influence of all nine contiguous cells on the update.

For systems of conservation laws in one dimension, $\mathbf{u}_t + \mathbf{f}_x = \mathbf{0}$, one may make use of the eigenstructure of the Jacobian matrix $A = \partial\mathbf{f}/\partial\mathbf{u}$. Then from the matrix R of right eigenvectors and the diagonal matrix of eigenvalues, given by $AR = R\Lambda$, one can introduce the characteristic variables $\mathbf{w}$ and the equations they satisfy as

$$\partial\mathbf{w} = R^{-1}\partial\mathbf{u}, \quad \mathbf{w}_t + \Lambda\mathbf{w}_x = \mathbf{0}. \tag{30}$$

Then the formal extension of the characteristic Galerkin scheme (10) – (12) is given by

$$(\mathbf{U}^{n+1} - \mathbf{U}^n, \phi_i) + \int R(\mathbf{U}^n)\left[\int_x^{x+\Lambda(U^n)\Delta t} \phi_i(s)ds\right] R^{-1}(\mathbf{U}^n)d\mathbf{U}^n = \mathbf{0}. \tag{31}$$

A scheme based on a particular continuous piecewise linear parameterisation of $[\mathbf{U}, x]$ has been given in [11], where it is compared with the similar Osher-Solomon scheme [20].

On the other hand, extending these ideas to systems of equations in several dimensions needs completely new ideas, particularly as regards approximating the evolution operator. The discussion of these lies beyond the scope of the present paper, but see [15].

4 Error analysis of evolution Galerkin methods

There is a widely recognised difficulty in establishing the first order accuracy on a nonuniform mesh of the simple conservative upwind scheme, i.e. (13) for $a(u) \geq 0$. This is because the usually defined truncation error is not $O(h)$: for, since U_i^n is the average of U^n over a cell centred at x_i, it is natural to choose u_i^n as either the average of the exact solution $u(x, t_n)$ over the cell, or the value at the cell centre x_i, and to define the truncation error for $a(u) \geq 0$ as

$$\frac{u_i^{n+1} - u_i^n}{\Delta t} + \frac{f(u_i^n) - f(u_{i-1}^n)}{h_i}. \tag{32}$$

However, since $h_i = x_{i+\frac{1}{2}} - x_{i-\frac{1}{2}} \neq x_i - x_{i-1}$, this is not $O(h)$ in general. What is then normally done is to split (32) into two parts and invoke a *supraconvergence* argument as in [14] — see also [19] and [27]. We shall show below that this is not necessary if a more appropriate definition of truncation error is used.

In order to establish the full accuracy of the recovered solution, we start with the second form of the scheme in (23), namely $\tilde{U}^{n+1} = R_h P_h E_\Delta \tilde{U}^n$, and decompose the error $e^n = u(\cdot, t_n) - \tilde{U}^n$ into two parts, the *projection error* η^n and the *evolution error* ξ^n. These are defined by introducing a projection of the exact solution onto the *trial space* spanned by $\{\phi_i\}$, which we can assume to be the piecewise constant space; namely, we will later choose a projection Q_h and set

$$u^n = Q_h u(\cdot, t_n) \text{ and } \tilde{u}^n = R_h Q_h u(\cdot, t_n). \tag{33}$$

Thence we have

$$\begin{aligned} e^n &\equiv u(\cdot, t_n) - \tilde{U}^n \\ &= [u(\cdot, t_n) - R_h Q_h u(\cdot, t_n)] + [R_h Q_h u(\cdot, t_n) - \tilde{U}^n] \\ &\equiv \eta^n + \xi^n, \end{aligned} \tag{34}$$

defining the two parts of the error.

The projection error depends only on the choices of Q_h and R_h, but the evolution error is accumulated through the action of the scheme. Introducing the action of the scheme on $\tilde{u}^n$ as well as on $\tilde{U}^n$, we have

$$\begin{aligned} \xi^{n+1} &\equiv \tilde{u}^{n+1} - \tilde{U}^{n+1} \\ &= [\tilde{u}^{n+1} - R_h P_h E_\Delta \tilde{u}^n] + [R_h P_h E_\Delta \tilde{u}^n - R_h P_h E_\Delta \tilde{U}^n]. \end{aligned} \tag{35}$$

Now the first term here defines the truncation error.

$$T^n = \frac{1}{\Delta t_n}[\tilde{u}^{n+1} - R_h P_h E_\Delta \tilde{u}^n], \tag{36}$$

which will depend on the choice of Q_h. To bound the second term requires a stability property for the scheme: for simplicity here, we assume *strong stability* in some chosen norm, namely that

$$\left\| R_h P_h E_\Delta \tilde{u}^n - R_h P_h E_\Delta \tilde{U}^n \right\| \leq \left\| \tilde{u}^n - \tilde{U}^n \right\| \equiv \|\xi^n\|, \tag{37}$$

which is to hold for any pair of functions $\tilde{u}^n, \tilde{U}^n$ in the space of recovered functions; typical choices of norm are L_∞, L_2 or L_1. Then from (35, (36) and (37) it follows that

$$\|\xi^{n+1}\| \leq \Delta t_n \|T^n\| + \|\xi^n\|, \tag{38}$$

and hence we obtain the error bound from (34) and (38),

$$\|e^n\| \leq \|\eta^n\| + \|\xi^0\| + \sum_0^{n-1} \Delta t_m \|T^m\|. \tag{39}$$

It should be noted that the most severe assumptions made to obtain this result are buried in the strong stability assumption (37). It not only places a restriction on the approximation E_Δ to the evolution operator, but also on the recovery operator R_h. The transport collapse operator, which we have already noted is equalivalent to the E_Δ used in the characteristic Galerkin method (12), is contractive in the L_1 norm, which is a good reason for adopting this norm in (37). Also the projection P_h is obviously contractive; but unless $R_h \equiv I$, i.e. no recovery is applied, R_h will not be contractive. Indeed, from the assumption (20) which says that $P_h R_h = I$ on span$\{\phi_i\}$, it is an approximate inverse to P_h. Thus (37) places a restriction on how far the recovery process can be applied; in practical terms this means that the oscillations damped by the projection P_h onto piecewise constants must not be too greatly enhanced by R_h — a requirement that is usually a very strong design criterion in practical recovery or reconstruction algorithms.

Now let us apply this general analysis to some particular cases. We consider only the scalar conservation law $u_t + f_x = 0$ assuming that $a(u) \geq 0$, and start with the unrecovered "first order" scheme on an arbitrary mesh. We use the finite volume formulation (16) to calculate the truncation error, so we can substract from it the result of integrating the exact conservation law over $\Omega_i^n = (x_{i-\frac{1}{2}}, x_{i+\frac{1}{2}}) \times (t_n, t_{n+1})$ to obtain

$$\begin{aligned} T_i^n &= \frac{(Q_h u - P_h u)_i^{n+1} - (Q_h u - P_h u)_i^n}{\Delta t_n} \\ &+ \frac{(f(u_i^n) - \bar{f}_{i+\frac{1}{2}}^{n+\frac{1}{2}}) - (f(u_{i-1}^n) - \bar{f}_{i-\frac{1}{2}}^{n+\frac{1}{2}})}{h_i}, \end{aligned} \tag{40}$$

where the exact cell edge fluxes are given by

$$\bar{f}_{i\pm\frac{1}{2}}^{n+\frac{1}{2}} = \frac{1}{\Delta t_n} \int_{t_n}^{t_{n+1}} f(u(x_{i\pm\frac{1}{2}}, t))dt. \tag{41}$$

In the first line here we have emphasised the difference between the cell average of the exact solution $(P_h u)_i^n$ and the projection used in the definition

(36), $u_i^n \equiv (Q_h u)_i^n$. Now from the second line it is clear that, to obtain a first order truncation error on an arbitrary mesh, $f(u_i^n)$ should be evaluated at the outflow edge $x_{i+\frac{1}{2}}$. Hence we choose

$$u_i^n \equiv (Q_h u)_i^n = u(x_{i+\frac{1}{2}}, t_n). \tag{42}$$

Both terms in (40) are then clearly $O(h)$ for a sufficiently smooth solution. Indeed, it is easy to show that (see [16])

$$\begin{aligned} |T_i^n| &\leq \frac{1}{\Delta t_n} \|u_{xt}\|_{L_1(\Omega_i^n)} + \frac{1}{h_i} \|u_{tt}\|_{L_1(\Omega_i^n)} \\ &= O(h_i + \Delta t_n). \end{aligned} \tag{43}$$

Thus by this simple and natural choice we demonstrate the first order accuracy of the scheme on a completely arbitrary mesh.

The argument is extended in [16] to cover recovery by discontinuous linears and continuous quadratics. In general, for the use of a recovery polynomial of degree p to obtain an accuracy of order h^{p+1}, we define Q_h at the outflow edge to match P_h to this order; for example, for piecewise liners we define $u_i^n \equiv (Q_h u)_i^n = (u - \frac{1}{2} h_i u_x)(x_{i+\frac{1}{2}}, t_n)$. A similar construction can also be used on a rectangular or triangular mesh for the two-dimensional problem, or though a sonic point where $f'(u)$ changes sign.

Typical of the results is that if discontinuous linear recovery is used, then the three-point scheme resulting from a downwind slope recovery is second order accurate on any mesh, if it is stable; but use of adaptive slope recovery, to ensure stability, will give second order accuracy only on a smoothly varying mesh.

5 Conclusion

The methods described here are designed to solve transient hyperbolic systems, where accurate prediction of the time variation is crucial. Thus the Lagrange Galerkin schemes had their earliest application in such areas as estuarial flows, modelled by the incompressible Navier-Stokes equations, and oil reservoir modelling, using the porous media flow equations; in both cases the flow velocity can be used to define the trajectory (4) and in the definition of E_Δ in (5). Usually, piecewise linear elements on either structured or unstructured meshes are used, with no recovery stage. Thus the basically third order scheme (19) for the advection lies at the heart of the schemes, but other terms in the equations limit the accuracy to second order. One of the chief difficulties is the choice of appropriate quadrature methods to maintain both accuracy and stability.

The compressible flow problems, that are of primary interest to most people at this meeting, at the present time usually involve only steady flows; and when they are unsteady, the time variation is often slow and implicit or semi-implicit methods are called for. Thus the characteristic schemes described above are of less direct interest. However, the use of the finite volume formulation, with a piecewise constant primary approximation followed by recovery to second or third order, links the methods very clearly with the cell centre finite volume methods most commonly used. They provide some new insights — especially in more than one dimension — and may yet be useful for transient problems in this field.

The most recent and important field of application for Lagrangian methods has been in weather forecasting and pollutant dispersal in the atmosphere. Following Staniforth [23], most applications use ***semi-Lagrangian*** methods, in which the trajectories from a set of mesh points are traced back and an interpolation performed to pick up information at the earlier time level. Arguments regarding efficient implementation are usually invoked to preclude the use of Galerkin projections; and maintenance of conservation does not play as key a role as in the shock modelling of aerodynamic flows. However, these semi-Lagrangian methods still fall within the framework of spline recovery described in section 2.4. For, if we regard a delta function as a spline of order $s = 0$, its inner product with a cubic spline ($s = 4$) gives the same result as the inner product of two linear splines, or of a piecewise constant with a quadratic spline: that is, the $\frac{1}{6}, \frac{2}{3}, \frac{1}{6}$ coefficients on the left of (21) are just the three nonzero nodal values of the cubic B-spline; and solving (21) gives the coefficients in a cubic B-spline expansion which reproduces the nodal values $\{U_i^n\}$. Then (22) gives the semi-Lagrangian update of the nodal values.

References

1 J. P. Benqué, G. Labadie, and J. Ronat, *A new finite element method for the Navier-Stokes equations coupled with a temperature equation*, Proc. 4th Int. Symp. on Finite Element Methods in Flow Problems, T. Kawai, ed., North-Holland, 1982, pp. 295-302.

2 Y. Brenier, *Average multivalued solutions for scalar conservation laws*, SIAM J. Numer. Anal., 21 (1984), pp. 1013-1037.

3 P.N. Childs and K.W. Morton, *Characteristic Galerkin methods for scalar conservation laws in one dimension*, SIAM J. Numer. Anal., 27 (1990), pp. 553-594.

4 P. Collela, *Multidimensional upwind methods for hyperbolic conservation laws*, J. Comput. Phys., 87 (1990), pp. 171-200.

5 P. Colella and P. Woodward, *The piecewise parabolic method (PPM) for gasdynamical simulations*, J. Comput. Phys., 54 (1984), pp. 174-201.

6 I. Dawkins, *Development of Practical Evolution Galerkin Algorithms on Unstructured Meshes*, Ph.D. Thesis, Oxford, 1997.

7 J. Douglas and T.F. Russell, *Numerical methods for convection-dominated diffusion problems based on combining the method of characteristics with finite element or finite difference procedures*, SIAM J. Numer. Anal., 19 (1982), pp. 871-885.

8 B. Engquist and S. Osher, *One-sided difference approximations for nonlinear conservation laws*, Math. Comp., 36 (1981), pp. 321-352.

9 P. Lesaint, *Numerical solution of the equation of continuity*, in Topics in Numerical Analysis III, J.J.H. Miller, ed., Academic Press, 1977, pp. 199-222.

10 R. J. Le Veque, *High-Resolution conservative algorithms for advection in compressible flow*, SIAM J. Numer. Anal., 33 (1996), pp. 627-666.

11 P. Lin and K.W. Morton, *An upwind finite element method for nonlinear hyperbolic systems of conservation laws*, Tech. Report NA93/10, Oxford University Computing Laboratory, Wolfson Building, Parks Road, Oxford, OX1 3QD, 1993.

12 P. Lin, K.W. Morton and E. Süli, *Euler characteristic Galerkin scheme with recovery*, M^2AN, 27 (1993), pp. 863-894.

13 P. Lin, K.W. Morton and E. Süli , *Characteristic Galerkin schemes for conservation laws in two and three space dimensions*, SIAM J. Numer. Anal., 34 (1997), pp. 779-796.

14 T.A. Manteuffel and A.B. White Jr., *The numerical solution of second-order boundary value problems on nonuniform meshes*, Math. Comp., 47 (1986), pp. 511-535.

15 K.W. Morton, *Approximation of multidimensional hyperbolic partial differential equations*, Proc. State of the Art in Numerical Analysis Conf., I.S. Duff and G.A. Watson, eds., OUP, 1996, pp. 473-502.

16 K.W. Morton, *Finite Volume Methods for Evolutionary Problems*, to appear SIAM J. Numer. Anal.

17 K.W. Morton and P.N. Childs, *Characteristic Galerkin methods for hyperbolic systems*, in Nonlinear Hyperbolic Equations – Theory, Computation Methods and Applications, J. Ballmann and R. Jeltsch, eds., Notes on Numerical Fluid Mechanics vol. 24, Vieweg, 1989, pp. 435-455.

18 K.W. Morton, A. Priestley, and E. Süli. *Stability analysis of the Lagrange-Galerkin method with non-exact integration.* M^2AN, 22 (1988), pp. 625-653.

19 K.W. Morton and E. Süli, *Evolution-Galerkin methods and their supraconvergence*, Numer. Math., 71 (1995), pp. 331-355.

20 S. Osher and F. Solomon. *Upwind difference schemes for hyperbolic systems of conservation laws.* Math. Comp., 38 (1982), pp. 339–374.

21 O. Pironneau, *On the transport-diffusion algorithm and its application to the Navier-Stokes equations*, Numer. Math., 38 (1982), pp. 309-332.

22 Yu. B. Radvogin and N. A. Zaitsev *Multidimensional minimum stencil supported 2nd order accurate upwind schemes for solving hyperbolic and Euler systems*, Keldysh Inst. of Appld. Math. Report 22, 1996.

23 A. Staniforth and J. Côté. *Semi-Lagrangian integration schemes for atmospheric models — A review.* Mon. Wea. Rev., 119 (1991), pp. 2206–2223.

24 V. Thomée and B. Wendroff. *Convergence estimates for Galerkin methods for variable coefficient initial value problems.* SIAM J. Numer. Anal., 11 (1974), pp. 1059–1068.

25 B. van Leer. *Towards the ultimate conservative difference scheme. V A second-order sequel to Godunov's method.* J. Comput. Phys., 32 (1979), pp. 101–136.

26 R.F. Warming, P. Kutler, and H. Lomax. *Second- and third-order non-centred difference schemes for nonlinear hyperbolic equations.* AIAA J., 11 (1973), pp. 189–196.

27 B. Wendroff and A.B. White Jr., *Some supraconvergent schemes for hyperbolic equations on irregular grids*, 2nd Int. Conf. on Nonlinear Hyperbolic Problems, Aachen, J. Ballmann and R. Jeltsch, eds., Vieweg, 1989, pp. 671-677.

Fluctuation Distribution Schemes on Adjustable Meshes for Scalar Hyperbolic Equations

M.J.Baines
Department of Mathematics
University of Reading
P O Box 220, Reading, RG6 6AX, UK

Paper given at the Meeting to Honour the 60th Birthday of Phil Roe, Arcachon, France, July 1998

Abstract. A problem with the convergence of fluctuation distribution schemes for steady hyperbolic equations on unstructured triangular meshes is that the fluctuations are not driven to zero.

One way of dealing with the problem is to allow the mesh to adjust, using the extra degrees of freedom to counteract the difficulty and improve the approximation. The method then becomes an approximate method of characteristics.

Iterative procedures for the solution of the resulting coupled equations are discussed, including steepest descent least squares and a procedure coupling least squares with multidimensional upwinding. An upwind least squares method is also proposed.

1 Preface

I first met Phil Roe when he visited Reading University about 20 years ago to show Bill Morton the results of his researches during the seventies. At that time he worked at the Royal Aircraft Establishment at Bedford, about 50 miles north of London. Later I had the opportunity to work with him while on three months leave from Reading University visiting RAE Farnborough. There were regular flights between the two establishments in old aircraft flown by enthusiastic pilots in world war II style. At our initial meeting he appeared from the fog having made the unlikely trip from Bedford in the ancient Dakota which they'd actually managed to start on that particular day!

By that time Phil had already constructed his Approximate Riemann Solver and the current issue was the design of limiters. Peter Sweby was soon to become a joint student of ours. I looked forward to communications from Phil since they often took an unexpected form. On one occasion I received a poem (reproduced below) which focussed on the then contentious issue of how to treat fluctuations on quadrilaterals.

Twixt boxes and arrows are hotly disputed
Great arguments no sooner proved than refuted
That their mutual ambitions should soon see the light
Let boxes and arrows combine in the fight

Both fluxes and crosses must each be computed
To each a particular talent is suited
Let arrows do fluxes and boxes do crosses
(There's horses for courses and courses for hosses!)

P.L.Roe (1982)

Another communication took the form of a witty cartoon (reproduced here) on the same theme.

While we kept in touch (not least through the Conferences and Workshops organised by Bill Morton's Institute for Computational Fluid Dynamics at Reading and Oxford) our paths diverged, with Phil moving to the College of Aeronautics at Cranfield and then on to higher things in Michigan (including the editorship of JCP), while my own interests developed into moving grids. So I was intrigued to read after many years his recent Barriers and Challenges paper [2] which adapted the grid as well as the solution on the way to convergence. Typically he had cut through a lot of unnecessary jargon to achieve a striking result. And now we are collaborating again. He is really a mathematician as well as an engineer with a fascination and respect for the beauty and power of algebra. Moreover he values elegance in all the things he does, whether in writing papers, grouping ideas or making up (inspired) limericks. I would say that there are three things which for me epitomise Phil Roe: good sense, good manners, and a nicely surreal sense of humour.

Let me turn now to a piece of work inspired by Phil which is an analysis of fluctuation-signal type algorithms when the grid is allowed to move.

2 Introduction

Phil Roe was the first to suggest the fluctuation-distribution framework [1] for the approximate solution of steady first order hyperbolic PDEs in multi-dimensions. In this approach nonzero *fluctuations,* or residuals, on cells are distributed by *signals*, implemented by adding weighted fractions of the fluctuations to the values of the solution at the corners of the cell. The cumulative update to the solution at a node is of the sum of the weighted contributions from all cells with that node as target.

The steps of the procedure are carried out repeatedly, updating the solution values until the total increments at every node become zero, at which point the process is said to have converged. However, for many types of

Arrows
are
Dead!

discretisation, even though the total increments at a node are zero. the individual cell fluctuations do not vanish but only their weighted sums. So even at convergence the fluctuation in a cell, and hence the PDE residual, is not zero, leading to an unsatisfactory solution.

One way to alleviate the difficulty is to increase the number of degrees of freedom available by including the mesh locations as additional variables [2],[3],[4],[5]. As a consequence, when the total increments become zero the individual fluctuations in a cell are closer to zero and give a much better approximation to the PDE. The approach then resembles an approximate method of characteristics.

In this paper we shall consider a steady scalar hyperbolic scalar equation on an unstructured triangular mesh for which this situation occurs.

3 Fluctuations

The scalar conservation law

$$div\underline{F} = 0 \tag{3.1}$$

with $\underline{F} = u\underline{a}$ becomes

$$div\,(u\underline{a}) = 0, \tag{3.2}$$

equivalent to the integral form

$$\oint_S u\underline{a}.\underline{\widehat{n}}dS = 0, \tag{3.3}$$

where $\underline{\widehat{n}}$ is the inward unit normal to the arbitrary closed surface S in a domain Ω, say. Taking the velocity field $\underline{a}$ to be divergence-free, equation (3.2) reduces to the advection equation

$$\underline{a}.\nabla u = 0. \tag{3.4}$$

Let the domain in which equation (3.4) holds be divided into cells and let u be approximated in each cell by a finite-dimensional function U. Then on each cell we may define the fluctuation to be

$$\phi = -\int \underline{a}.\nabla U d\Omega \tag{3.5}$$

(see [1],[2]) which is the integral of the residual error incurred in replacing u by U in equation (3.4).

Consider now the two-dimensional domain Ω which is the union of triangular cells T_e and let U be linear in each triangle of the form

$$U = \sum_{ei} U_{ei}\psi_{ei}(\underline{x}) \tag{3.6}$$

where the $\psi_{ei}(\underline{x})$ are linear basis functions and the suffix ei refers to the corners of the triangle. Then from equation (3.5) the fluctuation in triangle T_e is

$$\phi_e = -\int_{T_e} \underline{a}.\nabla U \, d\Omega = -\int_{T_e} \underline{a}.\nabla \left(\sum_{ei} U_{ei}\psi_{ei}(\underline{x}) \right) d\Omega \tag{3.7}$$

$$= -\sum_{ei} U_{ei} \int_{T_e} \underline{a}.\nabla \left(\psi_{ei}(\underline{x})\right) d\Omega \tag{3.8}$$

which may be written

$$\phi_e = \sum_{ei} k_{ei} U_{ei} \tag{3.9}$$

where

$$k_{ei} = -\int_{T_e} \underline{a}.\nabla \left(\psi_{ei}(\underline{x})\right) d\Omega = -\frac{1}{2}\underline{a}_e.\underline{n}_{ei}, \tag{3.10}$$

$\underline{n}_{ei}$ being the inward normal to the side of the triangle e opposite node i multiplied by the length of that side (see fig.1).

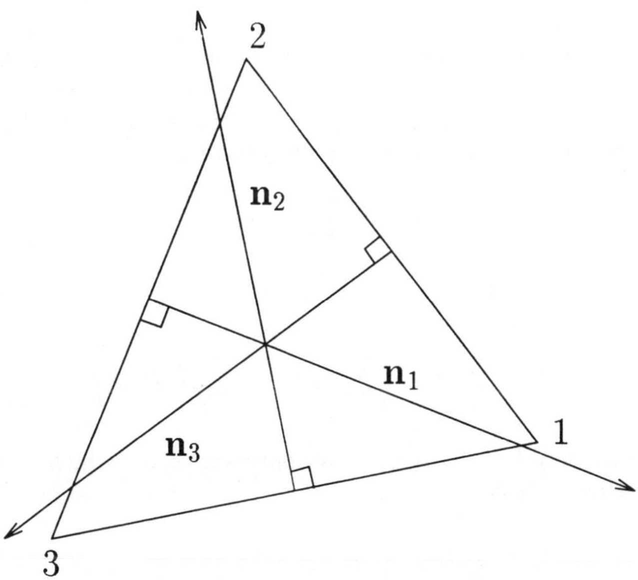

Fig. 1. The normals.

The vector $\underline{a}_e$ is the average field velocity

$$\underline{a}_e = \frac{1}{S_e} \int_{T_e} \underline{a} d\Omega \tag{3.11}$$

where S_e is the area of the triangle T_e.

In equation (3.10) the coefficients k_{ei} depend on $\underline{a}_e$ and the mesh coordinates. However, since

$$\sum_{ei} \psi_{ei}(\underline{x}) = 1 \tag{3.12}$$

we may deduce from equation (3.10) that

$$\sum_{ei} k_{ei} = 0. \tag{3.13}$$

Suppose that the suffix eI denote inflow nodes and from now on reserve the suffix ei for non-inflow nodes. Then equation (3.9) may be written

$$\phi_e = \sum_{ei} k_{ei} U_{ei} + \sum_{eI} k_{eI} U_{eI}, \tag{3.14}$$

where the U_{eI} are prescribed by the inflow conditions.

4 Signals

Define $\{T_j\}$ to be the set of triangles surrounding node j (see fig 2). In the fluctuation-signal algorithm each ϕ_e is calculated from equation (3.14) and added to the value of U_j at node j with predetermined weights w_{je}. The cumulative signal or total update at node j is then

$$\delta U_j = \sum_{\{T_j\}} w_{je} \phi_e, \tag{4.1}$$

i.e. using equation (3.14),

$$\delta U_j = \sum_{\{T_j\}} w_{je} \left(\sum_{ei} k_{ei} U_{ei} + \sum_{eI} k_{eI} U_{eI} \right). \tag{4.2}$$

The procedure may be repeated until convergence, in which case $\delta U_j = 0$ and the U values satisfy

$$\sum_{\{T_j\}} w_{je} \left(\sum_{ei} k_{ei} U_{ei} + \sum_{eI} k_{eI} U_{eI} \right) = 0. \tag{4.3}$$

We refer to three examples of weights w_{je}. First, suppose that the weights w_{je} arise from a least squares minimisation of ϕ over the whole domain (see [2]). Then the weights are proportional to the k_{je} and their sum is zero by equation (3.13). The method is non-conservative in the usual sense, but the

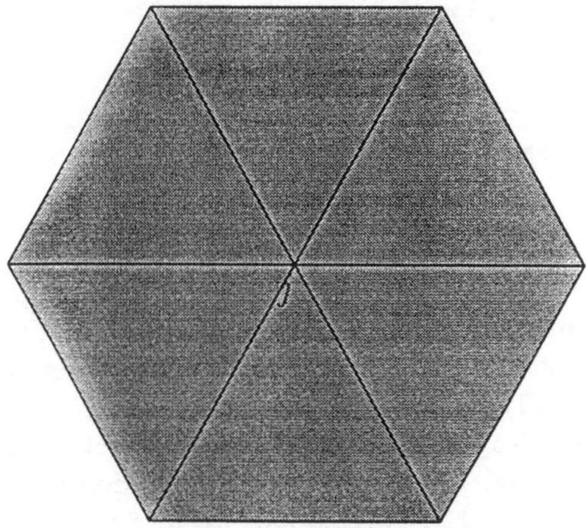

Fig. 2. A patch of triangles T_j (shaded) surrounding node j.

main advantage is the existence of the least squares functional which can be used as a monitor to measure convergence and in designing iterative procedures for reaching the solution. Secondly, let the constant fraction one-third be added to a multiple of the least squares weights, giving a two-dimensional Lax-Wendroff scheme [6]. The sum of the weights is now unity, enforcing conservation, and the method is second-order accurate, but there is no objective functional. In the third example the weights are those associated with the PSI method of Multidimensional Upwinding [6],[7], which is a conservative positive upwind scheme. In this scheme the weights are all non-negative but are solution-dependent in general.

5 Null Space

Equation (3.14) may be written in the matrix-vector form

$$\Phi = K\mathbf{U} + K_I\mathbf{U}_I \tag{5.1}$$

where Φ and $\mathbf{U}$ are vectors of the values of ϕ_e and U_j taken over all cells and nodes, respectively. The function U is assumed to be continuous and the matrices K, K_I are therefore assembled matrices containing combinations of the coefficients k_{ei} as elements. Equation (4.1) may now be written in the matrix-vector form

$$\delta\mathbf{U} = W\Phi \tag{5.2}$$

where W is the matrix of weights w_{ei}. From equation (5.1)

$$\delta\mathbf{U} = W\left(K\mathbf{U} + K_I\mathbf{U}_\mathbf{I}\right). \tag{5.3}$$

At convergence equation (5.2) gives

$$W\Phi = 0 \tag{5.4}$$

while from (5.3) $\mathbf{U}$ satisfies the matrix equation

$$WK\mathbf{U} = -WK_I\mathbf{U_I}. \tag{5.5}$$

Convergence of the fluctuation-signal approach is equivalent to solving the (possibly nonlinear) equation (5.5) for $\mathbf{U}$ (corresponding to $\delta\mathbf{U} = \mathbf{0}$). However, from equation (5.4), unless the matrix W is square and nonsingular $\delta\mathbf{U} = \mathbf{0}$ does not imply that $\Phi = 0$.

The three examples quoted in the previous section have matrix forms characterised as follows. Least squares minimisation of Φ, which results in weights proportional to k_{je}, corresponds to $W = K^t$ and equation (5.5) becomes

$$K^tK\mathbf{U} = -K^tK_I\mathbf{U_I} \tag{5.6}$$

while equation (5.4) becomes $K^t\Phi = 0$ so that Φ lies in the null space of K^t. The Lax-Wendroff method corresponds to $W = \frac{1}{3}I + \alpha K^t$ where α is a constant: Φ lies in the null space of $\frac{1}{3}I + \alpha K^t$. For the PSI method W depends on $\mathbf{U}$ but is ≥ 0: the null space cannot include vectors with only positive components (unlike the previous two methods).

We now describe a simple configuration in which this situation arises. Let the domain be the unit square with $n+1$ nodes along each side. The interior of the square is discretised with $2n^2$ triangular cells and $(n+1)^2$ nodes by dividing each square of a uniform grid by a constantly orientated diagonal (see fig.3). Suppose further that there are just two inflow sides so that the number of non-inflow nodes is n^2. Then the dimensions of the vectors Φ and $\mathbf{U}$ are $2n^2$ and n^2 respectively, while those of the matrices K and W are $2n^2 \times n^2$ and $n^2 \times 2n^2$. It follows that the matrix W possesses a null space of dimension n^2 and that at convergence equation (5.4) merely tells us that Φ belongs to that null space. For example, if the weights in equation (4.1) sum to zero then any row of W has a zero sum and the vector $\Phi = (1, 1, ..., 1)$ lies in the null space of W.

In general W will always possess a null space and as a result $\delta\mathbf{U} = \mathbf{0}$ will not automatically imply that Φ is small.

6 Adjustable Nodes

One way of tackling the problem of the null space is to increase the number of degrees of freedom by regarding the mesh locations as additional variables, in a way similar to that exploited elsewhere [3],[4],[5]. A natural construction in the present context is to use the fluctuations to generate signals for adjusting the mesh as well as the solution and this can be implemented as follows.

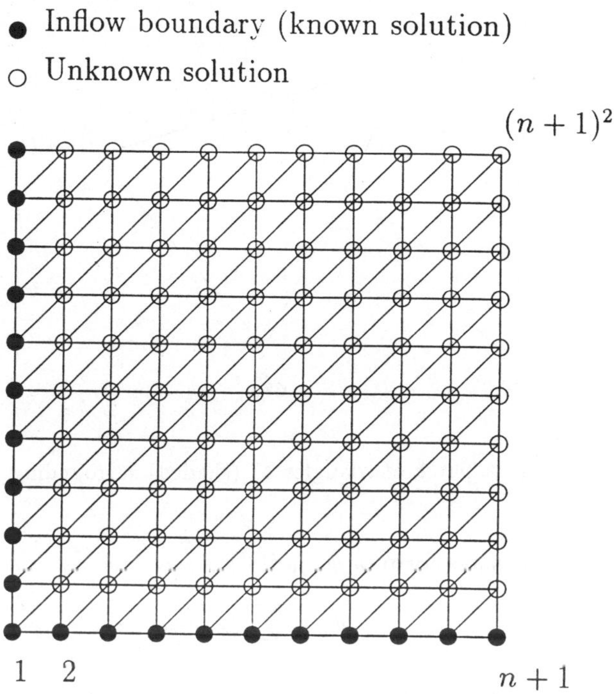

Fig. 3. A square domain with $2n^2$ triangular cells and n^2 unknowns.

6.1 Fluctuations

First we derive a nodal based form of the fluctuation (3.9), which may be obtained by writing

$$\phi_e = \sum_{ei} k_{ei} U_{ei} = -\frac{1}{2} \sum_{ei} U_{ei} \underline{a}_e . \underline{n}_{ei} \tag{6.1}$$

$$= -\frac{1}{2} \sum_{ei} U_{ei} \underline{a}_e . \left(-(Y_{ei2} - Y_{ei1})\underline{i} + (X_{ei2} - X_{ei1})\underline{j}\right) \tag{6.2}$$

where $\underline{i}$,$\underline{j}$ are unit vectors in the x, y directions and $ei1, ei2$ are the vertices of T_e taken anticlockwise from ei (fig 3). Writing $\underline{a}_e = (a_e, b_e)$, (6.2) becomes

$$\frac{1}{2} \sum_{ei} U_{ei} \left((Y_{ei2} - Y_{ei1})a_e - (X_{ei2} - X_{ei1})b_e\right) \tag{6.3}$$

$$= \frac{1}{2} \sum_{ei} \left(-Y_{ei}(U_{ei2} - U_{ei1})a_e + X_{ei}(U_{ei2} - U_{ei1})b_e\right) \tag{6.4}$$

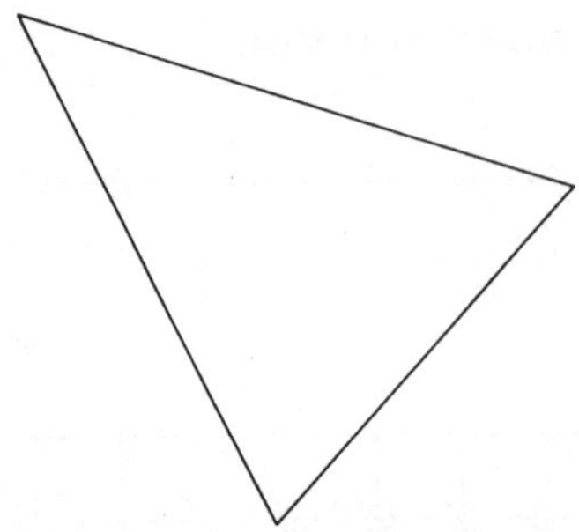

Fig. 4. A patch of triangles T_j (shaded) surrounding node j.

so that an alternative form of ϕ_e is ([8])

$$\phi_e = \sum_{ei} \underline{l}_{ei} \cdot \underline{X}_{ei} \tag{6.5}$$

(cf. (3.10)), where

$$\underline{l}_{ei} = \frac{1}{2} \Delta U_{ei} \begin{pmatrix} b_e \\ -a_e \end{pmatrix} \tag{6.6}$$

(cf. (3.12)), and $\Delta U_{ei} = U_{ei2} - U_{ei1}$ is the difference in the U values across the side opposite corner ei, taken anticlockwise around the triangle (see fig 4). Clearly

$$\sum_{ei} \underline{l}_{ei} = 0. \tag{6.7}$$

As in equation (3.14), separating out the inflow points leads to the form

$$\phi_e = \sum_{ei} \underline{l}_{ei} \cdot \underline{X}_{ei} + \sum_{eI} \underline{l}_{eI} \cdot \underline{X}_{eI}. \tag{6.8}$$

Inflow nodes are separated out here on the grounds that the $\underline{X}_{eI}$ coordinates will also naturally be partially prescribed at inflow. In some problems the $\underline{X}_{ei}$ may also be prescribed at *outflow* in order to constrain the nodes to remain on the physical boundaries of the domain.

6.2 Signals

Nodal positions may now be updated by signals in the same way as for U. Denoting the weights for these signals by $\underline{\gamma}_{je}$ the analogue to equation (4.1) is

$$\delta \underline{X}_j = \sum_{\{T_j\}} \underline{\gamma}_{je} \phi_e = \sum_{\{T_j\}} \underline{\gamma}_{je} \left(\sum_{ei} \underline{l}_{ei} \cdot \underline{X}_{ei} + \sum_{eI} \underline{l}_{eI} \cdot \underline{X}_{eI} \right). \qquad (6.9)$$

At convergence

$$\sum_{\{T_j\}} \underline{\gamma}_{je} \left(\sum_{ei} \underline{l}_{ei} \cdot \underline{X}_{ei} + \sum_{eI} \underline{l}_{eI} \cdot \underline{X}_{eI} \right) = 0 \qquad (6.10)$$

(cf. (4.3)).

If the weights are derived from a least squares approach they are, as in the case of U updates, proportional to $\underline{l}_{je}$.

6.3 Null Space

In matrix-vector form equation (6.8) may be written as

$$\Phi = \underline{L}.\underline{\mathbf{X}} + \underline{L}_I.\underline{\mathbf{X}}_I \qquad (6.11)$$

where $\underline{\mathbf{X}}$ is a vector of all the 2-vectors $\underline{X}_j$ taken over the nodes. The piecewise linear function X is assumed to be continuous and the matrices $\underline{L}$,$\underline{L}_I$ are therefore assembled matrices containing combinations of the coefficients l_{ei} as elements.

If $\underline{\Gamma}$ is defined to be the matrix of weights $\underline{\gamma}_{je}$, the matrix-vector forms of equations (6.9) and (6.10) are

$$\delta\underline{\mathbf{X}} = \underline{\Gamma}\Phi = \underline{\Gamma}(\underline{L}\underline{\mathbf{X}} + \underline{L}_I\underline{\mathbf{X}}_I) \qquad (6.12)$$

and

$$\underline{\Gamma}\underline{L}\underline{\mathbf{X}} = -\underline{\Gamma}\underline{L}_I\underline{\mathbf{X}}_I. \qquad (6.13)$$

In the least squares method $\underline{\Gamma}$ is proportional to $\underline{L}^t$ and equation (6.13) becomes

$$\underline{L}^t\underline{L}\underline{\mathbf{X}} = -\underline{L}^t\underline{L}_I\underline{\mathbf{X}}_I \qquad (6.14)$$

(cf. (5.6)).

Considering again the illustration of a square domain with two inflow sides discussed in the previous section, the vector $\underline{\mathbf{X}}$ will have dimension $2n^2$ and both matrices $\underline{L}$ and $\underline{\Gamma}$ have dimensions $2n^2 \times 2n^2$. Hence $\underline{\Gamma}$ is square and provided that it is non-singular, $\delta\underline{\mathbf{X}}= \mathbf{0}$ implies that $\Phi = \mathbf{0}$, as desired.

One way to understand the situation is as follows. For a given solution U, by adjusting $\mathbf{X}$ we may construct an approximate "characteristic mesh" as a triangulation of points which lie on the characteristics of the equation (3.4). On the resulting mesh a piecewise linear approximation to U on each characteristic is approximately constant and therefore within each triangle

$$\underline{a}.\nabla U = 0 \qquad (6.15)$$

giving a zero Φ.

However such an approximate triangulation is far from unique and it is clear that the equation

$$\phi_{ei} = \sum_{ei} \underline{l}_{ei} \cdot \underline{X}_{ei} = 0 \tag{6.16}$$

does not uniquely determine $\underline{\mathbf{X}}$.

7 Constrained Nodal Movement

The above analysis of the situation is still valid when the nodal coordinates are constrained to move perpendicular to the approximate characteristics which suggests that we may achieve the same result when only one of the nodal coordinates is allowed to vary. A single varying nodal coordinate combined with variable U is still sufficient to allow (a) the sides of the triangles to align with the characteristics and (b) U to become constant along them: equation (6.15) holds and Φ_e vanishes as required.

Suppose then that $\underline{X}_j$ is constrained to move only in a chosen direction $\underline{\widehat{N}}_j$ and denote the corresponding coordinate by N_j. We may take $\underline{\widehat{N}}_j$ to be in the direction perpendicular to the characteristic velocity field $\underline{a}_j$ at node j. Then equation (6.9) reduces to

$$\delta N_j = \sum_{\{T_j\}} \widetilde{\gamma}_{ei} \phi_e = \sum_{\{T_j\}} \widetilde{\gamma}_{ei} \left(\sum_{ei} \widetilde{l}_{ei} N_e + \sum_{eI} \widetilde{l}_{eI} N_e \right) \tag{7.1}$$

where $\widetilde{\gamma}_{je} = \underline{\widehat{N}}_j \cdot \underline{\gamma}_{je}$ and $\widetilde{l}_{ei} = \underline{\widehat{N}}_e \cdot \underline{l}_{ei}$ with $\underline{\widehat{N}}_e$ perpendicular to $\underline{a}_e$, which at convergence becomes

$$\sum_{\{T_j\}} \widetilde{\gamma}_{je} \left(\sum_{ei} \widetilde{l}_{ei} N_e + \sum_{eI} \widetilde{l}_{eI} N_e \right) = 0. \tag{7.2}$$

Note that as in equation (6.7)

$$\sum \widetilde{l}_{ei} = 0. \tag{7.3}$$

The matrix-vector equation (6.12) then becomes

$$\delta \mathbf{N} = \widetilde{\Gamma} \Phi = \widetilde{\Gamma} \left(\widetilde{L} \mathbf{N} + \widetilde{L}_I \mathbf{N}_I \right) \tag{7.4}$$

where $\widetilde{L}, \widetilde{L}_I$ are matrices of the coefficients $\widetilde{l}_{ei}, \widetilde{l}_{eI}$ and $\widetilde{\Gamma}$ contains the coefficients $\widetilde{\gamma}_{je}$. At convergence

$$\widetilde{\Gamma} \Phi = \mathbf{0}, \qquad \widetilde{\Gamma} \widetilde{\mathbf{L}} \mathbf{N} = -\widetilde{\Gamma} \widetilde{\mathbf{L}}_{\mathbf{I}} \mathbf{N}_{\mathbf{I}}. \tag{7.5}$$

In the least squares case $\widetilde{\Gamma}$ is proportional to $\widetilde{L}^t$ and equation (7.5) becomes

$$\widetilde{L}^t \widetilde{L} \underline{\mathbf{N}} = -\widetilde{L}^t \widetilde{L}_I \underline{\mathbf{N}}_I. \tag{7.6}$$

8 Coupled Solutions

In the earlier illustration of simple advection across a square domain (see fig 3)the vector $\mathbf{N}$ is of dimension n^2 and the matrices L and $\widetilde{\Gamma}$ now have dimensions $2n^2 \times n^2$ and $n^2 \times 2n^2$, respectively. By itself equation (7.4) has the same drawbacks as equation (5.3) with $\widetilde{\Gamma}$ possessing a null space and $\delta\mathbf{N} = \mathbf{0}$ not implying $\Phi = \mathbf{0}$. However, taken together with equation (5.3) we obtain at convergence the square systems

$$\begin{pmatrix} W \\ \widetilde{\Gamma} \end{pmatrix} \Phi = \mathbf{0} \tag{8.1}$$

for Φ and

$$\begin{pmatrix} WK & 0 \\ 0 & \widetilde{\Gamma}\widetilde{L} \end{pmatrix} \begin{pmatrix} \mathbf{U} \\ \mathbf{N} \end{pmatrix} = - \begin{pmatrix} WK_I & 0 \\ 0 & \widetilde{\Gamma}\widetilde{L}_I \end{pmatrix} \begin{pmatrix} \mathbf{U_I} \\ \mathbf{N_I} \end{pmatrix} \tag{8.2}$$

for $\mathbf{U}$ and $\mathbf{N}$. Provided that the left hand side matrices are non-singular we can always solve for a pair $\mathbf{U}$, $\mathbf{N}$ for which $\Phi = \mathbf{0}$.

In three dimensions a similar argument shows that the inclusion of *two* nodal coordinates (rather than the full three available) gives rise to a square system and hence, given nonsingularity, to vanishing Φ at convergence, which again accords with the method of characteristics. Indeed, in d dimensions using simplexes, the number of variable nodal coordinates to be included should be $d-1$.

We consider some examples.

8.1 Least Squares Weights

If the fluctuation distribution scheme is derived from a least-squares minimisation of Φ over both $\mathbf{U}$ and $\mathbf{N}$, the weights in W are proportional to k_{je} as before while those in $\widetilde{\Gamma}$ are now proportional to

$$\widetilde{l}_{je} = \underline{\widehat{N}}_j . \underline{l}_{je} = \frac{1}{2} \Delta U_{je} (a_j a_{je} + b_j b_{je})/(a_j^2 + b_j^2)^{1/2} = \frac{1}{2} \left|\underline{a}\right| \Delta U_{je} \tag{8.3}$$

for a constant velocity field $\underline{a}$ (see equation (7.1)).

A necessary condition for nonsingularity in equation (8.1) is that the null spaces of W and $\widetilde{\Gamma}$ be disjoint. However, from equations (3.13) and (7.3), the vector $\Phi = (1, 1, ..., 1)$ lies in both of these null spaces, so in this special case we have not elminated the null space entirely.

Nevertheless the fact that Φ is small on an approximate characteristic mesh ensures a good solution to the differential equation when $||\Phi||_2$ is minimised. Even though the null space may not have been totally eradicated, all the Φ components (including those in the null space) are small. Moreover this argument still holds when the least squares minimisation is taken over nodal coordinates alone: given any consistent weights for $\mathbf{U}$, including those possessing a null space, least squares minimisation of $||\Phi||_2$ over $\mathbf{X}$ (or $\mathbf{N}$) alone still ensures that the effect of the null space is nullified.

We therefore may either

- choose a unified least squares minimisation of $||\Phi||_2$ over $\mathbf{U}$ and $\mathbf{X}$ (or $\mathbf{N}$), leading to a method with an objective functional that can be monitored, *or*
- choose a least squares minimisation of $||\Phi||_2$ over $\mathbf{X}$ (or $\mathbf{N}$) alone, combined with some other choice of weights W for $\mathbf{U}$ with other properties but lacking an objective functional. In the examples cited earlier the Lax-Wendroff or PSI schemes can be incorporated with the latter option.

We have already cited the former, unified least squares, method as an example in the text. We shall discuss the latter approach in Section 10 and also extend the idea to an upwind least squares procedure.

9 Iterative Solution Methods

We now consider iterative methods for the solution of equation (8.2) for $\mathbf{U}$ and $\mathbf{N}$. Since K depends upon $\mathbf{N}$ and L depends upon $\mathbf{U}$ in general, the two component equations in (8.2) are coupled. However, if $\mathbf{U}$ and $\mathbf{N}$ are frozen in WK, the set of equations (5.5) form a sparse linear system for $\mathbf{U}$. Similarly, if $\mathbf{N}$ and $\mathbf{U}$ are frozen in $\widetilde{\Gamma} L$, the equations (7.5) form a sparse linear system for $\mathbf{N}$. This linearisation suggests an iterative solution procedure in which steps for equation (5.5) are alternated with those for equation (7.5).

9.1 A Jacobi Iteration

For example, the relaxed Jacobi method for the independent solutions of the linearised equations (5.5) and (7.5) is

$$diag(WK)\delta\mathbf{U} = -\tau W\Phi = -\tau\left(WK\mathbf{U} + WK_I\mathbf{U}_\mathbf{I}\right) \tag{9.1}$$

and

$$diag(\widetilde{\Gamma}\widetilde{L})\delta\mathbf{N} = -\sigma\widetilde{\Gamma}\Phi = -\sigma\left(\widetilde{\Gamma}\widetilde{L}\mathbf{N} + \widetilde{\Gamma}\widetilde{L}_I\mathbf{N}_\mathbf{I}\right), \tag{9.2}$$

where τ and σ are positive relaxation factors. In component form, for each non-inflow node j these equations (excluding inflow terms) read

$$\left(\sum_{\{T_j\}} w_{je}k_{ej}\right)\delta U_j = -\tau\left(\left(\sum_{\{T_j\}} w_{je}k_{ej}\right)U_j + \sum_{\{T_j\}} w_{je}\sum_{ei} k_{ei}U_{ei}\right) \tag{9.3}$$

using equation (3.13) and

$$\left(\sum_{\{T_j\}} \widetilde{\gamma}_{je}\widetilde{l}_{ej}\right) \delta N_j = -\sigma \left(\left(\sum_{\{T_j\}} \widetilde{\gamma}_{je}\widetilde{l}_{ej}\right) N_j + \sum_{\{T_j\}} \widetilde{\gamma}_{je} \sum_{ei} \widetilde{l}_{ei} N_{ei} \right) \tag{9.4}$$

using equation (6.7), denoting by ei the index of those nonzero terms in the j'th equation of equation (4.3) which are different from j. In the iterations described here we may interleave these two linearised steps.

9.2 Steepest Descent Methods

In the case of a unified L_2 minimisation of the norm of Φ we may construct iterative algorithms directly using the techniques of optimisation. For example, steepest descent methods applied to the least squares equations (5.6) and (7.5) generate the iterations

$$\delta \mathbf{U} = -\tau' K^t (K\mathbf{U} + K_I \mathbf{U_I}) \tag{9.5}$$

and

$$\delta \mathbf{N} = -\sigma' \widetilde{L}^t (\widetilde{L}\mathbf{N} + \widetilde{L}_I \mathbf{N_I}) \tag{9.6}$$

where τ', σ' are relaxation coefficients. Optimal choices of τ' and σ' consistent with the linearisation are

$$\tau' = -\frac{\mathbf{R}^t\mathbf{R}}{\mathbf{R}^t K^t K \mathbf{R}} \qquad \sigma' = -\frac{\mathbf{S}^t\mathbf{S}}{\mathbf{S}^t \widetilde{L}^t \widetilde{L} \mathbf{S}} \tag{9.7}$$

where

$$\mathbf{R} = K^t (K\mathbf{U} + K_I \mathbf{U_I}) \tag{9.8}$$

and

$$\mathbf{S} = \widetilde{L}^t (\widetilde{L}\mathbf{N} + \widetilde{L}_I \mathbf{N_I}) \tag{9.9}$$

are the residuals associated with the normal equations (5.6) and (7.5).

Equations (9.5) and (9.6) are identical with the relaxed Jacobi iterations (9.3) and (9.4) under the choices

$$\tau = \tau' \sum_{\{T_j\}} k_{ei}^2 \qquad \sigma = \sigma' \sum_{\{T_j\}} \widetilde{l}_{ei}^2 \tag{9.10}$$

and may be rewritten, using equations (3.13) and (7.3), as

$$\delta U_j = \frac{\sum_{\{T_j\}} k_{je} \left(\sum_{ei} k_{ei} (U_{ei} - U_j)\right)}{\sum_{\{T_j\}} k_{je} \left(\sum_{ei} k_{ei}\right)} \tag{9.11}$$

and

$$\delta N_j = \frac{\sum_{\{T_j\}} \widetilde{l}_{je} \left(\sum_{ei} \widetilde{l}_{ei} (N_{ei} - N_j)\right)}{\sum_{\{T_j\}} \widetilde{l}_{je} \left(\sum_{ei} \widetilde{l}_{ei}\right)}, \tag{9.12}$$

each being an averaging process on U or N. Averaging processes have been exploited elsewhere [7],[9].

These algorithms clearly belong to the class of fluctuation-signal type schemes, even though the sum of the weights is zero and the iterations are non-conservative in the usual sense.

10 An Upwind Hybrid Method

As proposed in section 8 we may combine upwind iterations for **U** with least squares iterations for **N** in an obvious way by interleaving iterations for **U** from equation (9.3) with the least squares procedure for **N** from equations (9.6). In this way we may combine the benefits of a conservative and positive iteration for **U**, which follows the flow of information along characteristics, with an iteration for **N** which drives $||\Phi||$ down to zero and thereby counteracts the effect of the null space.

We now consider weights based on physically based iterations together with their properties and link them to least squares iterations.

Recall that the fluctuation-signal mechanism (with appropriate weights) may be regarded as originating in a scheme for solving the time-dependent equation

$$u_t + div\,(u\underline{a}) = 0 \tag{10.1}$$

(cf. equation (3.2)), and hence if used as an iteration to solve equation (3.2) may be interpreted as proceeding to the steady state limit through physical states. Such a physical approach to the limit suggests that the flow of information along characteristics should be respected and the initial approximation should consist of inflow conditions, which in turn suggests the use of upwind weights for U satisfying

$$\begin{aligned} w_{ei} &= 0 \quad k_{ei} \leq 0 \\ w_{ei} &\geq 0 \quad k_{ei} > 0. \end{aligned} \tag{10.2}$$

It is also desirable (although not essential) that the weights chosen should be such that the iteration for U is conservative. This means that appropriately scaled weights should satisfy

$$\sum w_{ei} = 1 \tag{10.3}$$

(which rules out least squares weights).

This type of iteration is associated with Multidimensional Upwinding methods [5] where by careful choice of the weights the schemes can be made both conservative and positive, as in the PSI scheme [6],[9].

Similar arguments can be applied to the weights γ (see below).

We may then combine such an iteration with a least squares iteration for N. In this way we obtain the benefits of the multidimensional upwind schemes as well as driving $||\Phi||$down to zero, thus nullifying the effect of the null space.

We may go further by trying to incorporate upwinding into the iteration for N. In doing so we give up the formal minimisation of $||\Phi||$ but obtain much faster convergence to a very similar optimal grid.

11 Upwind Least Squares Methods

The least squares approach embeds the original equation in a higher order equation. The correct solution is picked out from the larger set of solutions by an outflow condition which is the original differential equation applied at the outflow boundary. It may be therefore be argued that the weights should still exhibit an upwind bias.

One way of achieving such a bias is to carry out the minimisation of $||\Phi||$ over only ***downwind*** nodal values and admit temporary discontinuities in U. The updates resulting from the minimisation still reduce $||\Phi||$ but at the expense of generating discontinuities. This step may be followed by a projection step which resets all upwind values of U in a cell in such a way as to restore continuity of U. The two steps are then repeated until convergence is achieved. The second step is not a descent step and will not reduce $||\Phi||$ in general. At convergence we nevertheless obtain a continuous U which minimises $||\Phi||$ because the gradient is zero.

The algorithm has a strong upwind bias which reflects the nature of the original problem and its dependence on characteristics. In fact the second of the two steps is automatically implied if the updates in the first step are carried out in the usual least squares manner but with upwinded updates suppressed. With an appropriate scaling the first step is then equivalent to the LDA scheme of Multidimensional Upwinding [2],[11].

The generalisation to adaptive meshes is straightforward and is again motivated by the flow of information. The norm $||\Phi||$ is minimised over only ***downwind*** values of N and when the local updates are carried out N becomes discontinuous at the nodes. The projection step then resets all upwind N values in a cell so as to make N continuous again. At convergence we obtain a continuous N which has driven $||\Phi||$ even further down towards zero.

The method may be called Adaptive Discontinuous Least Squares (ADLS).

As in the previous section this procedure for N may be interleaved with iterations from a standard upwinding scheme for U, such as PSI, providing an algorithm which takes full advantage of the hyperbolic nature of the problem as well as speeding up convergence and counteracting the effect of the null space.

12 Numerical Results

The canonical result for the unified least squares method is due to Phil Roe in [2]. The problem is that of the circular advection equation,

$$yu_x - xu_y = 0, \tag{12.1}$$

in a rectangular region $|x| < 1, 0 < y < 1$ with initial data $u = 0$ except for $u = 1$ at two adjacent gridpoints on the inflow side. The method is to minimise the residual

$$F = \sum_{cells} \frac{\Phi^2}{S} \tag{12.2}$$

over the nodal values and nodal coordinates. The iteration is performed using steepest descent least squares (for both the solution and the grid) and the result is shown in fig. 3. In [2] the comparison between this result and that on a fixed grid is shown via graphs of the outflow profile. A different comparison may be made by shading the cells by the value of the local residual ($||\Phi||$). In fig. 4 this shading is shown on grid plots for both the fixed and optimal grid solutions, as well as at two stages during the iteration, demonstrating the way in which the grid movement capability drives the residual down to very low values. Fig. 5 shows the corresponding convergence profile.

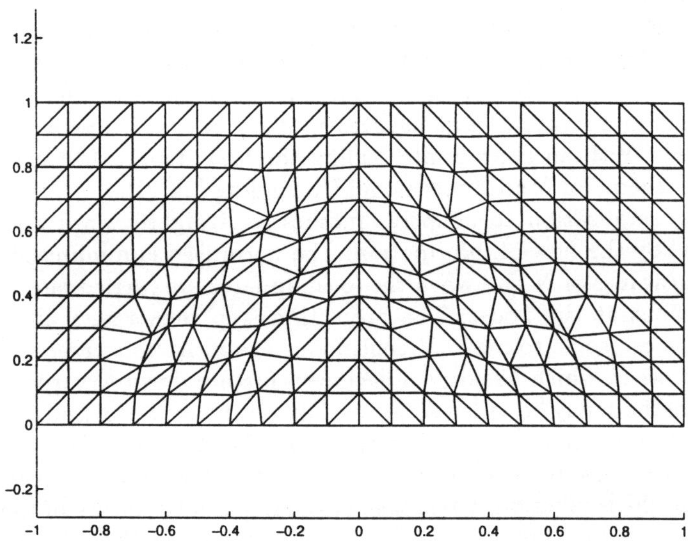

Fig. 5. Resulting grid when using steepest descent least squares.

One of the drawbacks of the method is slow convergence. This is likely to be partly due to the way in which the least squares method updates upwind nodes as well as downwind nodes and partly due to lack of conservation. The iteration may be considerably accelerated by changing to the upwind least squares method of the previous section (for both the solution and the grid). However we find that some grid smoothing is needed. At the end of a period of alternately interleaving solution and node iterations the grid is subjected to the simple smoothing

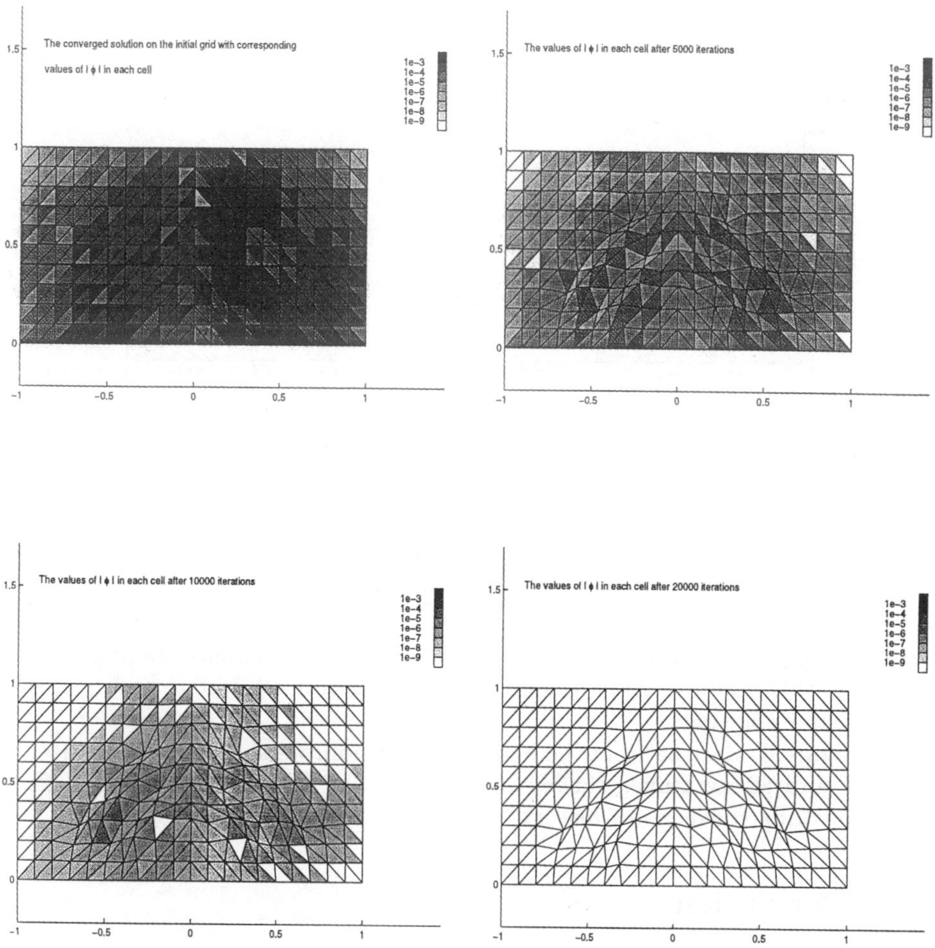

Fig. 6. Results using steepest descent least squares.

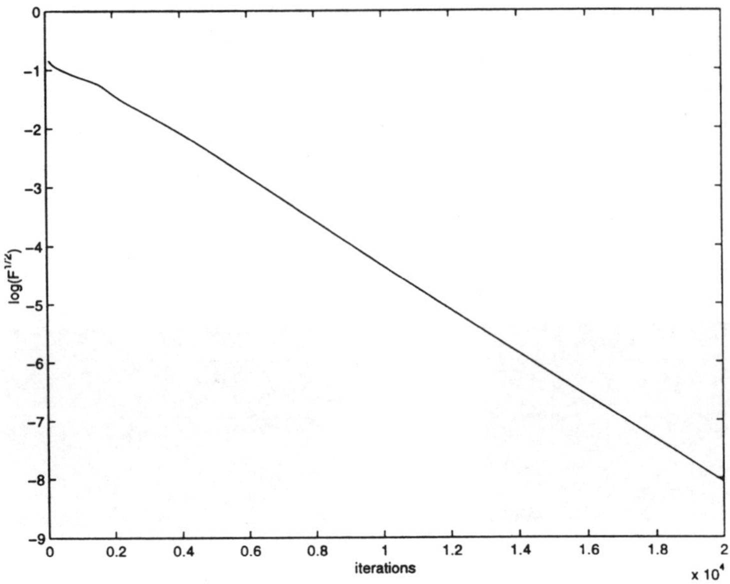

Fig. 7. Convergence history when using steepest descent least squares.

$$\mathbf{X}_j^{new} = \frac{1}{N} \sum_{ji=1}^{N} \mathbf{X}_{ji} \tag{12.3}$$

where ji goes round the N outer nodes of the patch of cells surrounding node j, unless this results in mesh tangling. The result is much faster convergence to the same solution.

In figure 6, starting from the fixed grid solution, we show grid plots after 0,20,40 and 80 iterations, shaded by the value of $||\Phi||$, and in figure 7 the corresponding convergence history.

13 Conclusion

In this paper difficulties experienced with fluctuation distribution methods on fixed unstructured grids in generating zero fluctuations/residuals has been addressed by including the mesh locations as variables. This allows the grid to adjust in order to drive the least squares residual of the fluctuation down closer to zero. The null space invoked by fixed grid methods is counteracted by the extra degrees of freedom provided by the moving nodes.

For a scalar advection problem in 2-D the procedure may be used in association with least squares minimisation of the average residual, giving a unified scheme for the solution and the grid, as in [2]. The resulting scheme

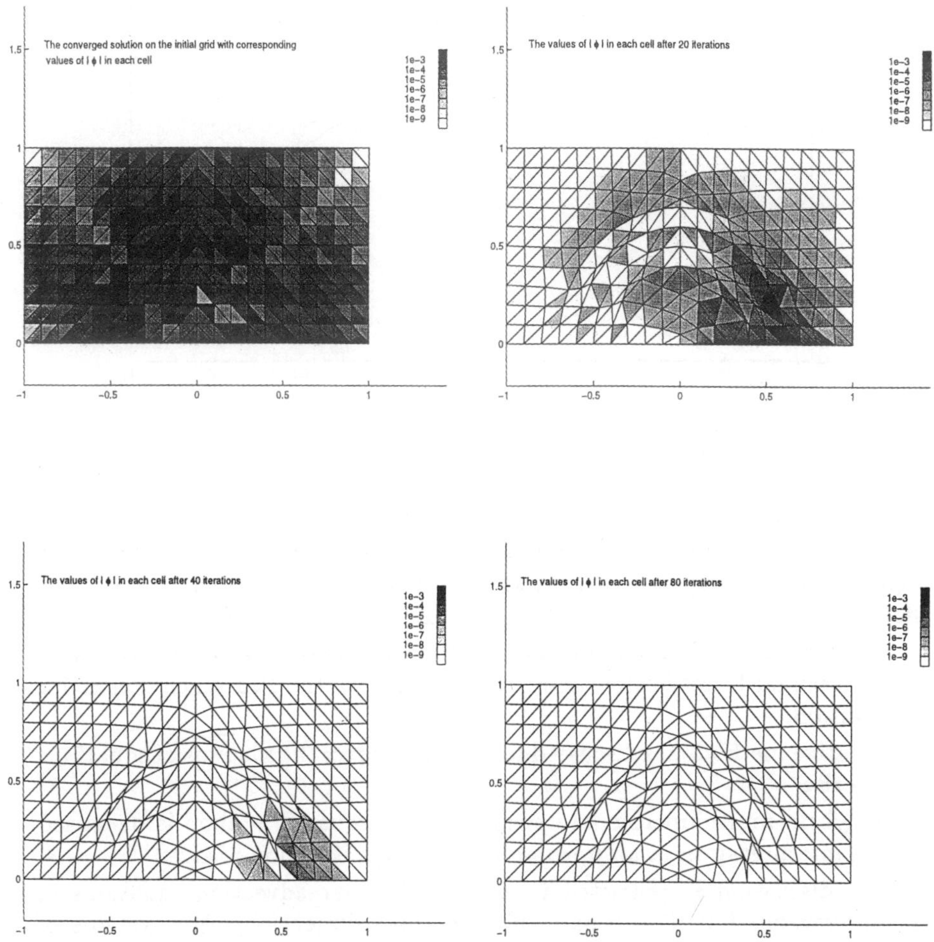

Fig. 8. Results using upwind least squares.

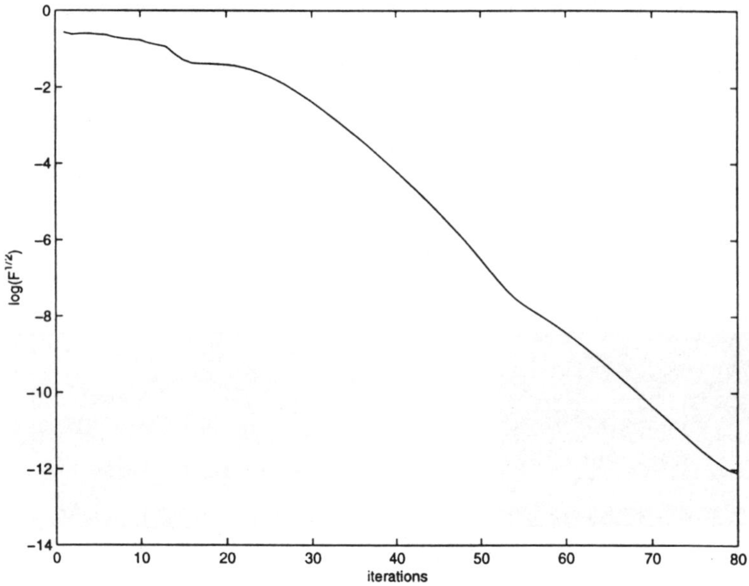

Fig. 9. Convergence history when using upwind least squares.

is an approximate method of characteristics on a triangular grid. The nodes need only move perpendicular to the characteristics to provide the desired effect.

The argument still holds when least squares is used only for the grid, with a more standard, physically based, scheme such as the PSI scheme used for the solution.

Iterative methods are discussed for the solution of the systems of equations which arise.

A hybrid multidimensional upwinding-least squares method is discussed and an interpretation of an upwind least squares iteration is given.

The discussion is restricted to a class of scalar advection equations in two dimensions but is also applicable to three dimensions. For systems of equations, where there is no single family of characteristics, the approach must rely on mesh movement in the mean.

14 Acknowledgements

The author has benefitted greatly from discussions with Phil Roe, Yves Tourigny, Matthew Hubbard, Paul Houston and Stephen Leary.

15 References

[1] P.L.Roe (1983). Fluctuation and Signals: a framework for Numerical Evolution Problems. In Proceedings of IMA Conference on Numerical Methods for Fluid Dynamics, Reading, UK March 1982 (Morton and Baines (eds.)), p1, Academic Press.

[2] P.L.Roe (1996). Compounded of Many Simples, in Proceedings of Workshop on Barriers and Challenges in CFD, ICASE, NASA Langley, August 1996, (Ventakrishnan, Salas and Chakravarthy (eds.)), p241, Kluwer, 1998.

[3] Y.Tourigny and M.J.Baines (1997). Analysis of an Algorithm for Generating Locally Optimal Meshes for L_2 Approximation by Discontinuous Piecewise Polynomials, Math. Comp., 66,623-650.

[4] Y.Tourigny and F.Hulsemann (1997). A New Moving Mesh Algorithm for the Finite Element Solution of Variational Problems, SIAM.J.Num.An.(to appear).

[5] M.J.Baines (1997). On Variational Techniques and Least Squares Methods with Adjustable Nodes. Numerical Analysis Report 2/97, Department of Mathematics, University of Reading.

[6] H.Deconinck, P.L.Roe and R.Struijs (1993). A Multidimensional Generalisation of Roe's Flux Difference Splitter for the Euler Equations. Computers and Fluids, 22, 215.

[7] M.E.Hubbard (1994). Multidimensional Upwinding. PhD thesis, Department of Mathematics, University of Reading.

[8] M.J.Baines (1997). A Note on Duality for a Scalar Hyperbolic Equation. Numerical Analysis Report 8/97, Department of Mathematics, University of Reading.

[9] M.J.Baines and M.E.Hubbard (1998). Multidimensional Upwinding and Grid Adaptation. In Numerical Methods for Wave Propagation Problems (E.F.Toro and J.F.Clarke (eds.)), Kluwer.

[10] N.R.C.Birkett and M.Rudgyard (1998). In Proceedings of 1998 ICFD Conference on Numerical Methods for Fluid Dynamics (M.J.Baines (ed.)), ICFD, Oxford University Computing Laboratory.

Superconvergent lift estimates through adjoint error analysis

M.B. Giles and N.A. Pierce

Oxford University Computing Laboratory
Oxford OX1 3QD, United Kingdom
email: *giles@comlab.ox.ac.uk*

Key Words: ERROR ANALYSIS, FUNCTIONALS, ADJOINT EQUATIONS

Abstract. This paper introduces a new idea, using adjoint error analysis to obtain approximate values for integral quantities, such as lift and drag, which are twice the order of accuracy of the flow solution. The theory is presented for both linear and nonlinear applications and numerical results confirm the effectiveness of the technique for the one-dimensional Poisson equation and the quasi-1D Euler equations.

1 Introduction

In engineering applications of CFD, there are usually a few integral quantities of primary concern, such as lift and drag on an aircraft, total mass flux through a turbomachine, or total heat flux into a turbine blade. The rest of the flow solution is often needed only for qualitative purposes, for example to see if there is a bad flow separation. In this paper we show how the order of accuracy of an important integral quantity can be greatly improved, usually doubled, compared to the accuracy of the flow solution on which the estimate is based. This is accomplished through an error analysis using an approximate solution to the adjoint flow equations. These are the same adjoint equations that are solved to efficiently obtain the linear sensitivity of an objective function in design optimisation [Jameson (95), Jameson (97), Anderson (97), Elliott (97)], but in the present context, the adjoint variables reveal the contributions of flow solution approximation errors to the error in the computed integral. Correcting the leading order error produces a corrected value for the integral which is much more accurate.

This idea is closely related to the *a priori* and *a posteriori* analysis of the *superconvergence* of integral functionals arising from finite element computations in a variety of applications [Babuška (84), Barrett (87), Becker (96), Paraschivoiu (97), Giles (97b), Süli (97), Monk (98)]. However, with these methods the superconvergence arises naturally from Galerkin orthogonality without the addition of a correction term. Previous work by the present authors on doubling the order of accuracy of quasi-1D lift estimates obtained from a first order upwind method [Giles (98)] was based on a discrete truncation error viewpoint [Giles (97c)]. The new approach uses an analytic viewpoint which leads to a much simpler implementation when using more accurate discretisations. We are not aware of other work on the use of adjoint solutions to improve the accuracy of integral quantities through the evaluation of a correction term.

The paper begins by presenting the linear theory and numerical results for the one-dimensional Poisson equation. The nonlinear theory is then presented and applied to the quasi-1D Euler equations. Results are given for subsonic flow and transonic flow, with and without shocks. These demonstrate the effectiveness of the approach, and the paper concludes with a discussion of the challenges to be overcome in extending the technique to multi-dimensional applications.

2 Linear theory

Let u be the solution of the linear differential equation

$$Lu = f,$$

on the domain Ω, subject to homogeneous boundary conditions for which the problem is well-posed. The adjoint differential operator L^* and associated homogeneous boundary conditions are defined by the identity

$$(v, Lu) = (L^*v, u),$$

for all u, v satisfying the respective boundary conditions. Here the notation $(.,.)$ denotes an integral inner product over the domain Ω.

If we are concerned with the value of the functional $J=(g,u)$, where g is a given function defined on Ω, an equivalent dual formulation of the problem is to evaluate the functional $J=(v,f)$, where v satisfies the adjoint equation

$$L^*v = g,$$

subject to the homogeneous adjoint boundary conditions. The equivalence of the two forms of the problem follows immediately from the definition of the adjoint operator.

$$(v, f) = (v, Lu) = (L^*v, u) = (g, u).$$

Suppose that u_h and v_h are approximations to u and v, respectively, and satisfy the homogeneous boundary conditions. The subscript h denotes that the approximate solutions are derived by interpolating the results of a numerical computation using a grid with average spacing h. The functions f_h and g_h are defined by

$$Lu_h = f_h, \quad L^*v_h = g_h.$$

It is assumed that v_h and v_h are sufficiently smooth that f_h and g_h lie in $L_2(\Omega)$. If u_h and v_h were equal to u and v, then f_h and g_h would be equal to

f and g. Thus, the *residual errors* $f_h - f$ and $g_h - g$ are a computable indication of the extent to which u_h and v_h are not the true solutions.

Now, using the definitions and identities given above, we obtain the following expression for the functional:

$$\begin{aligned}(g,u) &= (g,u_h) - (g_h, u_h - u) + (g_h - g, u_h - u) \\ &= (g,u_h) - (L^* v_h, u_h - u) + (g_h - g, u_h - u) \\ &= (g,u_h) - (v_h, L(u_h - u)) + (g_h - g, u_h - u) \\ &= (g,u_h) - (v_h, f_h - f) + (g_h - g, u_h - u).\end{aligned}$$

The first term in the final expression is the value of the functional obtained from the approximate solution u_h. The second term is an inner product of the residual error $f_h - f$ and the approximate adjoint solution v_h. The adjoint solution gives the weighting of the contribution of the local residual error to the overall error in the computed functional. Therefore, by evaluating and subtracting this adjoint error term we obtain a more accurate value for the functional.

The third term is the remaining error after making the adjoint correction. If $g_h - g$ is of the same order of magnitude as $v_h - v$ then the remaining error has a bound which is proportional to the product $\|u_h - u\| \, \|v_h - v\|$ (using L_2 norms), and thus the corrected functional value is superconvergent. If the solution errors $u_h - u$ and $v_h - v$ are both $O(h^p)$ so that halving the grid spacing leads to a 2^p reduction in the errors, then the error in the functional is $O(h^{2p})$.

For simplicity of presentation, we have assumed above that the primal problem has homogeneous boundary conditions, and that the functional is simply an inner product over the whole domain and does not have a boundary integral term. More generally, inhomogeneous boundary conditions and boundary integrals in the functional are both permissible. Inhomogeneous boundary conditions for the primal problem lead to a boundary integral term for the adjoint formulation, and similarly a boundary integral in the primal form of the functional leads to inhomogeneous adjoint boundary conditions. Although the analysis is slightly more complicated, the final form of the adjoint error correction is exactly the same as before, provided the approximate solutions u_h and v_h still exactly satisfy the inhomogeneous boundary conditions. If they do not, then there is an additional correction term to take account of this error.

3 Linear example

The example is the one-dimensional Poisson equation,

$$\frac{d^2u}{dx^2} = f,$$

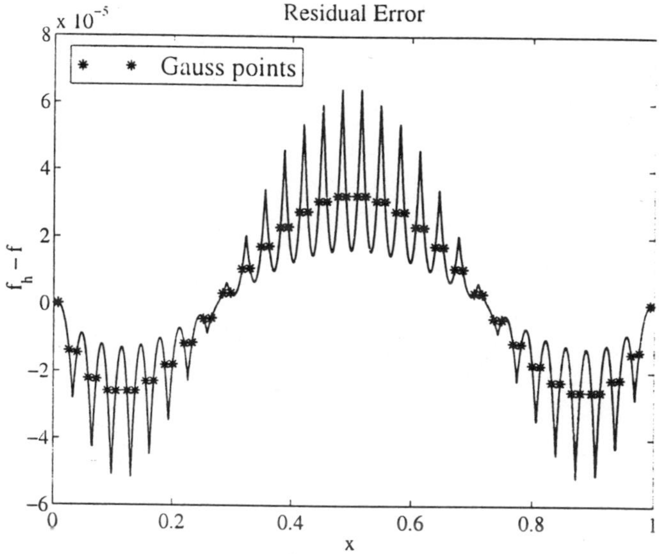

Fig. 1. Residual error for 1D Poisson equation.

on the unit interval $[0, 1]$ subject to homogeneous boundary conditions $u(0) = u(1) = 0$. This is approximated numerically on a uniform grid, with spacing h, using a simple second order finite difference discretisation,

$$h^{-2}\delta_x^2 u_j = f(x_j).$$

The approximate solution $u_h(x)$ is then defined by interpolation with a cubic spline through the nodal values u_j.

The dual problem is also a Poisson equation,

$$\frac{d^2v}{dx^2} = g,$$

subject to the same homogeneous boundary conditions, and the approximate adjoint solution v_h is obtained in exactly the same manner.

Numerical results have been obtained for the case

$$f = x^3(1-x)^3, \qquad g = \sin(\pi x).$$

Figure 1 shows the residual error $f_h - f$ when $h = \frac{1}{32}$, as well as the values at the two Gaussian quadrature points on each sub-interval which are used in the numerical integration of the inner product $(v_h, f_h - f)$. Since u_h is a cubic spline, $f_h \equiv \frac{d^2u_h}{dx^2}$ is continuous and piecewise linear. The best piecewise linear approximation to f has an approximation error whose dominant term is quadratic on each sub-interval; this explains the scalloped shape of the residual error. Figure 2 shows the approximate adjoint solution v_h, illustrating that the residual error in the center of the domain contributes most significantly to the overall error in the functional.

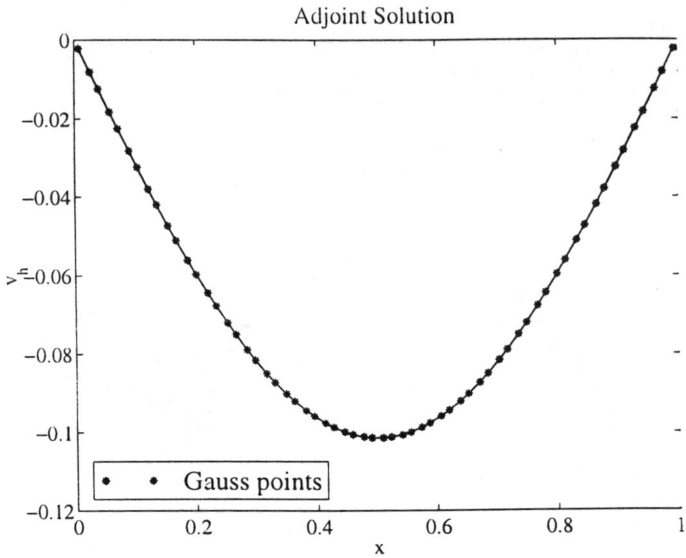

Fig. 2. Adjoint solution for 1D Poisson equation.

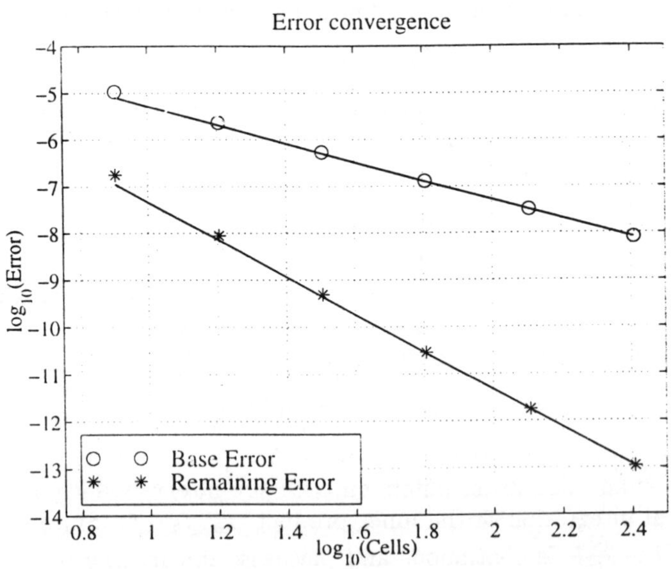

Fig. 3. Error convergence for 1D Poisson equation.

Figure 3 is a log-log plot of two quantities versus the number of cells: the error in the base value of the functional (g, u_h) and the error remaining after subtracting the adjoint error correction term $(v_h, f_h - f)$. The superimposed lines have slopes of -2 and -4, confirming that the base solution is second order accurate while the corrected functional is fourth order accurate. It is also worth noting that on a grid with 16 cells, which might be a reasonable choice for practical computations, the error in the corrected value is over 200 times smaller than in the uncorrected value.

4 Nonlinear theory

Let u be the solution of the nonlinear differential equation

$$N(u) = f,$$

on the domaon Ω subject to certain boundary conditions, and let the functional of interest, $J(u)$, be an integral over the domain of a nonlinear algebraic function of u. The linear differential operator L_u is defined to be the Fréchet derivative [Collatz (66)] of N,

$$L_u\, \tilde{u} \equiv \lim_{\epsilon \to 0} \frac{N(u + \epsilon \tilde{u}) - N(u)}{\epsilon},$$

and, similarly, the function $g(u)$ is defined by

$$(g(u), \tilde{u}) \equiv \lim_{\epsilon \to 0} \frac{J(u + \epsilon \tilde{u}) - J(u)}{\epsilon}.$$

The linear adjoint problem is

$$L_u^* v = g,$$

subject to the appropriate homogeneous adjoint boundary conditions [Giles (97)]. Now consider approximate solutions u_h, v_h which have again been obtained by interpolating the results of a finite volume calculation. The quantities f_h, g_h are defined by

$$N(u_h) = f_h, \qquad L_{u_h}^* v_h = g_h.$$

Note the use of $L_{u_h}^*$, the Fréchet derivative based on u_h which is known, instead of L_u^* based on u which is not known. In addition, the analysis requires averaged Fréchet derivatives $\overline{L}_{(u,u_h)}$ and $\overline{g}(u, u_h)$ defined by

$$\overline{L}_{(u,u_h)} = \int_0^1 L|_{u+\theta(u_h - u)}\, d\theta,$$

$$\overline{g}(u, u_h) = \int_0^1 g(u + \theta(u_h - u)) d\theta,$$

so that

$$N(u_h)-N(u) = \int_0^1 \frac{\partial}{\partial\theta} N(u+\theta(u_h-u))\, d\theta$$
$$= \overline{L}_{(u,u_h)}\,(u_h-u),$$

and similarly

$$J(u_h)-J(u) = (\overline{g}(u,u_h), u_h-u).$$

Using the above definitions, we obtain the following result:

$$\begin{aligned}
J(u) &= J(u_h) - (\overline{g}(u,u_h), u_h-u) \\
&= J(u_h) - (g_h, u_h-u) + (g_h-\overline{g}(u,u_h), u_h-u) \\
&= J(u_h) - (L^*_{u_h} v_h, u_h-u) + (g_h-\overline{g}(u,u_h), u_h-u) \\
&= J(u_h) - (v_h, L_{u_h}(u_h-u)) + (g_h-\overline{g}(u,u_h), u_h-u) \\
&= J(u_h) - (v_h, \overline{L}_{(u,u_h)}(u_h-u)) + (g_h-\overline{g}(u,u_h), u_h-u) \\
&\qquad - (v_h, (L_{u_h}-\overline{L}_{(u,u_h)})(u_h-u)) \\
&= J(u_h) - (v_h, N(u_h)-N(u)) + (g_h-\overline{g}(u,u_h), u_h-u) \\
&\qquad - (v_h, (L_{u_h}-\overline{L}_{(u,u_h)})(u_h-u)) \\
&= J(u_h) - (v_h, f_h-f) + (g_h-\overline{g}(u,u_h), u_h-u) \\
&\qquad - (v_h, (L_{u_h}-\overline{L}_{(u,u_h)})(u_h-u))
\end{aligned}$$

The first term in the final result is the functional evaluated using the approximate solution u_h. The second term is the adjoint error correction term which is again an inner product of the residual error and the approximate adjoint solution. Since both of these are known, this second term can be computed and subtracted from the first to form a corrected value for the functional.

The last two terms, which cannot be computed since the analytic solution u is not known, form the remaining error in the corrected functional. If the solution error for the nonlinear primal problem and the linear adjoint problem are of the same order, and they are both sufficiently smooth that the corresponding residual errors are also of the same order, then the order of accuracy of the functional approximation after making the adjoint correction is twice the order of accuracy of the the primal and adjoint solutions on which it is based.

5 Quasi-1D Euler equations

The steady quasi-1D Euler equations in conservative form are

$$\frac{d}{dx}(AF) - \frac{dA}{dx}P = 0,$$

where $A(x)$ is the cross-sectional area of the duct and U, F and P are defined as

$$U = \begin{pmatrix} \rho \\ \rho q \\ \rho E \end{pmatrix}, \qquad F = \begin{pmatrix} \rho q \\ \rho q^2 + p \\ \rho q H \end{pmatrix}, \qquad P = \begin{pmatrix} 0 \\ p \\ 0 \end{pmatrix}.$$

Here ρ is the density, q is the velocity, p is the pressure, E is the total internal energy and H is the stagnation enthalpy. The system is closed by the equation of state for an ideal gas,

$$H = E + \frac{p}{\rho} = \frac{\gamma p}{(\gamma - 1)\rho} + \tfrac{1}{2}q^2,$$

where γ is the ratio of specific heats.

Numerical results have been obtained using a standard second order finite volume method with characteristic smoothing on a uniform computational grid. Except when there is a shock, the approximate solution $u_h(x)$ is constructed from the discrete nodal values u_j by cubic spline interpolation of the three components of U. All other variables are then calculated from these. Evaluation of the residual error $f_h - f$ requires first derivatives of flow quantities; these are obtained by differentiating the cubic spline representation.

The linear adjoint problem is approximated by the 'continuous' method, which involves linearising the nonlinear flow equations, constructing the analytic adjoint equations, and then forming a discrete approximation to these on the same uniform grid as the flow solution [Jameson (95), Jameson (97), Anderson (97)]. An alternative approach which could have been used is the 'discrete' method in which one takes the discretised nonlinear flow equations, linearises them and then uses the transpose of the linear matrix as the discrete adjoint operator [Elliott (97)]. Previous research has shown that both approaches produce consistent approximations to the analytic adjoint solution which has been determined in closed form for the quasi-1D Euler equations [Giles (98)].

Results have been obtained for three test cases: a subsonic flow, a shock-free transonic flow with subsonic inflow and supersonic outflow, and a shocked flow with supersonic inflow and subsonic outflow. The Mach number distributions for these three cases are shown in Figure 4. In each case the functional of interest is the integral of pressure along the duct; this serves as a prototype for the lift in airfoil and aircraft calculations.

5.1 Subsonic flow

Figure 5 shows the error convergence for a subsonic flow in a converging-diverging duct. The base error, which is the error before applying the adjoint correction, is second order, as indicated by the superimposed line of slope -2. This is as expected given the second order truncation error in approximating the nonlinear flow equations. The other superimposed line of slope -4 shows that the error remaining after the adjoint correction is fourth order.

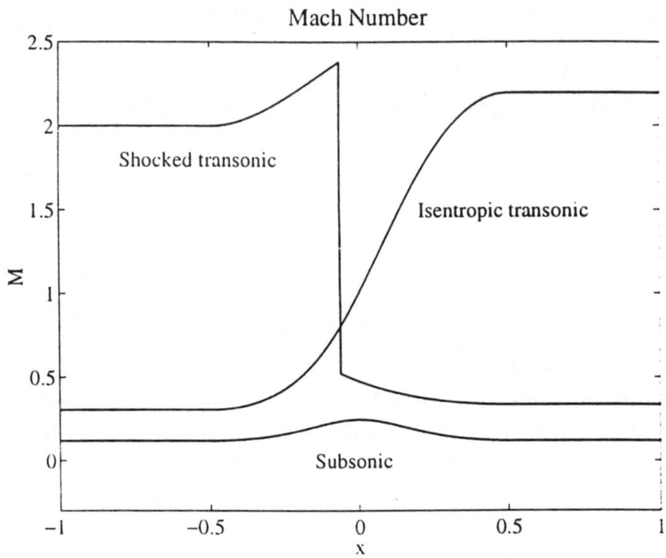

Fig. 4. Mach number distributions for quasi-1D test cases.

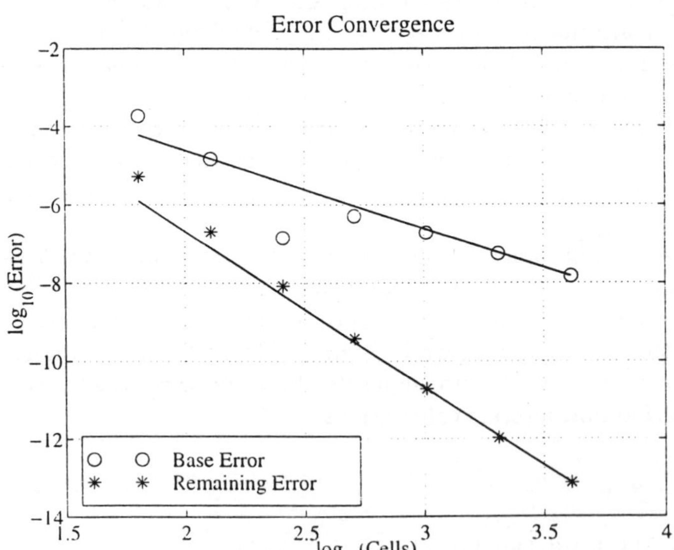

Fig. 5. Error convergence for quasi-1D subsonic flow.

5.2 Isentropic transonic flow

Figure 6 shows the error convergence for a transonic flow in a converging-diverging duct with the throat located at $x = 0$. The flow is subsonic upstream of the throat and supersonic downstream of the throat. Again the results show that the base error is second order while the remaining error after the adjoint correction is fourth order.

The accuracy of the corrected functional in this case is a little puzzling because the adjoint solution has a logarithmic singularity at the throat [Giles (98)], as shown in Figure 7. Therefore, $v_h - v$ is $O(1)$ in a small region of size $O(h)$ on either side of the throat. Based on this, one would expect that the remaining error might be $O(h^3)$ since the numerical results confirm that the residual error for the nonlinear equations is $O(h^2)$. The explanation for the fourth order convergence must lie in a leading order cancellation within the two remaining error integrals, but we do not yet have a complete understanding of this phenomenon.

5.3 Shocked transonic flow

The final example is for flow in a diverging duct, where a shock separates supersonic upstream and subsonic downstream regions. Previous research has proved that the analytic adjoint solution is continuous and has zero gradient at the shock, so the adjoint variables pose no special difficulty in this case [Giles (98)]. The challenge is the reconstruction of the approximate solution $u_h(x)$ from the nodal quantities u_j coming from the finite volume calculation.

The analytic solution is discontinuous at the shock, and satisfies the Rankine-Hugoniot shock jump relations which require that there is no discontinuity in the nonlinear flux F. The discrete solution has a slightly smeared shock, and so if one interpolates the conservative variables U it is clear that locally in a neighborhood of size $O(h)$ the error in the reconstructed solution $u_h(x)$ will be $O(1)$.

To recover a discontinuous approximate solution $u_h(x)$ we instead use the fact that F is known to be continuous at the shock and therefore choose to interpolate the nodal values of F. From these one can deduce the conservation variables U by solving a quadratic equation, one branch of which gives a subsonic flow solution, the other being supersonic. Therefore, given a shock position, one can reconstruct a supersonic solution on the upstream side, a subsonic solution on the downstream side, and automatically satisfy the Rankine-Hugoniot shock jump conditions at the shock itself. To determine the shock position, we rely on prior research [Giles (96)] which shows that the integrated pressure along the duct is correct to second order when using a finite volume method which is conservative and second order accurate in smooth flow regions. Therefore, we iteratively adjust the position of the shock until the reconstructed solution has the same base functional value

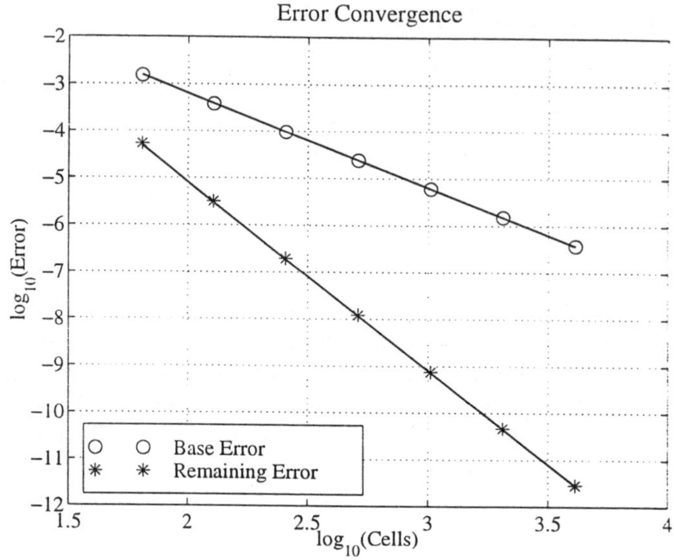

Fig. 6. Error convergence for quasi-1D shock-free transonic flow.

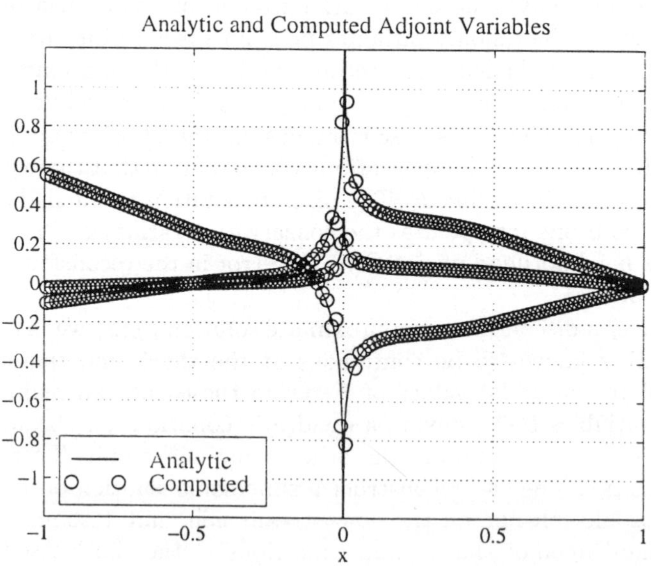

Fig. 7. Adjoint solution for quasi-1D shock-free transonic flow.

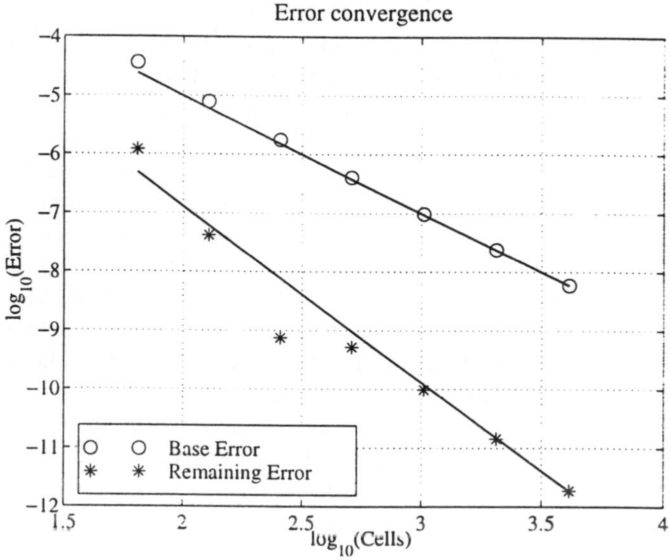

Fig. 8. Error convergence for quasi-1D shocked flow.

(i.e. without the adjoint correction) as the original numerical approximation, thereby obtaining the correct shock position to second order.

Figure 8 shows the error convergence. As expected, the base error is again second order. Because there is still an $O(h)$ error in the approximate solution $u_h(x)$ in the neighbourhood of the shock, the corrected error is now third order, not fourth. However, in future work we hope to recover overall fourth order accuracy, based on the average cell size, by using local grid adaptation at the shock.

6 Concluding remarks

In this paper we have outlined a means of calculating improved estimates of integral quantities such as lift and drag from CFD calculations, by evaluating an adjoint correction term which is an inner product of the residual error in approximating the flow equations and an approximate solution to the corresponding adjoint equations. The numerical results demonstrate the effectiveness of the technique applied to a second order finite volume approximation of the quasi-1D Euler equations. When the flow is smooth, the error in the integrated pressure is fourth order; when there is a shock, it is third order.

The theory is equally applicable to the Euler and Navier-Stokes equations in multiple dimensions. However, there are three important issues to be addressed before similar results can be obtained for airfoil and aircraft applications of engineering interest. The first is the treatment of curved surfaces; to achieve fourth order accuracy for corrected functional such as lfit and drag, it is likely that smooth curved surfaces will need to be approximated in a way which ensures continuity in the surface normal. as opposed to the use of simple linear (or bi-linear) facets. The second issue is the resolution of singularities; the adjoint flow solution in two dimensional airfoil applications has an inverse square root singularity along the incoming stagnation streamline [Giles (97)] and this will need to be well resolved. The final issue concerns unstructured grid calculations which are needed for complex applications. The approximate solution u_h needs to be sufficiently smooth that the error in ∇u_h is of the same order as the error in u_h itself. To achieve this on unstructured grids where the solution error has a significant high-frequency content may require the use of multi-dimensional smoothed cubic splines.

Another interesting direction for future research is *a posteriori* estimation of the error remaining after making the adjoint correction. The goal of such research would be to develop a mathematical framework on which one could base efficent grid refinement indicators, and thereby obtain the value of a functional to the desired level of accuracy and at a minimum computational cost.

Acknowledgments

This research was supported by EPSRC under grant GR/K91149.

References

W.K. Anderson and V. Venkatakrishnan. Aerodynamic design optimization on unstructured grids with a continuous adjoint formulation. AIAA Paper 97-0643, 1997.

J.W. Barrett and C.M. Elliott. Total flux estimates for a finite-element approximation of elliptic equations. *IMA J. Numer. Anal.*, 7:129–148, 1987.

I. Babuška and A. Miller. The post-processing approach in the finite element method – Part 1: calculation of displacements, stresses and other higher derivatives of the displacements, *Intern. J. Numer. Methods Engrg.*, 20:1085–1109, 1984.

R. Becker and R. Rannacher. Weighted *a posteriori* error control in finite element methods. Technical report No. 96-1, Universitat Heidelberg, 1996.

L. Collatz. *Functional analysis and numerical mathematics.* Academic Press, 1966.

J. Elliott and J. Peraire. Practical 3D aerodynamic design and optimization using unstructured meshes. *AIAA J.*, 35(9):1479–1485, 1997.

M.B. Giles. Analysis of the accuracy of shock-capturing in the steady quasi-1D Euler equations. *Comput. Fluid Dynamics J.*, 5(2):247–258, 1996.

M.B. Giles and N.A. Pierce. Adjoint equations in CFD: duality, boundary conditions and solution behaviour. AIAA Paper 97-1850, 1997.

M.B. GILES, M.G. LARSON, J.M. LEVENSTAM, AND E. SÜLI. Adaptive error control for finite element approximations of the lift and drag in viscous flow. Technical Report NA97/06, Oxford University Computing Laboratory, 1997.

M.B. GILES On adjoint equations for error analysis and optimal grid adaptation in CFD, In *Computing the Future II: Advances and Prospects in Computational Aerodynamics*, M. Hafez, editor, World Scientific, 1998.

M.B. GILES AND N.A. PIERCE. On the properties of solutions of the adjoint Euler equations. In *Numerical Methods for Fluid Dynamics VI*, M.J. Baines, editor, ICFD, 1998.

A. JAMESON. Optimum aerodynamic design using control theory. *Comput. Fluid Dynam. Rev.*, pages 495–528, 1995.

A. JAMESON, N. PIERCE, AND L. MARTINELLI, Optimum aerodynamic design using the Navier–Stokes equations. AIAA Paper 97-0101, 1997.

P. MONK AND E. SÜLI. The adaptive computation of far field patterns by *a posteriori* error estimation of linear functionals. Technical Report NA98/02, Oxford University Computing Laboratory, 1998.

M. PARASCHIVOIU, J. PERAIRE, AND A. PATERA. A posteriori finite element bounds for linear-functional outputs of elliptic partial differential equations. *Comput. Methods Appl. Mech. Engrg.*, 150(1-4):289–312, 1997.

E. SÜLI. *A posteriori* error analysis and adaptivity for finite element approximations of hyperbolic problems. Technical Report NA97/21, Oxford University Computing Laboratory, 1997.

Somewhere between the Lax-Wendroff and Roe schemes for calculating multidimensional compressible flows

Alain LERAT, Christophe CORRE and Ying HUANG

SINUMEF Laboratory, ENSAM
151, boulevard de l'Hopital
75013 Paris, FRANCE
email: lerat/corre/huang@paris.ensam.fr

Key Words: ACCURATE SCHEME, COMPRESSIBLE FLOWS

Abstract. A scheme located between the Lax-Wendroff and Roe schemes is presented for the multidimensional compressible Euler and Navier-Stokes equations. The scheme is second-order accurate at steady-state with a low internal dissipation. It is easy to code and involves no tuning parameter. Accuracy and efficiency are demonstrated for several flow problems including a turbulent transonic flow over a wing.

1 Introduction

The paper is devoted to the construction of an efficient multidimensional scheme with low internal dissipation for solving steady aerodynamic problems. We look for a truly second-order accurate scheme at steady-state which involves no tuning parameter and therefore no artificial viscosity or other correction. We also avoid limiters and switch-like ingredients that could prevent the scheme from converging to the steady-state by inducing limiting cycles.

To construct the scheme, we start from the Lax-Wendroff type scheme in [1]-[2] (see also the review paper [3]) and modify the dissipative operator to come closer to a Roe type scheme [4]. This is simply done by introducing a characteristic time-step in the Lax-Wendroff dissipative operator as in [5]. The challenge is to maintain the second-order accuracy at steady-state and to ensure a correct dissipation for a multidimensional problem without any tuning.

The paper is organized as follows. For a hyperbolic conservation law in two space-dimension, Section 2 presents a first attempt to modify the Lax-Wendroff scheme and shows that a direct application of the characteristic time-step idea fails. The idea is refined in Section 3 and a new scheme between Lax-Wendroff and Roe methods is obtained. The detailed spatial discretization and the properties of the fully discrete scheme are described in Section 4. The scheme is extended to hyperbolic systems of conservation laws in Section 5. to any number of space-dimensions in Section 6 and to the multidimensional Navier-Stokes equations in Section 7. Then implicit treatments are presented in Section 8 to allow the use of large CFL numbers and thus to increase the convergence rate to the steady-state. Finally, applications to the calculation of inviscid and viscous compressible flows are presented in Section 9.

2 A first attempt to modify the Lax-Wendroff scheme

Consider an initial-value problem for the two-dimensional scalar equation :

$$w_t + f_x + g_y = 0 \tag{1}$$

where the flux components $f = f(w)$ and $g = g(w)$ are smooth functions of the conservative variable $w = w(x, y, t)$. The Lax-Wendroff time-discretization for Eq. (1) can be written as :

$$\frac{w^{n+1} - w^n}{\Delta t} + f_x^n + g_y^n = P^n \tag{2}$$

with

$$P = \frac{1}{2}[\Delta t A(f_x + g_y)]_x + \frac{1}{2}[\Delta t B(f_x + g_y)]_y \tag{3}$$

where $w^n = w(x, y, n\Delta t)$, $A = \dfrac{df}{dw}$ and $B = \dfrac{dg}{dw}$. Space discretization is purely centered.

In addition to ensuring the second-order accuracy in time, the operator P corrects the destabilizing effect of the Euler forward approximation to w_t. To check that the operator P does produce a dissipative effect, we rewrite it as :

$$P = \frac{1}{2}[\Delta t(A^2 w_x + AB w_y)]_x + \frac{1}{2}[\Delta t(BA w_x + B^2 w_y)]_y$$

and consider the associated quadratic form :

$$Q = \frac{\Delta t}{2}(A^2\xi^2 + 2AB\xi\eta + B^2\eta^2) = \frac{1}{2}X^t M X$$

with

$$X = \begin{bmatrix} \xi \\ \eta \end{bmatrix}, \qquad M = \Delta t \begin{bmatrix} A^2 & AB \\ AB & B^2 \end{bmatrix}.$$

Dissipativity in a broad sense of the operator P means that the quadratic form Q is non-negative definite, *i.e.* the eigenvalues of the symmetric matrix M are positive or null. Here, these eigenvalues are precisely $\Delta t(A^2 + B^2)$ and 0.

With a suitable centered spatial discretization on a regular Cartesian mesh, the scheme (2)-(3) is linearly L_2-stable and 4th-order dissipative in the sense of Kreiss if $CFL < 1/d$, for any number d of space-dimension (see [3]). Here,

$d = 2$ and $CFL = \Delta t[(\frac{A}{\delta x})^2 + (\frac{B}{\delta y})^2]^{1/2}$, where δx and δy are the space increments.

This paper is concerned with the approximation of steady solutions of Eq. (1) using a scheme in form (2) as a time-dependent method to reach a steady-state. The steady solutions of scheme (2) satisfy :

$$f_x + g_y = P \tag{4}$$

A shortcoming of the Lax-Wendroff scheme is that the solutions of (4) may depend on the time-step Δt. A way to suppress this dependence is to replace the first and second time-steps appearing in the operator P respectively by the characteristic time-steps Δt_1^c and Δt_2^c defined as :

$$\begin{aligned} \Delta t_1^c|A| &= \delta x \\ \Delta t_2^c|B| &= \delta y. \end{aligned} \tag{5}$$

The modified scheme stays in form (2) with the new right-hand side :

$$P' = \frac{\delta x}{2}[sgn(A)(f_x + g_y)]_x + \frac{\delta y}{2}[sgn(B)(f_x + g_y)]_y.$$

Note that this modification comes to introduce some kind of upwinding.
A simplified version of P' without crossed derivatives reads :

$$P_R = \frac{\delta x}{2}[sgn(A)f_x]_x + \frac{\delta y}{2}[sgn(B)g_y]_y = \frac{\delta x}{2}[|A|w_x]_x + \frac{\delta y}{2}[|B|w_y]_y$$

It reduces to a simple 2-D extension of the Roe scheme, direction by direction. The second-order accuracy of the Lax-Wendroff scheme is lost by replacing P by P' or P_R. However using P' keeps the *second-order accuracy at steady state*, while the scheme with P_R is always first-order accurate. To show this property we note that, when $w^{n+1} = w^n = w$, the fully-discrete scheme with P' approximates :

$$f_x + g_y = P' + O(h^2), \tag{6}$$

where $O(h) = O(\delta x) = O(\delta y)$. From the expression of P' this yields :

$$f_x + g_y = O(h), \tag{7}$$

at least. Inserting (7) into the operator P' leads to :

$$P' = O(h^2)$$

and finally, from Eq. (6) :

$$f_x + g_y = O(h^2). \tag{8}$$

Consequently :

$$P' = O(h^3).$$

On the contrary, the simplified version cannot give better than :

$$P_R = O(h),$$

so that the fully discrete scheme with P_R stays first-order.
Unfortunately, the fully-discrete scheme with P' is linearly unstable. The reason lies in the *non-dissipativity* of the operator P'. This operator is associated with the quadratic form :

$$Q' = \frac{1}{2} X^t M' X$$

with

$$M' = \begin{bmatrix} \delta x|A| & C \\ C & \delta y|B| \end{bmatrix},$$

where

$$C = \frac{1}{2} sgn(A\, B)(\delta x|B| + \delta y|A|).$$

Eigenvalues of M' are non-negative if and only if :

$$(\delta x|B| - \delta y|A|)^2 \leq 0.$$

Therefore the operator P' is not dissipative, except if

$$\frac{|A|}{\delta x} = \frac{|B|}{\delta y}, \tag{9}$$

that is for the calculation of an advection along one of the mesh diagonals, which is a very special case.
The first-order Roe type scheme associated with P_R is linearly stable under a CFL condition. The corresponding quadratic form is :

$$Q_R = \frac{1}{2} X^t M_R X$$

with

$$M_R = \begin{bmatrix} \delta x|A| & 0 \\ 0 & \delta y|B| \end{bmatrix}$$

Clearly, the operator P_R is always dissipative.

3 The new scheme between Lax-Wendroff and Roe

We now look for an operator P meeting all the following requirements :

- R_1 : The operator P can be discretized on a 3×3-point stencil
- R_2 : The operator P is independent of Δt
- R_3 : The operator P leads to second-order accuracy at steady-state
- R_4 : The operator P is dissipative

First, note that the definition (5) and the condition (9) yields $\Delta t_1^c = \Delta t_2^c$, *i.e.* the equality of the two characteristic time-steps. Generally speaking, consider the Lax-Wendroff operator P with two arbitrary time-steps Δt_1 and Δt_2, that is :

$$P'' = \frac{1}{2}[\Delta t_1 A(f_x + g_y)]_x + \frac{1}{2}[\Delta t_2 B(f_x + g_y)]_y.$$

Dissipativity of P'' is tantamount to

$$[(\Delta t_1 - \Delta t_2)AB]^2 \leq 0$$

and thus to $\Delta t_1 = \Delta t_2$. In other words, stable modifications can only be obtained by using a *single* time-step in the 2-D Lax-Wendroff operator.
To ensure dissipativity, we now use a unique time-step Δt^c and relax the constraints (5) as :

$$\begin{aligned} \Delta t^c|A| &= \delta x\, \Phi \\ \Delta t^c|B| &= \delta y\, \Psi \end{aligned} \tag{10}$$

The coefficients Φ and Ψ appears as CFL numbers in x and y-directions respectively. Consider the Lax-Wendroff operator P given by (3) and replace Δt in P by Δt^c defined by (10). The new scheme is in form (2) with the right-hand side :

$$P^c = \frac{\delta x}{2}[\Phi sgn(A)(f_x + g_y)]_x + \frac{\delta y}{2}[\Psi sgn(B)(f_x + g_y)]_y. \tag{11}$$

Clearly, the operator P^c meets the requirements R_1, R_2 and R_3 for any choice of the coefficients Φ and Ψ.
Let us now find suitable coefficients Φ and Ψ. First, to define a unique Δt^c from the two relations (10), the following compatibility condition should hold:

$$\frac{|A|}{\delta x}\Psi = \frac{|B|}{\delta y}\Phi. \tag{12}$$

From the relations (10), we also deduce :

$$\Phi \geq 0 \text{ and } \Psi \geq 0 \tag{13}$$

Conditions (12) and (13) ensure the dissipativity of the operator P^c (requirement R_4). To end the determination of Φ and Ψ, we choose to keep the operator P^c as close as possible to the operator P' associated to the basic characteristic time steps (5). Since P' is not dissipative, this is a way to minimize the numerical dissipation due to P^c. Note that P' is defined by $\Phi = \Psi = 1$. More precisely, we minimize the quantity

$$|\Phi - 1| + |\Psi - 1| \tag{14}$$

under the constraints (12) and (13). This optimization problem is solved by considering Φ and Ψ as functions of the advection direction α :

$$\Phi = \Phi(\alpha), \qquad \Psi = \Psi(\alpha)$$

where

$$\alpha = \frac{\delta x|B|}{\delta y|A|}. \tag{15}$$

In terms of the parameter α, the constraints (12)-(13) read :

$$\Psi(\alpha) = \alpha\Phi(\alpha)$$
$$\Phi(\alpha) \geq 0$$

Quantity (14) is optimized for each value of α, *i.e.*, for a given $\alpha > 0$, we look for $z = \Phi(\alpha) \geq 0$ minimizing

$$I(z) = |z-1| + |\alpha z - 1|.$$

The optimal function Φ is found to be

$$\Phi(\alpha) = \begin{cases} 1 \text{ if } 0 < \alpha \leq 1 \\ \frac{1}{\alpha} \text{ if } \alpha > 1. \end{cases}$$

Note that $\alpha < 1$ expresses that $|A|/\delta x$ is greater than $|B|/\delta y$, *i.e.* that the advection direction is between the x-direction and a mesh diagonal; in other words, the x-direction is dominant. Similarly, $\alpha > 1$ means that the y-direction is dominant.
Now observe that the optimal solution can also be written as

$$\Phi = \min(1, \frac{1}{\alpha}) = \min(1, \frac{\delta y|A|}{\delta x|B|}),$$

$$\Psi = \min(1, \alpha) = \min(1, \frac{\delta x|B|}{\delta y|A|}),$$

so that, from (10), it corresponds to :

$$\Delta t^c = \min(\frac{\delta x}{|A|}, \frac{\delta y}{|B|}) = \min(\Delta t_1^c, \Delta t_2^c).$$

The 2-D characteristic time-step is therefore the *smallest of the two 1-D characteristic time-steps.*
Note also that for $\alpha < 1$, the operator P^c reads :

$$P^c = \frac{\delta x}{2}[sgn(A)(f_x + g_y)]_x + \frac{\delta y}{2}[\alpha sgn(B)(f_x + g_y)]_y.$$

Compared to the non-dissipative operator P', one observes a reduction by the factor α of the second term.

4 Fully discrete formulation

Let us now present the spatial discretization of the proposed scheme (2),(11), that is of :

$$\frac{1}{\Delta t}(w^{n+1} - w^n) + [f - \frac{\delta x}{2}\Phi'(f_x + g_y)]_x^n + [g - \frac{\delta y}{2}\Psi'(f_x + g_y)]_y^n = 0, \quad (16)$$

where

$$\begin{aligned} &\Phi' = sgn(A)\,\Phi, \ \Psi' = sgn(B)\,\Psi \\ &\Phi = \min(1, \frac{1}{\alpha}), \ \Psi = \min(1, \alpha), \ \alpha = \frac{\delta x\,|B|}{\delta y\,|A|}. \end{aligned} \quad (17)$$

To ensure linear stability and dissipation (in the sense of Kreiss) for the fully discrete scheme in several space dimensions, it is important to use a special discretization of (16) involving a predictor step for each space-direction as shown in [3] for the pure Lax-Wendroff method.
Consider a regular Cartesian mesh $x_j = j\delta x$, $y_k = k\delta y$. For any mesh function $v_{j,k}$, we define the basic difference and average operators :

$$\begin{aligned} &(\delta_1 v)_{j+\frac{1}{2},k} = v_{j+1,k} - v_{j,k} && (\delta_2 v)_{j,k+\frac{1}{2}} = v_{j,k+1} - v_{j,k} \\ &(\mu_1 v)_{j+\frac{1}{2},k} = \tfrac{1}{2}(v_{j+1,k} + v_{j,k}) && (\mu_2 v)_{j,k+\frac{1}{2}} = \tfrac{1}{2}(v_{j,k+1} + v_{j,k}). \end{aligned}$$

The fully discrete scheme is obtained by calculating a centered approximation of $\delta x\,(f_x + g_y)^n$ at the cell interface $j + \frac{1}{2}, k$ (predictor π_1), a centered approximation of $\delta y\,(f_x + g_y)^n$ at the cell interface $j, k + \frac{1}{2}$ (predictor π_2), then the numerical fluxes at the cell interfaces and finally w^{n+1} at the cell centers j, k :

a) Predictors :

$$\begin{aligned} (\pi_1)^n_{j+\frac{1}{2},k} &= (\delta_1 f + \tfrac{\delta x}{\delta y}\mu_1\delta_2\mu_2 g)^n_{j+\frac{1}{2},k} \\ (\pi_2)^n_{j,k+\frac{1}{2}} &= (\tfrac{\delta y}{\delta x}\mu_2\delta_1\mu_1 f + \delta_2 g)^n_{j,k+\frac{1}{2}} \end{aligned} \quad (18)$$

b) Numerical fluxes :

$$\begin{aligned} (h_1)^n_{j+\frac{1}{2},k} &= (\mu_1 f - \tfrac{1}{2}\Phi'\pi_1)^n_{j+\frac{1}{2},k} \\ (h_2)^n_{j,k+\frac{1}{2}} &= (\mu_2 g - \tfrac{1}{2}\Psi'\pi_2)^n_{j,k+\frac{1}{2}} \end{aligned} \quad (19)$$

where Φ' and Ψ' at cell interfaces are defined by (17) using averages of A and B at $j + \frac{1}{2}, k$ and $j, k + \frac{1}{2}$.

c) New cell-values :

$$w^{n+1}_{j,k} = w^n_{j,k} - \Delta t(\frac{\delta_1 h_1}{\delta x} + \frac{\delta_2 h_2}{\delta y})^n_{j,k}. \quad (20)$$

The only difference between the Lax-Wendroff type scheme and the above scheme lies in the replacement of $\Delta t\,A$ by $\delta x\,\Phi'$ and $\Delta t\,B$ by $\delta y\,\Psi'$ in the numerical fluxes (19). This introduces some kind of upwinding that we are going to illustrate. Suppose for instance that, at the cell interfaces $j+\frac{1}{2}, k$ and $j, k+\frac{1}{2}$, we have $A>0$, $B>0$ and $\alpha<1$. Then :

$$\Phi'_{j+\frac{1}{2},k} = 1 \ , \ \Psi'_{j,k+\frac{1}{2}} = \alpha_{j,k+\frac{1}{2}}$$

and the numerical fluxes reduce to :

$$\begin{aligned}
(h_1)_{j+\frac{1}{2},k} &= f_{j,k} \\
&\quad -\frac{1}{8}\frac{\delta x}{\delta y}(g_{j,k+1} - g_{j,k-1} + g_{j+1,k+1} - g_{j+1,k-1}) \\
(h_2)_{j,k+\frac{1}{2}} &= \alpha_{j,k+\frac{1}{2}} g_{j,k} + (1-\alpha_{j,k+\frac{1}{2}})\frac{g_{j,k}+g_{j,k+1}}{2} \\
&\quad -\frac{1}{8}\frac{\delta y}{\delta x}\alpha_{j,k+\frac{1}{2}}(f_{j+1,k} - f_{j-1,k} + f_{j+1,k+1} - f_{j-1,k+1}).
\end{aligned}$$

In this situation (x-direction is dominant) :

- *the numerical flux in x-direction is upwinded and a correction involving a y-derivative is added*
- *the numerical flux in y-direction is a linear combination of an upwind and a centered flux, plus a x-derivative correction.*

On a regular Cartesian mesh, the scheme molecule is shown on Fig. 1. It depends on the advection direction and involves 8 points, except when the advection is along a mesh diagonal (7 points) or along the x and y-directions (2 points).

In any situation, the scheme (18)-(20) is a first-order accurate approximation to Eq. (1) but it is second-order accurate at steady-state, *i.e.*,

$$\left(\frac{\delta_1 h_1}{\delta x} + \frac{\delta_2 h_2}{\delta y}\right)_{j,k} = 0 \tag{21}$$

with (18) and (19), is a second-order approximation to

$$f_x + g_y = 0. \tag{22}$$

The L_2-stability of the scheme (18)-(20) can be studied for Eq. (1) and the linear fluxes $f = Aw$, $A = Const.$ and $g = Bw$, $B = Const.$. In this case and for $\alpha < 1$ (x-direction is dominant), the amplification factor of the scheme reads :

$$G = 1 - R - iS \tag{23}$$

where $i^2 = -1$ and

$$\begin{aligned}
R &= [1 - cos(\xi) + \tilde{\alpha}^2(1 - cos(\eta)) + \tilde{\alpha} sin(\xi) sin(\eta)]|\dot{A}| \\
S &= [sin(\xi) + \tilde{\alpha} sin(\eta)]\dot{A} \\
\tilde{\alpha} &= sgn(AB)\alpha = \frac{\delta x B}{\delta y A}, \quad \dot{A} = \frac{\Delta t}{\delta x} A,
\end{aligned}$$

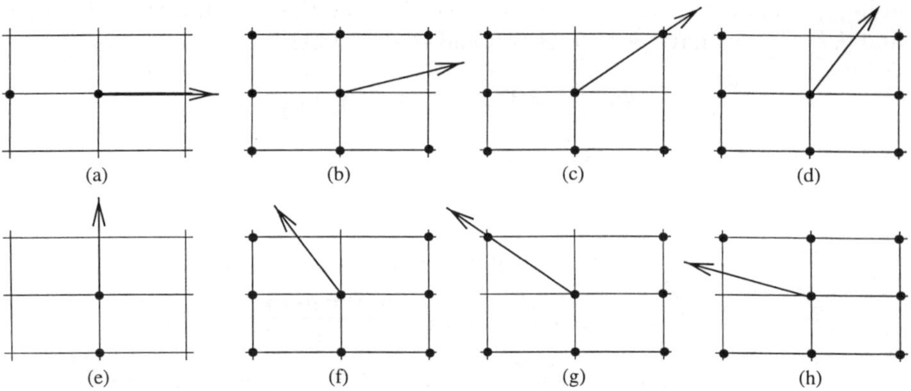

Fig. 1. Scheme molecule for a scalar equation with advection velocity (A, B) with $B \geq 0$ and (a) $A > 0$ and $B = 0$, (b) $A > 0$ and $\alpha < 1$, (c) $A > 0$ and $\alpha = 1$, (d) $A > 0$ and $\alpha > 1$, (e) $A = 0$ and $B > 0$, (f) $A < 0$ and $\alpha > 1$, (g) $A < 0$ and $\alpha = 1$, (h) $A < 0$ and $\alpha < 1$.

ξ and η being the reduced wave numbers in x and y directions.
A similar expression is obtained for $\alpha > 1$ (y-direction is dominant).
The stability domain of the scheme, that is the domain in which $|G| \leq 1$, in terms of $\dot{A}$ and $\dot{B} = \frac{\Delta t}{\delta y} B = \tilde{\alpha}\dot{A}$, is shown on Fig. 2. Practically, a sufficient stability condition is:

$$CFL = \Delta t[(\frac{A}{\delta x})^2 + (\frac{B}{\delta y})^2]^{1/2} \leq 0.64 \tag{24}$$

5 Extension to hyperbolic systems

Consider now Eq. (1) as a hyperbolic system of conservation laws. Quantities w, f, g are now vector-valued and A and B are matrix-valued. Extension of the present scheme to systems is made simply as follows.
The semi-discrete scheme is still in the form (16), but the coefficient Φ' (resp. Ψ') is now a matrix having the same eigenvectors as A (resp. B) and eigenvalues deduced from (17). More precisely, let T_A (resp. T_B) be the matrix whose column vectors are the right eigenvectors of A (resp. B) and $a^{(i)}$ (resp. $b^{(i)}$) the eigenvalues of A (resp. B). Owing to the hyperbolicity of (1), these eigenvalues are real and T_A is regular, so that :

$$A = T_A Diag[a^{(i)}]T_A^{-1} \quad , \quad B = T_B Diag[b^{(i)}]T_B^{-1},$$

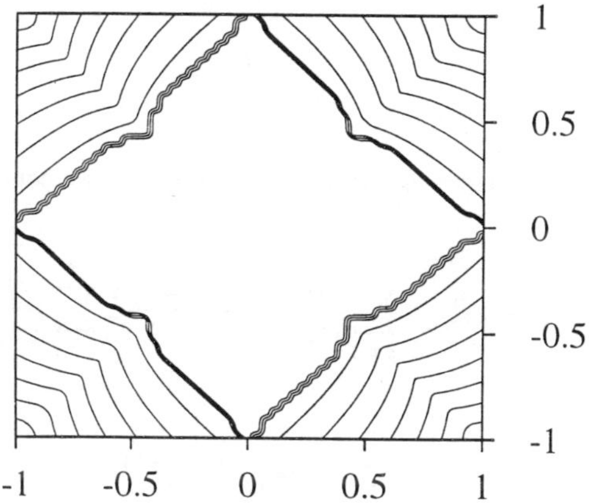

Fig. 2. Stability domain (inside the closed curve in bold line) of the present 2-D scheme in terms of $\dot{A} = \dfrac{\Delta t}{\delta x} A$ and $\dot{B} = \dfrac{\Delta t}{\delta y} B$.

where $Diag[d^{(i)}]$ denotes a diagonal matrix with diagonal entries $d^{(i)}$.
The matrix Φ' and Ψ' are defined by :

$$\Phi' = T_A Diag[\phi'^{(i)}] T_A^{-1} \quad , \quad \Psi' = T_B Diag[\psi'^{(i)}] T_B^{-1}, \tag{25}$$

with

$$\phi'^{(i)} = sgn(a^{(i)}) \phi^{(i)} \quad , \quad \psi'^{(i)} = sgn(b^{(i)}) \psi^{(i)} \tag{26}$$

$$\phi^{(i)} = min(1, \frac{\delta y |a^{(i)}|}{\delta x \, m(B)}) \quad , \quad \psi^{(i)} = min(1, \frac{\delta x |b^{(i)}|}{\delta y \, m(A)}) \tag{27}$$

where $m(A) = min_i(|a^{(i)}|)$ and $m(B) = min_i(|b^{(i)}|)$.
The fully discrete scheme is still in the form (18)-(20). At a cell interface $j+\frac{1}{2}, k$, the matrix $\Phi'_{j+\frac{1}{2},k}$ is computed from (25)-(27) by replacing A and B by their Roe averages $(A_R)_{j+\frac{1}{2},k}$ and $(B_R)_{j+\frac{1}{2},k}$ (see [4]). The same is done for $\Psi'_{j,k+\frac{1}{2}}$.

6 Extension to d dimension

For the hyperbolic system in d space-dimension :

$$\frac{\partial w}{\partial t} + \sum_{p=1}^{d} \frac{\partial f_p}{\partial x_p} = 0 \ , \tag{28}$$

with flux Jacobian matrices $A_p = df_p/dw$, the present scheme reads, in semi-discrete form :

$$\frac{1}{\Delta t}(w^{n+1} - w^n) + \sum_{p=1}^{d} \frac{\partial}{\partial x_p}(f_p - \frac{\delta x_p}{2}\Phi'_p \sum_{q=1}^{d} \frac{\partial f_q}{\partial x_q})^n = 0 \tag{29}$$

with

$$\Phi'_p = T_{A_p} Diag[\phi'^{(i)}_p] T^{-1}_{A_p} \tag{30}$$

$$\phi'^{(i)}_p = sgn(a^{(i)}_p)\phi^{(i)}_p \tag{31}$$

$$\phi^{(i)}_p = \min_{\substack{q=1,d \\ q\neq p}} (1, \frac{\delta x_q}{\delta x_p} \frac{|a^{(i)}_p|}{m(A_q)}). \tag{32}$$

To write down the fully discrete form of (29), we use the framework of [3]. Let $j = (j_1, j_2, ..., j_d)$ be a multi-integer associated with a point $x_j = (j_1\delta x_1, j_2\delta x_2, ..., j_d\delta x_d)$ of a regular Cartesian mesh. For any mesh function v_j, the difference and average operators over a mesh interval in the x_p-direction are defined as :

$$(\delta_p v)_{j+\frac{1}{2}e_p} = v_{j+e_p} - v_j \ , \ (\mu_p v)_{j+\frac{1}{2}e_p} = \frac{1}{2}(v_{j+e_p} + v_j)$$

where e_p is a multi-integer with components e_{pq} equal to 0 if $q \neq p$ and to 1 if $q = p$.
The fully discrete scheme uses d predictors $(p = 1, 2, ..., d)$:

$$(\pi_p)^n_{j+\frac{1}{2}e_p} = (\delta_p f_p + \delta x_p \sum_{\substack{q=1 \\ q\neq p}}^{d} \mu_p \mu_q \frac{\delta_q f_q}{\delta x_q})^n_{j+\frac{1}{2}e_p}. \tag{33}$$

The numerical fluxes are defined by :

$$(h_p)^n_{j+\frac{1}{2}e_p} = (\mu_p f_p - \frac{1}{2}\Phi'_p \pi_p)^n_{j+\frac{1}{2}e_p} \tag{34}$$

and the new cell-values are :

$$w^{n+1}_j = w^n_j - \Delta t (\sum_{p=1}^{d} \frac{\delta_p h_p}{\delta x_p})^n_j. \tag{35}$$

The matrix $(\Phi'_p)_{j+\frac{1}{2}e_p}$ is computed by using Roe averages of the matrices A_p at the cell interface $j+\frac{1}{2}e_p$.
The molecule of the above scheme has $1+2d^2$ points (3 points in 1-D, 9 points in 2-D, 19 points in 3-D) at most. This number can be reduced locally, for instance if a supersonic flow is calculated.
The scheme stability domain in 3-D for a linear scalar problem is shown on Fig.3. A sufficient criterion is:

$$CFL = \Delta t[\sum_{p=1}^{3} \frac{|A_p|^2}{\delta x_p^2}]^{1/2} \leq 0.49. \tag{36}$$

Note that the scheme can be easily extended to curvilinear meshes in a cell-centered finite-volume formulation.

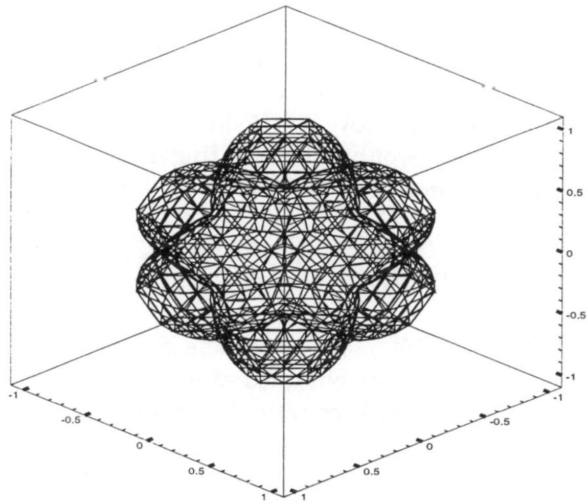

Fig. 3. Stability domain (inside the closed surface) of the present 3-D scheme in terms of $\dot{A}_p = A_p \dfrac{\Delta t}{\delta x_p}$, $p = 1, 2, 3$.

7 Extension to the Navier-Stokes equations

The multidimensional system (28) is now considered with flux components f_p including an inviscid or Euler part f_p^E and a viscous part f_p^V :

$$f_p = f_p^E(w) - f_p^V(w, w_{(1)}, w_{(2)}, \dots, w_{(d)}) \tag{37}$$

where $w_{(q)} = \partial w / \partial x_q$.
The construction of the scheme follows the same lines as for an hyperbolic system [6]. The semi-discrete scheme reads formally as (29) where f_p is the total flux (37). The calculation of Φ'_p is based on the eigenvectors of the Euler Jacobian matrix $A_p^E = df_p^E/dw$ and the eigenvalues $\phi_p^{(i)}$ may be modified as follows in order to include viscous effects :

$$\Phi'_p = T_{A_p^E} Diag[\phi_p'^{(i)}] T_{A_p^E}^{-1} \tag{38}$$

$$\phi_p'^{(i)} = sgn((a^E)_p^{(i)}) \phi_p^{(i)} \tag{39}$$

$$\phi_p^{(i)} = \min_{\substack{q=1,d \\ q \neq p}} (1, \frac{\delta x_q}{\delta x_p} \frac{|(a^E)_p^{(i)}|}{m(A_q^E)}, \frac{1}{2} Re_{p,p}^{(i)}, \frac{1}{2} \frac{\delta x_q}{\delta x_p} Re_{p,q}^{(i)}). \tag{40}$$

with the mesh Reynolds numbers $Re_{p,q}^{(i)} = \delta x_q |(a^E)_p^{(i)}| / \rho_q^V$, where the viscous coefficient ρ_q^V is the spectral radius of the Jacobian matrix $A_q^V = \partial f_q^V / \partial w_{(q)}$.
The choice of eigenvalues (40) yields a lower amount of numerical dissipation in flow regions where the viscous effects are dominant (mesh Reynolds numbers lower than 2), but may prove a little less robust than the choice based on Euler characteristic speeds only. The resulting discrete scheme is in any case a true second-order approximation of the steady Navier-Stokes equations.

8 Implicit treatment

The robustness and efficiency of the scheme are increased by adding to its explicit stage an implicit stage obtained after linearization and some simplifications. For the Navier-Stokes equations in d space-dimension, this implicit stage can be written as :

$$\Delta w_j^n + \sum_{p=1}^{d} \frac{\Delta t}{\delta x_p} [\mu_p (A_p^E \delta_p \Delta w) - \delta_p (\frac{A_p^V}{\delta x_p} \delta_p \Delta w) - \frac{1}{2} \delta_p (\Phi_p |A_p^E| \delta_p \Delta w)]_j^n = \Delta w_j^{expl} \tag{41}$$

where $\Delta w_j^n = w_j^{n+1} - w_j^n$, j is still a multi-integer and the eigenvalues of Φ_p are defined by (40).
The implicit scheme (41) is unconditionally linearly stable and provides fast convergence rates to steady state when used with large CFL numbers but, when applied to the computation of multidimensional flows on meshes containing large numbers of points, the linear system associated with (41) at each time-step must be approximately solved in order to avoid excessive CPU cost and memory requirements. For a steady problem, a line-relaxation algorithm can be used to solve this linear system without requiring the iterative procedure to converge. If we assume, for brevity's sake only, constant Jacobian

matrices, a typical step of the relaxation process at inner iteration l can be written under the form :

$$D\Delta w_j^{(l)} + L_p \Delta w_{j-e_p}^{(l)} + U_p \Delta w_{j+e_p}^{(l)} = \Delta w_j^{expl} - \sum_{\substack{q=1 \\ q\neq p}}^{d} (L_q \Delta w_{j-e_q}^{(l-1)} + U_q \Delta w_{j+e_q}^{(l-1)})$$

with

$$L_p = -\frac{\Delta t}{\delta x_p}[\frac{(A_p^E + \Phi_p |A_p^E|)}{2} + \frac{A_p^V}{\delta x_p}]\,;\; U_p = -\frac{\Delta t}{\delta x_p}[\frac{(-A_p^E + \Phi_p |A_p^E|)}{2} + \frac{A_p^V}{\delta x_p}]\,;$$

$$D = I + \sum_{q=1}^{d} \frac{\Delta t}{\delta x_q}(\Phi_q |A_q^E| + \frac{2A_q^V}{\delta x_q}) \tag{42}$$

where $\Delta w^{(l)} = w^{(l)} - w^n$. A maximal efficiency is obtained when alternating the line-relaxation process : p is successively taken equal to $1, \ldots, d$ [7]. Note that in order to ensure the stability, the matrices Φ_p multiplying $|A_p^E|$ in the implicit stage are actually taken equal to identity turning (41) into an implicit stage of Harten type with a viscous part. An alternate-line symmetric Gauss-Siedel method is used for all the following flow problems.

9 Applications

The properties of the scheme presented in the preceding sections are now illustrated through the computation of a linear advection problem, an inviscid transonic flow over an airfoil, a boundary layer flow and a turbulent transonic flow over a wing.

9.1 Circular advection

We consider a linear advection with rotation around the point ($x = \frac{1}{2}, y = 0$) over the square domain $[0, 1] \times [0, 1]$, similarly as in [8] but here we transport a Gaussian profile. The initial solution is set to zero everywhere except on the boundary part : $0 \le x \le \frac{1}{2}, y = 0$. More precisely, we look for the steady solution of the following initial boundary-value problem :

$$\begin{cases} \dfrac{\partial w}{\partial t} + y\dfrac{\partial w}{\partial x} + (\dfrac{1}{2} - x)\dfrac{\partial w}{\partial y} = 0, & 0 < x < 1,\ 0 < y < 1. \\ w(x, y, 0) = 0, \quad 0 < x < 1,\ 0 < y < 1 & \\ w(x, 0, t) = exp(-75(x - \dfrac{1}{4})^2), & 0 \le x \le \dfrac{1}{2},\ t \ge 0 \\ w(x, 1, t) = 0, \quad 0 \le x \le 1,\ t \ge 0 & \\ w(0, y, t) = 0, \quad 0 \le y \le 1,\ t \ge 0 & \end{cases}$$

On the boundary parts : $\frac{1}{2} < x \le 1,\ y = 0$ and $x = 1,\ 0 \le y \le 1$, no boundary condition is required (outflow boundary). Numerically, the boundary values are obtained from extrapolation of the interior values.

The exact steady solution is $w = Const.$ on any circle of center $(\frac{1}{2}, 0)$.
The numerical solution is shown in Fig.4. It has been computed by the present implicit scheme on a 40×40 uniform Cartesian mesh using CFL=1000. The iso-lines of the numerical solution are nearly perfect circles and the solution after an half turn, on the outflow boundary : $\frac{1}{2} < x \leq 1, \quad y = 0$, is in excellent agreement with the exact solution.

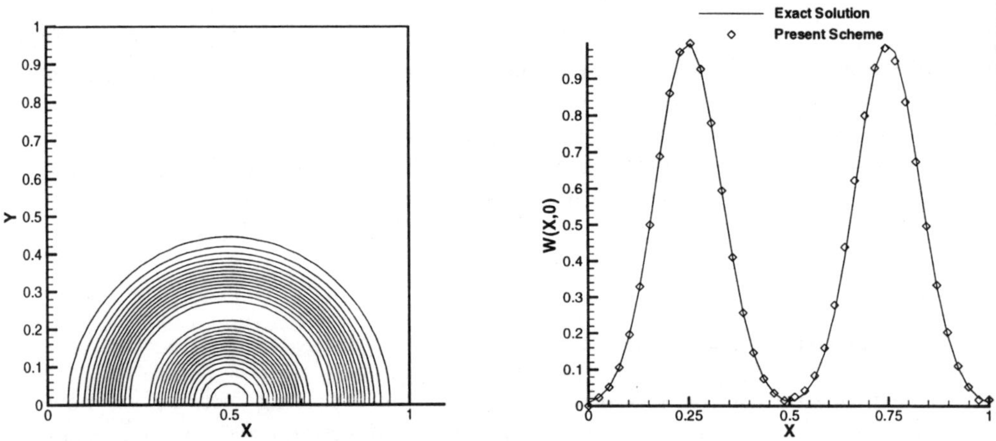

Fig. 4. Circular advection. Left : iso-lines of the solution w. Right : exact and numerical solution profile on line $y = 0$.

9.2 Inviscid transonic flow over an airfoil

Then, we consider a steady inviscid transonic flow over the NACA0012 airfoil at Mach number 0.85 and zero angle of attack. The flow is symmetric and computed in the upper half domain on a C-mesh composed of 124×32 cells. On the airfoil, the slip condition is prescribed and the pressure is deduced from a conservative integral form of the momentum equation projected on the normal to the wall.

The computation starts from an uniform flow and is run with a local time-step associated with a constant and uniform CFL number. The L_2 residual is reduced by 5 orders of magnitude in 360 iterations by the implicit method using alternate line-relaxation at CFL=500 (see Fig.5).

The steady numerical solution of the Euler equations is shown on Fig.5. and Fig.6. The flowfield is oscillation free and the shock profile is spread over two mesh-cells only. The entropy error is low : its relative value with respect to the freestream entropy is 2×10^{-3}. Let us remind that the solution is independent of the CFL number and does not require any parameter tuning (no correction in the method).

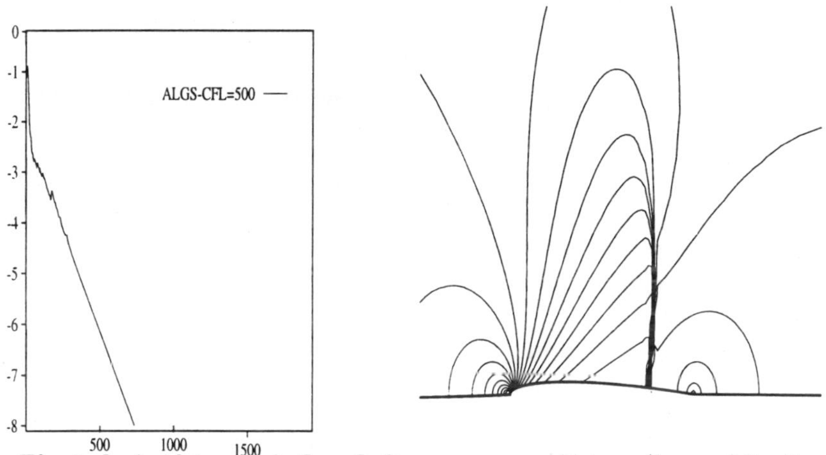

Fig. 5. Inviscid transonic flow. Left : convergence history (log_{10} of the L_2- residual in terms of time iterations). Right : pressure contours .

9.3 Boundary layer flow

The full Navier-Stokes equations are now used to compute the boundary layer over a flate plate in subsonic regime ($M_\infty = 0.5$). The computational domain is the square : $-1 \leq x \leq 1,\ 0 \leq y \leq 2$. The plate is located at $y = 0$, between $x = 0$ and $x = 1$. The Reynolds number based on the plate length is $Re = 1000$. The calculation is made on a Cartesian mesh having 80×30 cells. It is started from an uniform flow and run with CFL=1000. The residual is reduced by 5 orders of magnitude in 50 iterations and the solution agrees perfectly with the Blasius solution (see Fig.7).

9.4 Turbulent transonic flow over a wing

The transonic flow over the ONERA M6 wing at $M_\infty = 0.84$, incidence $\alpha = 3.06°$ and Reynolds number $Re = 2.6 \times 10^6$ has been computed by the implicit scheme using the algebraic turbulent model of Baldwin-Lomax. Calculations

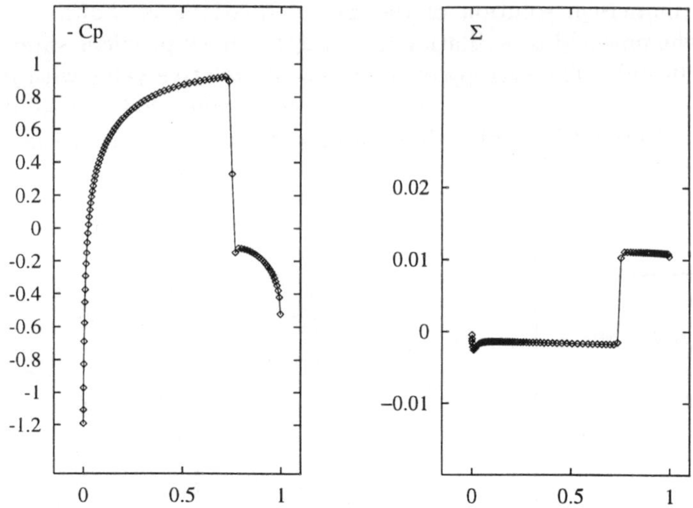

Fig. 6. Inviscid transonic flow. Solution on the airfoil. Left : pressure coefficient. Right : entropy deviation.

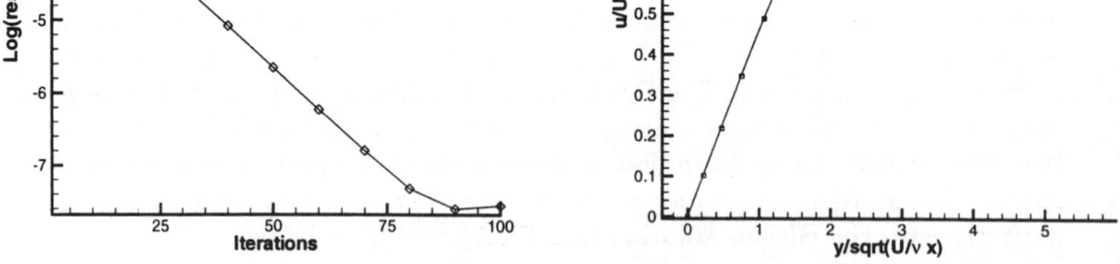

Fig. 7. Boundary layer flow. Left : convergence history. Right : analytic (Blasius) and numerical u-velocity profile.

have been carried on two grids. Each grid is made of two subdomains : one above the wing and the other below. The upper and lower subdomains have the same number of mesh cells. For the first grid, this number is $98 \times 27 \times 34$ (98 along the half wing, 27 in the spanwise direction and 34 in the normal direction). For the finer grid, it is $98 \times 51 \times 34$. The method requires 15 MW of storage and needs 45 mn of CPU time on a Cray C90 computer to decrease the residual by 4 orders of magnitude on the first grid (see Fig.8,left). Wall-pressure distribution in the wing section (Fig.8, right) compares well with the experiments reported in [9]. Refining the grid in the spanwise direction slightly improves the numerical results but the shock structure on the upper wing surface is already well represented on the coarser grid (see Fig.9.). The same conclusion has been found after a further grid refinement in the normal direction ($98 \times 51 \times 66$ in each subdonmain).

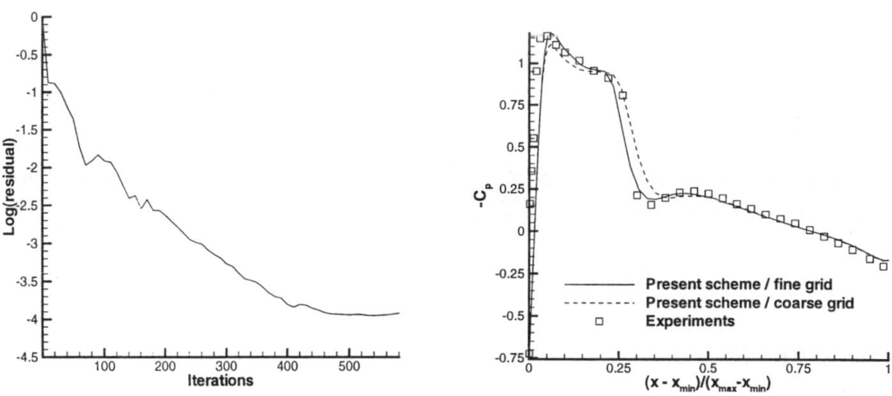

Fig. 8. 3-D turbulent transonic flow. Left : convergence history. Right : pressure coefficient on the upper surface near the wing tip (90% spanwise).

10 Conclusions

A scheme collecting together the advantages of the Lax-Wendroff and Roe schemes has been presented for the steady compressible Euler and Navier-Stokes equations. This scheme is really second-order accurate and easy to code for three-dimensional problems. It ensures a low but sufficient numerical dissipation without any tuning parameter or limiter and its solution is

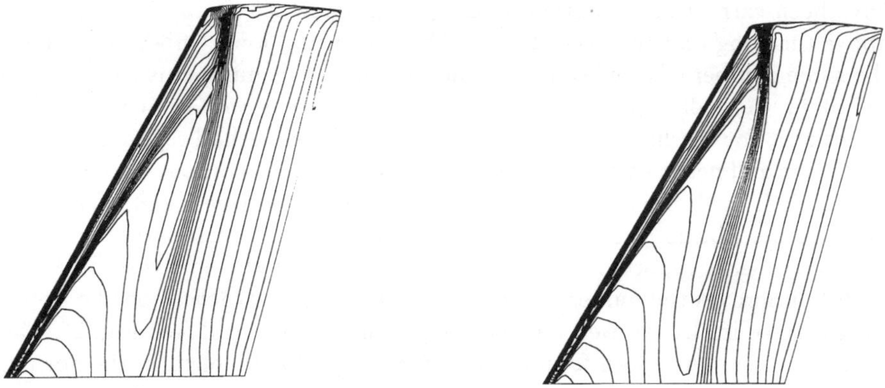

Fig. 9. 3-D turbulent transonic flow. Pressure contours on the upper surface. Left : coarse grid. Right : fine grid.

independent of the CFL number used to converge to the steady state. For 3-D transonic flow problems, the scheme has proved its efficiency, robustness and its weak dependence on the mesh used.

Acknowledgements : The authors warmly thank V. Couaillier (ONERA) for providing the mesh for the ONERA M6 wing calculations.

References

[1] LERAT A., Implicit method of second-order accuracy for the Euler equations. *AIAA Journal*, 23, pp. 33-40 (1985).

[2] LERAT A. and SIDES J.. Efficient solution of the steady Euler equations with a centered implicit method. in *Numerical Methods for Fluid Dynamics 3, edited by K.W.Morton and M.J.Baines, Clarendon-Oxford*, pp. 65-86 (1988).

[3] LERAT A., Multidimensional centered schemes of the Lax-Wendroff type. *CFD Review 1995, J. Wiley*, pp. 124-140 (1995).

[4] ROE P.L., Approximate Riemann solvers, parameter vectors and difference schemes. *J. Comp. Phys.*, 43, pp.357-372 (1981).

[5] HUANG Y. and LERAT A., Second-order upwinding through a characteristic time-step matrix for compressible flow calculations. *J. Comp. Phys.*, 142, pp.445-472 (1998).

[6] CORRE C., and LERAT A., Efficient calculation of 3-D turbulent transonic flows. *Lecture Notes in Physics,* 515, pp. 19-24 (1998).
[7] CORRE C., KHALFALLAH K., LERAT A., Line-relaxation methods for a class of centred schemes. *CFD Journal,* vol.5, pp. 213-246 (1996).
[8] DECONINCK H., STRUIJS R., BOURGEOIS G. and ROE P.L., Compact advection schemes on unstructured grids. VKI Lecture Series 1993-04, in *High Resolution (upwind and TVD) Methods for the Compressible Flow Equations,* edited by VKI (1994).
[9] SCHMITT V., CHARPIN F., Pressure distributions on the ONERA-M6-Wing at transonic Mach numbers. *AGARD-AR-138* (1979).

FLUX SCHEMES FOR SOLVING NONLINEAR SYSTEMS OF CONSERVATION LAWS

Jean-Michel GHIDAGLIA

Centre de Mathématiques et de Leurs Applications
CNRS URA 1611 et Ecole Normale Supérieure de Cachan
61 avenue du Président Wilson
94235 Cachan Cedex, FRANCE
email : jmg@cmla.ens-cachan.fr

Key Words: FINITE DIFFERENCES, HYPERBOLIC SYSTEMS OF CONSERVATION LAWS, FINITE VOLUMES, SOURCE TERMS

Abstract. We introduce a distinction among the classical three point finite difference schemes for nonlinear systems of conservation laws : those which are flux schemes and those which are not. We discuss at the continuous level why we think that these schemes are of interest. We also study some of their properties and their extensions to the multidimensional (finite volumes) case.

1 Introduction

Let us consider a system of m-conservation laws ($m \geqslant 1$)

$$\frac{\partial v}{\partial t} + \frac{\partial \varphi}{\partial x} = 0, \quad x \in \mathbb{R}, \quad t > 0. \tag{1}$$

Here the flux φ is related, at the continuous level, to the so-called conservative variable v by a nonlinear function $f : G \longrightarrow \mathbb{R}^m$ (G is an open set in $\mathbb{R}^m$) :

$$\varphi = f(v). \tag{2}$$

In view of the discretization of (1) and given two positive real numbers Δt and Δx, it is classical to associate to this system of equations a discrete dynamical system $(v_j^n)_{j\in\mathbb{Z}} \longrightarrow (v_j^{n+1})_{j\in\mathbb{Z}}$ where

$$v_j^{n+1} = v_j^n - \lambda(g_{j+\frac{1}{2}}^n - g_{j-\frac{1}{2}}^n), \tag{3}$$

and $\lambda = \Delta t/\Delta x$. The vector $g_{j+\frac{1}{2}}^n$ is the *numerical flux*, and for a three point explicit numerical scheme we have :

$$g_{j+\frac{1}{2}}^n = g(\lambda; v_j^n, v_{j+1}^n), \tag{4}$$

where the function $g(\lambda; ., .)$ is to be specified.

Consistency, which is necessary if we want that solutions to (3) "converge" to solutions to (1), simply amounts imposing the condition

$$g(\lambda; v, v) = f(v), \quad \forall v \in G, \tag{5}$$

and we shall assume this identity from now on.
Let us for the sake of simplicity in the notations omit the dependence on λ in g and set also

$$g(v, w) = \frac{f(v) + f(w)}{2} + h(v, w)\,. \tag{6}$$

so that (5) simply reads $h(v, v) = 0$. The function h represents the non centered part in the numerical flux and as is well-known, it is the part that contributes to the stability of the scheme. Since $h(v, v) = 0$, we expect that $h(v, w)$ is the product of a $m \times m$ matrix $M(v, w)$ by the vector $w - v$. Indeed all the consistent numerical flux are of this form (see Proposition 1) :

$$h(v, w) = \frac{1}{2} M(v, w)(w - v)\,. \tag{7}$$

We would like in this article to consider a subclass of schemes that we term as *flux schemes* according to the following definition.

Definition 1 *The numerical flux g corresponds to a flux scheme when there exists a $m \times m$ matrix $S(v, w)$ such that*

$$h(v, w) = \frac{1}{2} S(v, w)(f(w) - f(v)). \tag{8}$$

As a result, when (8) holds true, we observe that the numerical flux is a linear combination (whose coefficients depends on v and w) of $f(v)$ and $f(w)$:

$$g(v, w) = \frac{(I - S(v, w))}{2} f(v) + \frac{(I + S(v, w))}{2} f(w)\,. \tag{9}$$

The goal of this article is to study some properties of schemes corresponding to such numerical fluxes.

Remark 1 *When we only have (7), we can write*

$$g(v, w) = \frac{f(v) + f(w)}{2} + \frac{1}{2} M(v, w)(w - v) \tag{10}$$

and M is usually termed as the viscosity matrix. We shall therefore term the corresponding scheme as a viscous scheme.

2 Some remarks at the continuous level

Let us first consider a smooth solution to (1) - (2). We denote by $A(v)$ the gradient of $f(v)$:

$$A_{ij}(v) = \frac{\partial f_i}{\partial v_j}(v), \quad 1 \leq i, j \leq m, \quad \forall v \in G. \tag{11}$$

We multiply (1) by $A(v)$ and observe that since $A(v)\frac{\partial v}{\partial t} = \frac{\partial \varphi}{\partial t}$, we have

$$\frac{\partial \varphi}{\partial t} + A(v)\frac{\partial \varphi}{\partial x} = 0. \tag{12}$$

We deduce from this computation that, in a certain sense, the flux φ satisfies a linear equation (of which coefficients depend on v).

Let us now prove that equation (12) remains valid across a shock. We consider an open set $U \subset \mathbb{R} \times \mathbb{R}_+^*$, which is divided in two open parts U^+ and U^- by a C^1 curve $\Sigma = \{(x(t), t), a < t < b\}$. A function v, which is smooth in $U \backslash \Sigma$, will be a weak solution to (1)-(2) in U, if and only if it is a classical solution in $U \backslash \Sigma$ and accros Σ, the Rankine-Hugoniot jump condition is fulfilled :

$$\begin{aligned} &[f(v)] = \sigma[v], \text{ where } \sigma(t) = \frac{dx}{dt}, \quad [w] \equiv w^+ - w^- \\ &\text{and } w^\pm = \lim\{w(x,t), x \to x(t), (x,t) \in U^\pm\}. \end{aligned} \tag{13}$$

We extend σ to the whole line $\{(y, t), y \in \mathbb{R}\}$ by setting

$$\sigma(y, t) = \sigma(t), \quad \forall y \in \mathbb{R}. \tag{14}$$

Since $[\varphi - \sigma v] = 0$, we can define across Σ the functions $A(v)\frac{\partial(\varphi - \sigma v)}{\partial x}$ and $(\varphi - \sigma v)_t$ as L^∞_{loc}-functions and then consider

$$\mathcal{N} \equiv (\varphi - \sigma v)_t + A(v)(\varphi - \sigma v)_x + \sigma_t v. \tag{15}$$

In U^+ or U^-, $\mathcal{N}$ is equal to zero since (1) implies

$$v_t + A(v)v_x = 0 \text{ in } U \backslash \Sigma. \tag{16}$$

Since $\mathcal{N}$ is a L^∞_{loc}-function, we deduce that $\mathcal{N}$ is almost everywhere equal to zero :

$$(\varphi - \sigma v)_t + A(v)(\varphi - \sigma v)_x = -\sigma_t v. \tag{17}$$

This gives a weak formulation to (12) across a shock.

Theorem 1 *Let v be a discontinuous solution to (1) - (2) with a finite number of shocks that do not interact. Then equation (12) holds true in the sense that $\mathcal{N}$ defined by (15) is almost everywhere equal to zero.*

This results is according to us a mathematical reason for which flux schemes are of interest. Indeed, equation (12) shows that a kind of "superposition principle" should be true at the dicrete level and flux schemes, as already noticed (see (9)), are reminiscent of such a principle.

Remark 2 *In the case of a shock interaction, it is not possible to give a sense to (17) since it may happen that σ undergoes a discontinuity at the point of interaction and then $\sigma_t v$ cannot be defined.*

3 A catalogue of numerical schemes

Let us consider a few numerical flux.

- We start with the Lax-Friedrichs scheme whose numerical flux reads as :

$$g^{LF}(v,w) = \frac{f(v)+f(w)}{2} - \frac{1}{2\lambda}(w-v). \tag{18}$$

- Next for the Lax-Wendroff scheme :

$$g^{LW}(v,w) = \frac{f(v)+f(w)}{2} - \frac{\lambda}{2}A(v,w)(f(w) \quad f(v))$$

where A maps $G \times G$ into $\mathcal{M}_m(\mathbb{R})$ and satisfies

$$A(v,w) = A(v). \tag{19}$$

- Then for the Godunov scheme :

$$g^{GOD}(v_l, v_r) = f(w_R(0, v_l, v_r)), \tag{20}$$

where $w_R(\xi; v_l, v_r)$ is the value at time $t = 1$ of the entropic solution (when it exists) to the Riemann problem (1) - (3) with initial data $v(x,0) = v_l$ for $x < 0$ and $v(x,0) = v_r$ for $x > 0$.
- In order to define Roe's scheme, we assume that we can construct a mapping A_R from $G \times G$ into $\mathcal{M}_m(\mathbb{R})$ such as :

$$A_R(v,w) \text{ has real eigenvalues and a corresponding set of eigenvectors that are complete in } \mathbb{R}^m\,, \tag{21}$$

$$A_R(v,w) = A(v)\,, \tag{22}$$

$$A_R(v,w)(w-v) = f(w) - f(v)\,. \tag{23}$$

Let us denote by $sgn(M)$ the sign of a matrix M. For example when M satisfies (21) we have :

$$sgn(M) = P^{-1}diag(sgn(\lambda_i))P, \text{ where } M = P^{-1}diag(\lambda_i)P\,, \tag{24}$$

and $sgn(\lambda) = 0$ if $\lambda = 0$ and $\lambda/|\lambda|$ otherwise. We also set $|M| = sgn(M)M$.

Then Roe's scheme corresponds to

$$g^{ROE}(v,w) = \frac{f(v)+f(w)}{2} - \frac{1}{2}|A_R(v,w)|(w-v)\,. \tag{25}$$

And we observe that thanks to (23),

$$g^{ROE}(v,w) = \frac{f(v)+f(w)}{2} - \frac{1}{2}sgn(A_R(v,w))(f(w)-f(v))\,. \tag{26}$$

Remark 3 *These two expressions show that Roe's scheme is both a viscous scheme (25) and a flux scheme (26).*

- Gallouet and Massella [3] have introduced a modification of Roe's method by taking (they call this scheme VFRoe, for Finite Volume Roe)

$$g^{VFRoe}(v,w) = \frac{f(v)+f(w)}{2} - \frac{1}{2}|A(\frac{v+w}{2})|(w-v)\,. \tag{27}$$

- Finally we end up with the scheme introduced in Ghidaglia, Kumbaro and Le Coq [4] (that we termed as the characteristic flux scheme). Let $\mu(v,w)$ denote a mean value between v and w which is independent of the continuous flux *e.g.* $\mu(v,w) = \frac{v+w}{2}$.
  By using (12) we construct in [4] the following numerical flux :

$$g^{CF}(v,w) = \frac{f(v)+f(w)}{2} - \frac{1}{2}sgn(A(\mu(v,w))(f(w)-f(v))\,. \tag{28}$$

  The motivation for the construction of such schemes comes from the case where we are considering complex systems (*e.g.* those occuring in two phase flows). In such a case Roe's solver is not available, and this numerical flux has shown very good results in 1D, 2D and 3D computations (see *e.g.* Boucker [2] and Ghidaglia, Kumbaro, Le Coq [4]).

Remark 4 *In the Characteristic Flux scheme (28), the matrix $sgn(A)$ which appears produces the upwinding (as in Roe's solver). In the context of two-phase flows, it is common that the convection operator that occurs is not hyperbolic. In such a case, the matrix A can have complex eigenvalues. In this case we take for $sgn(M)$ the matrix given by (24) where this time λ_i denotes the real part of the eigenvalues. See also Halaoua [5] for an extension of Roe's scheme in this context (complex eigenvalues).*

On the relation between flux schemes and viscous schemes

Let us observe that every flux scheme can be written as a viscous scheme. Indeed, denoting by $A_T(v,w)$ the Taylor matrix :

$$A_T(v,w) = \int_0^1 A(\theta v + (1-\theta)w)d\theta, \tag{29}$$

we see that any numerical flux corresponding to a flux scheme :

$$g(v,w) = \frac{f(v)+f(w)}{2} + \frac{1}{2}S(v,w)(f(w)-f(v)), \tag{30}$$

can be recast as a numerical flux corresponding to a viscous scheme :

$$g(v,w) = \frac{f(v)+f(w)}{2} + \frac{1}{2}M(v,w)(w-v), \tag{31}$$

with viscosity matrix

$$M(v,w) = S(v,w)A_T(v,w). \tag{32}$$

In general the converse is false. It is obvious that the Lax-Friedrichs scheme is not a flux scheme and we shall prove by a rather sophisticated, although interesting, way that Godunov's scheme is not a flux scheme either (see Proposition 2).

Proposition 1 *Any consistent scheme can be written as a viscous scheme.*

Proof. This follows from Taylor's expansion formula :

$$h(v,w) - h(v,v) = \left(\int_0^1 D_2h(v,\theta v + (1-\theta)w)d\theta\right)(w-v) \tag{33}$$

and therefore we can take

$$M(v,w) = 2\int_0^1 D_2h(v,\theta v + (1-\theta)w)d\theta. \tag{34}$$

Hence provided the numerical flux function g is C^1 (or $W^{1,\infty}$) we can write any three points scheme as a viscous scheme. However the expression of the associated viscosity matrix $M(v,w)$ can be out of reach. For example, we do not know whether Godunov's scheme has a "simple" viscosity matrix.

4 On the relation with source terms

In Alouges, Ghidaglia and Tajchman [1], we have considered the interaction between the numerical flux used and the dicretisation of source terms. We consider instead of (1), the equation :

$$\frac{\partial v}{\partial t} + \frac{\partial \varphi}{\partial x} = S, \quad x \in \mathbb{R}, \quad t > 0, \tag{35}$$

where S depends only on x. Now we discretize this equation by

$$v_j^{n+1} = v_j^n - \lambda(g_{j+\frac{1}{2}}^n - g_{j-\frac{1}{2}}^n) + \Delta t\, \Sigma_j^n, \tag{36}$$

where

$$g^n_{j+\frac{1}{2}} = g(\lambda; v^n_j, v^n_{j+1}) \,, \tag{37}$$

and g corresponds to a flux scheme :

$$g(v,w) = \frac{f(v)+f(w)}{2} - U(v,w)\frac{f(w)-f(v)}{2} \,. \tag{38}$$

As discussed in [1], if we simply take :

$$\Sigma^n_j = \Sigma_j = \frac{1}{\Delta x_j}\int_{x_{j-1/2}}^{x_{j+1/2}} S(x)dx \,, \tag{39}$$

then due to the fact that the numerical flux is not centered (upwind bias) one can observe large errors on any steady solution under investigation. We have therefore introduced the notion of enhanced consistency as follows. Denoting by

$$\phi = \phi^{\text{conv.}}(x) \equiv \int_0^x S(y)dy \,, \tag{40}$$

we say that Σ^n_j satisfies, with respect to the numerical flux g, the enhanced consistency property when we have : if at some time-step, v^n_j is such that

$$f(v^n_j) = \frac{1}{\Delta x_j}\int_{x_{j-1/2}}^{x_{j+1/2}} \phi^{\text{conv.}}(x)dx \,, \quad \forall j \in \mathbb{Z} \,, \tag{41}$$

then v^{n+1}_j given by (36) must be equal to v^n_j. This condition can be equivalently formulated as

$$\Sigma^n_j = \frac{f^n_{j+1/2} - f^n_{j-1/2}}{\Delta x_j} \,, \tag{42}$$

if v^n_j satisfies (41).
With these notations, we show in [1] the following results.

Theorem 2 *Let g be given in the form (38) and denote by $U^n_j = U(v^n_j, v^n_{j+1})$. The enhanced consistency will be satisfied if we discretize the forcing term S according to the following formula*

$$\Sigma^n_j = \frac{1}{2\Delta x_j}\left\{\int_{x_{j-1}}^{x_{j+1}} S(y)dy - U^n_j\int_{x_j}^{x_{j+1}} S(y)dy + U^n_{j-1}\int_{x_{j-1}}^{x_j} S(y)dy\right\} \,. \tag{43}$$

Corollary 1 *Assuming that S is given by a piecewise constant function : $S(x) = S_j$ for $x \in \,]x_{j-1/2}, x_{j+1/2}[$, the formula (43) reads*

$$\Sigma^n_j = \frac{I+U^n_{j-1}}{4}\frac{\Delta x_{j-1}}{\Delta x_j}S_{j-1} + \frac{I-U^n_j+U^n_{j-1}}{2}S_j + \frac{I-U^n_j}{4}\frac{\Delta x_{j+1}}{\Delta x_j}S_{j+1} \,. \tag{44}$$

In order to illustrate this formula, let us show what it means in the case of a single linear equation $v_t + c\,v_x = S$ where *e.g.* $c > 0$ and $v \in \mathbb{R}$. Here the usual upwind scheme amounts to take $U^n_j = 1$, *i.e.* $g(v,w) = c\,v$ and (44) reads as :

$$\Sigma^n_j = \frac{1}{2}\frac{\Delta x_{j-1}}{\Delta x_j}S_{j-1} + \frac{1}{2}S_j \,. \tag{45}$$

5 Relations with entropies

At the level of a general system (1) - (2), there is no universal notion of entropy and the minimal requirement is that a good notion of entropy should select one solution to the Riemann problem which is moreover stable in a suitable sense.

For physical systems (*i.e.* systems steaming from problems in continuous physics) usually one has an entropy function i.e. a strictly convex function $\eta : G \longrightarrow \mathbb{R}$, sufficiently smooth, to which one can associate an entropy flux $\psi : G \longrightarrow \mathbb{R}$ such that *smooth* solutions to (1) - (3) satisfies the additional conservation law

$$\eta_t + \psi_x = 0 \,, \tag{46}$$

which amounts simply to the algebraic relations :

$$A(v)\nabla_v \eta(v) = \nabla_v \psi(v). \tag{47}$$

Weak solutions to (1) - (2) will not satisfy (46) in general but

$$\eta_t + \psi_x \leqslant 0 \,, \tag{48}$$

in the sense of distributions.

If we consider a shock (see section 2) we know that (1) - (3) is equivalent to the Rankine-Hugoniot condition (13) :

$$[f(v)] = \sigma[v] \,, \tag{49}$$

while (48) amounts to

$$[\psi(v)] \leq \sigma[v] \,, \tag{50}$$

provided we take $U^+ = \{(y,t) \in U, y > x(t)\}$.

Let us now consider two states v_l (left) and v_r (right) (with $v_l \neq v_r$) such that there exists $\sigma \in \mathbb{R}$ for which

$$\varphi_l - \varphi_r = \sigma(v_l - v_r) \,, \tag{51}$$

where $\varphi_l \equiv f(v_l)$ and $\varphi_r \equiv f(v_r)$.

We choose the time step Δt and the space step Δx so that

$$\sigma \frac{\Delta t}{\Delta x} = \sigma\lambda \in \mathbb{Z}. \tag{52}$$

For example, when $\sigma = 0$ (a stationary shock), this condition holds for every Δt and Δx.

Then considering any flux scheme

$$v_j^{n+1} = v_j^n - \lambda(g_{j+\frac{1}{2}}^n - g_{j-\frac{1}{2}}^n), \quad g_{j+\frac{1}{2}}^n = g(\lambda; v_j^n, v_{j+1}^n), \tag{53}$$

$$g(v,w) = \frac{f(v)+f(w)}{2} + \frac{1}{2}S(v,w)(f(w)-f(v)); \tag{54}$$

and starting with the initial data

$$v_j^0 = v_l \text{ for } j \leq 0 \text{ and } v_j^0 = v_r \text{ for } j \geq 1; \tag{55}$$

we generate a pure shock wave

$$v_j^n = v_{j-(\sigma\lambda)n}^0, \quad \forall j \in \mathbb{Z}, \quad \forall n \in \mathbb{N}. \tag{56}$$

But now since we can exchange the subscript left and right in (51), we can also propagate a pure shock wave connecting from left to right v_r and v_l.

This shows immediately that flux-schemes are not entropic. Indeed, the entropy condition (50) reads here as

$$\psi(v_r) - \psi(v_l) < \sigma(\eta(v_r) - \eta(v_l)) \tag{57}$$

since in general equality cannot hold.

Hence *one* of the two shock waves that we just constructed is not entropic (observe that as Δt and Δx goes to zero with λ constant, the discrete shock waves converge to continuous ones).

A by-product of this construction is the following result.

Proposition 2 *Godunov scheme is not a flux scheme.*

This follows from the well-known fact (see *e.g.* [6]) that Godunov scheme is entropic.

6 Generalization to multidimensional schemes

The generalization of conservative finite differences in space dimension larger than 1 are the finite volume methods. Here again, numerical methods can be described by numerical fluxes. Let us discuss briefly this now. We only perform a space discretization of the operators so that we shall end up with a system of o.d.e.'s.
We consider a system of m balance equations : $(v = (v_1, \ldots, v_m) \in \mathbb{R}^m))$:

$$\frac{\partial v}{\partial t} + \nabla \cdot F(v) = 0, \tag{58}$$

here $\nabla \cdot F(v) = \sum_{j=1}^{nd} \frac{\partial F^j(v)}{\partial x_j}$, where F^j maps $\mathbb{R}^m$ into $\mathbb{R}^m$. This equation is posed in a nd-dimensional domain Ω ($nd = 1, 2$ or 3 in practice).

We assume that the computational domain Ω is decomposed in smaller volumes (the so-called control volumes) $K : \Omega = \cup_{K\in\mathcal{T}} K$ and we assume for the sake of simplicity in the exposition that $\Omega = \cup_{K\in\mathcal{T}} K$ is "conformal" *i.e.* that it is a finite element triangulation of Ω. In practice one can use triangles for $nd = 2$ and tetrahedrons for $nd = 3$. The cell-center finite volume approach for solving (58) consists in approximating the means

$$v_K(t) \equiv \frac{1}{vol(K)} \int_K v(x,t)\, dx\,, \tag{59}$$

where $vol(K)$ denotes the nd-dimensional volume of K and $area(A)$ stands for the $(nd-1)$-dimensional volume of an hypersurface A. Integrating (58) on K makes the normal fluxes, $F^{\nu}_{\partial K}$, appear

$$F^{\nu}_{\partial K}(t) = \int_{\partial K} F(v(\sigma,t)).\nu(\sigma)\, d\sigma, \tag{60}$$

where ∂K is the boundary of K, $\nu(\sigma)$ the unit external normal on ∂K and $d\sigma$ denotes the (nd-1)-volume element on this hypersurface. Indeed, we have according to (58) :

$$\frac{dv_K}{dt} + \frac{1}{vol(K)} F^{\nu}_{\partial K} = 0\,. \tag{61}$$

The heart of the matter in finite volume methods consists in providing a formula for the normal fluxes $F^{\nu}_{\partial K}$ in terms of the $\{v_L\}_{L\in\mathcal{T}}$. Assuming that the control volumes K are polyhedra, as is most often the case, the boundary ∂K is the union of hypersurfaces $K \cap L$ where L belongs to the set $\mathcal{N}(\mathcal{K})$, the set of $L \in \mathcal{T}$, $L \neq K$, such that $K \cap L$ has positive $(nd-1)$-measure. We can therefore decompose the normal flux as a sum :

$$F^{\nu}_{\partial K} = \sum_{L\in\mathcal{N}(\mathcal{K})} F_{K,L}\,, \tag{62}$$

where ($\nu_{K,L}$ points into L) :

$$F_{K,L} = \int_{K\cap L} F(v(\sigma,t)).\nu_{K,L}\, d\sigma\,. \tag{63}$$

Motivated by the case where (58) describes wave propagation phenomena, it is natural to look for an approximation of (63) in terms of $v_K(t)$ and $v_L(t)$:

$$F_{K,L} \approx area(K \cap L)\, \Phi(v_K, v_L; K, L)\,, \tag{64}$$

where Φ is the numerical flux to be described.
At this level of generality Φ must satisfy 2 structural properties :
(i) consistency : $\Phi(w,w;K,L) = F(w) \cdot \nu_{K,L}$,
(ii) conservation : $\Phi(v,w;K,L) = -\Phi(w,v;L,K)$.

In this case we introduce the following definition of flux schemes.

Definition 2 *The numerical flux Φ corresponds to a flux scheme when there exists a matrix $U(v,w;K,L)$ such that*

$$\Phi(v,w;K,L) = \frac{F(v)+F(w)}{2}\cdot\nu_{K,L} - U(v,w;K,L)\frac{F(w)-F(v)}{2}\cdot\nu_{K,L}\,. \quad (65)$$

Then any of the numerical flux schemes introduced in the 1D case can be generalized by taking for the matrix $U(v,w;K,L)$ the matrix which corresponds to $f(v) \equiv F(v)\cdot\nu_{K,L}$. For example the extension of formula (28) to the multidimensional case is given throught the following definition.

Definition 3 *The numerical flux of the "VFFC [1]" method is obtained by formula (65) when we take :*

$$U(v,w;K,L) = sgn(\frac{\partial F(v)\cdot\nu_{K,L}}{\partial v}|_{v=\mu(v,w;K,L)})\,, \quad (66)$$

where $\mu(v,w;K,L)$ is a mean between v_K and v_L which only depends on the geometry of K and L, e.g. :

$$\mu(v,w;K,L) = \frac{vol(K)v + vol(L)w}{vol(K)+vol(L)}\,. \quad (67)$$

Proposition 3 *Combining (61), (62), (64), (65), (66) and (67), the finite volume approximation of (58) is the following system of o.d.e.'s :*

$$\frac{dv_K}{dt} + \frac{1}{vol(K)}\sum_{L\in\mathcal{N}(K)} area(K\cap L)\,\Phi(v_K,v_L;K,L) = 0\,. \quad (68)$$

References

[1] ALOUGES F., GHIDAGLIA J.M. and TAJCHMAN M., On the interaction of upwinding and forcing for nonlinear hyperbolic systems of conservation laws, Prépublication du CMLA and article to appear, 1999.

[2] BOUCKER M., *Modélisation numérique multidimensionnelle d'écoulements diphasiques liquide-gaz en régimes transitoires et permanents : méthodes et applications*, Thèse, ENS-Cachan, France, 1998.

[3] GALLOUET T. and MASELLA J.M., Un schéma de Godounov approché, *C.R.Acad. Sc. Paris*, 1996, **323**, I, 77-84.

[4] GHIDAGLIA J.M., KUMBARO A. and LE COQ G., Une méthode volumes-finis à flux caractéristiques pour la résolution numérique des systèmes hyperboliques de lois de conservation, *C.R.Acad. Sc. Paris*, 1996, **322**, I, 981-988.

[5] HALAOUA K., Quelques solveurs pour les opérateurs de convection et leur application à la mécanique des fluides diphasiques, Thèse, ENS Cachan, June 1998.

[6] LEVEQUE R.J., *Numerical methods for conservation laws*, Birkhäuser, Basel, 1992.

[1] Volumes Finis à Flux Caractéristiques ([4])

A Lax–Wendroff type theorem for residual schemes

Rémi Abgrall, Katherine Mer and B. Nkonga

Mathématiques Appliquées de Bordeaux
351, Cours de la Liberation, 33405 Talence Cedex, FRANCE
email : abgrall@math.u-bordeaux.fr, mer@math.u-bordeaux.fr,
nkonga@math.u-bordeaux.fr

Key Words: COMPRESSIBLE FLOW SOLVERS, FLUCTUATION SPLITTING SCHEMES, UPWIND SOLVERS, LAX WENDROFF THEOREM

Abstract. In this paper, we are interested in the class of residual schemes for compressible fluid dynamics. We indicate sufficient conditions which ensure that a residual scheme that approximates a system of hyperbolic equations does converges to a weak solution that satisfies an entropy condition.This generalises a celebrated result by P.D. Lax and B. Wendroff. We give some examples, with numerical applications.

1 Introduction

It is well known that, given the Cauchy problem for a 1D hyperbolic system, that, under stability and consistency assumptions, a conservative scheme converges to a weak solution : this is Lax–Wendroff theorem [6]. This result extends to multidimensional problems for schemes formulated with general unstructured meshes [4].

Recently, P.L. Roe, H. Deconinck and coauthors [8], [9] have considered a new class of numerical schemes (residual upwind schemes) that allow the approximation of the Euler. These schemes do not have a finite volume formulation in general. They are related to Hughes' SUPG scheme. We aim at giving sufficent conditions that allow the convergence of the numerical solution to a weak solution. In practice, they are satisfied by the schemes of [8], [9]. The same technique enables to give conditions to ensure the convergence to a weak solution that satisfies an entropy inequality.

We consider the following system

$$\begin{array}{lll} \partial_t u + \text{ div } F(u) = 0 & \text{in} & R^2 \times R^+ \\ u(x,0) = u_0(x) & \text{in} & R^2 \end{array} \tag{1}$$

where $u : R^2 \times R^+ \to R^m$ and $F(u) = (f(u), g(u))$, $F \in (C^1(R^m))^{2m}$. We consider the 2D case only to simplify the text.

Let $(\mathcal{T}_h)_h$ be a family of triangulations of R^2. The vertices of (T_h) are denoted by x_i, $i = 1, \cdots, n_s$. We consider the family of dual cells, $(C_h)_h$ constructed from the centroids and the midpoints of the triangles. We denote by $h = \sup_j$ diam (T_j), for T_j triangle of $\mathcal{T}_h$. We denote by $|T|$ the area of T. By convention, $\sum_{i=1}^{3} \phi_i$ should be red as $\sum_{j=1}^{3} \phi_{i_j}$ where $\{i_1, i_2, i_3\}$ are the indices of the vertices of T. Moreover, we assume that $\mathcal{T}_h$ satisfies the following assumption.

Assumption 1 (H0) *There exist constants C_1 and C_2 such that*

$$C_1 \leq \sup_j \frac{h^2}{|T_j|} \leq C_2.$$

A triangle T' is a neighbor of T if T' and T have a common edge.

We define the following spaces

$$V^h = \{v_h \in C^0(R^2); v_{h|T} \text{ linear } , \forall\, T \in \mathcal{T}_h\}$$

$$X^h = \{v_h; v_{h|C} \text{ constant } , \forall\, C \in \mathcal{C}_h\}.$$

The notation $\pi_h v$ represents the P_1–Lagrange interpolation of a continuous function. Let $L_h : V^h \to X^h$ the mass lumping operator defined by $L_h(v) = \sum_i v(x_i)\chi_{ih}$ (χ_{ih} is the characteristic function of C_i). We denote by Λ_i the P^1 basis function associated to note x_i, and in the following, $\mathbf{n}_i^T = 2|T|\nabla \Lambda_{i|T}$, that is the inward normal to T opposite to the vertex x_i.

We consider a distributive scheme that writes

$$\begin{aligned} u_i^{n+1} &= u_i^n - \tfrac{\Delta t}{|C_i|} \textstyle\sum_{T, x_i \in T} \phi_i^T(u^n), \\ u_i^0 &= u(x_i, 0) \end{aligned} \tag{2}$$

where $u^n : R^2 \to R^m$ is the numerical approximation at time t_n on the vertices of $\mathcal{T}_h$, which components are $u_i^n \in X_h$. The residual ϕ_i^T and the numerical approximation satisfies the following assumptions

Assumption 2 (H1) *Let $\mathcal{T}_h$ be a triangulation that satisfies (H0). For any $C \in R$, there exists $C' \in R$ (that depends only on $\mathcal{T}_h$) such that $\forall v \in X^h$, $||v||_{L^\infty(R^2)} \leq C$ we have*

$$\forall T, \forall i, x_i \in T, ||\phi_i^T|| \leq C_2\, h \sum_{x_j \in T} ||v_j - v_i|| \,. \tag{3}$$

Remark 1.1 : This assumption is in fact a continuity assumption of the residuals in term of the local values of v. We ask that when v is uniform, $\phi_i^T = 0$. Our results extend easily when ϕ_i^T satisfies (H1) and if the number of variables entering into the definition of ϕ_i^T is bounded independantly of h and T. In practice, this is always true if the triangulation is regular (assumption (H0)). •

Assumption 3 (H2) *There exists an approximation F^h of the flux F such that*

(i) $\forall v \in X^h$, $\sum_{i=1}^{3} \phi_i^T(v) = \int_T \; div\, F^h(v(x))dx$,

(ii) $\forall v \in X^h$, $\forall T_1, T_2$ *neighboring triangles,*

$$F^h(v)_{|T_1} \cdot \mathbf{n} = F^h(v)_{|T_2} \cdot \mathbf{n} \;\; p.p. \;\; on \;\; T_1 \cap T_2$$

where $\mathbf{n}$ *is a normal to* $T_1 \cap T_2$.

(iii) *For any constant* C, *there exists* C' *such that* $\forall v \in X^h$; $||v||_{L^\infty(R^2)} \leq C$, *we have for* $T \in \mathcal{T}_h$ *and* $F_T^h = F_{|T}^h$, $|| \; div\, F_T^h(v)|| \leq \frac{C'}{h} \sum_{i,j=1,3} ||v_j - v_i||$ *a.e. on* T.

(iv) *For any sequence* $(v_h)_h$ *such that* (v_h) *is bounded in* $L^\infty(R^2 \times R^+)$ *independantly of* h, *one has*

$$\lim_h ||F^h(v_h) - F(v_h)||_{L^1_{loc}(R^2 \times R^+)^m} = 0.$$

Remark 1.2 : The property (H1) implies H2-(*ii*) when div F_T^h is a constant on $T \in \mathcal{T}_h$ and if the mesh is regular. •

We give later example of schemes that satisfy these assumptions. The main result is

Theorem 1.1 *Let be* $u_0 \in L^\infty(R^2)^m$, *and* u_h *the approximation given by (2). Assume the scheme satisfies assumptions (H1) and (H2). Assume there exists* C *that depend only on* C_1, C_2 *and* u_0 *and a function* $u \in L^2$ *such that*

$$\sup_h \sup_{x,y,t} |u_h(x,y,t)| \leq C$$

$$\lim_h ||u - u_h||_{L^2_{loc}(R^2 \times R^+)^m} = 0$$

Then u *is a weak solution of (1).*

2 Proof of Theorem 1.1

We start with a lemma inspired from [4].

Lemma 2.1 *Let* $T > 0$ *and* N *the integer part of* $\frac{T}{\Delta t}$ *Let* $\mathcal{Q}$ *be any bounded open subset of* R^2. *Let* $(v_h)_h$ *be a sequence such that* $v_h(\;.\;, t_n) \in X_h$*for any* $n \leq N$. *Assume there exists a constant* C *independant of* h *and* $v \in L^2_{loc}(\mathcal{Q} \times [0,T])$ *such that*

$$\sup_h \sup_{x,y,t} |v_h(x,y,t)| \leq C, \quad \lim_h ||v_h - v||_{L^2(\mathcal{Q}\times[0,T])} = 0.$$

Then

1. $\lim_h \sum_{n=0}^N \Delta t \sum_{T\subset \mathcal{Q}} |T| \sum_{x_i,x_j\in T} ||v_i - v_j|| = 0.$
2. $\tilde{v}_h^n = \pi_h v_h^n$ *satisfies* $\lim_h h\ ||\nabla \tilde{v}_h||_{L^2(\mathcal{Q}\times[0,T])} = 0.$

Proof : We prove the result for real valued function. The extension to vector valued functions is obvious.

First claim : For any triangle T, we define

$$v^h_{|T} = \sum_{x_i\in T} v_i^n \chi_{C_i\cap T}$$

$$w^h_{|T} = \sum_{x_i\in T} v_{\sigma(i)}^n \chi_{C_i\cap T}.$$

where σ is a cycle of $\{1,2,3\}$. From this, we define the following two functions v and w on $R^2 \times R^+$. They are bounded independantly of h. Moreover, we have

$$|T|\sum_{i\neq j} |v_i^n - v_j^n| = 3\int_T |v^h - w^h|\, dx.$$

The "3" comes from the definition of the dual cells.

We have

$$\begin{aligned}\sum_{n=0}^N \Delta t \sum_{T\subset\mathcal{Q}} |T| \sum_{x_i,x_j\in T} |v_i - v_j| &= \int_0^{t_N} \int_{\cup T\subset\mathcal{Q}} |v^h - w^h|\, dxdt \\ &\leq \int_0^T \int_{\cup T\subset\mathcal{Q}} |v^h - w^h|\, dxdt.\end{aligned}$$

The sequence (v^h) is bounded : there exists $v' \in L^\infty(\mathcal{Q}\times[0,T])$ such that $v^h \to v'$ for the weak star topology. Similarly, there exists $w \in L^\infty(\mathcal{Q}\times[0,T])$, $w^h \to w$ for the weak star topology. Since $v_h \to v$ in L^2_{loc}, $v' = v$ because $\mathcal{Q}\times[0,T]$ is bounded and $C_0^\infty(\mathcal{Q}\times[0,T])$ is dense in $L^1(\mathcal{Q}\times[0,T])$.

We show that $w = v$. Let $\phi \in C_0^\infty(R^2\times R^+)$ We have

$$\begin{aligned}\int_0^T\int_{\mathcal{Q}} v_h\phi dxdt &= \int_0^T \sum_{T,T\cap\mathcal{Q}\neq\emptyset} \sum_{x_i\in T} v_i^h \int_{C_i\cap T}\phi(x,t)dxdt \\ &= \sum_{x_i\in T} w_i^h \int_{C_{\sigma^{-1}(i)}\cap T}\phi(x,t)dxdt \\ &= \int_0^T\int_{\mathcal{Q}} w_h\phi dxdt \\ &+ \sum_{x_i\in T} w_i^h\left(\int_{C_{\sigma^{-1}(i)}\cap T}\phi(x,t)dxdt - \int_{C_i\cap T}\phi(x,t)dxdt\right)\end{aligned}$$

Since

$$\int_{C_i\cap T}\phi dxdt = |C_i\cap T|\phi(x^i), \int_{C_{\sigma^{-1}(i)}\cap T}\phi dxdt = |C_i\cap T|\phi(x'^i)$$

for $x^i \in C_i\cap T$ et $x'^i \in C_{\sigma^{-1}(i)\cap T}$ well chosen. Since $\nabla\phi$ and v^h are bounded on $\mathcal{Q}\times[0,T]$, since $|C_i\cap T| = |C_{\sigma^{-1}(i)}\cap T|$ (see remark 2), ICI we have

$$\left|\int_0^T\int_{\mathcal{Q}} v_h\phi dxdt - \int_0^T\int_{\mathcal{Q}} w_h\phi dxdt\right| \leq Ch$$

where C is independant of h. Hence, $v = v'$.

By the same method, we see that (v_h^2) and (w_h^2) have the same weak star limit. Let us show it is v^2. By density of $C_0^\infty(\mathcal{Q}\times[0,T])$ in $L^1(\mathcal{Q}\times[0,T])$ and because v_h^2 is bounded independantly of, we can take test functions ϕ that are in $C_0^\infty(\mathcal{Q}\times[0,T])$. ϕ is bounded in $\mathcal{Q}\times[0,T]$, so

$$\int_{\mathcal{Q}\times[0,T]} |v - v_h|^2 \phi dxdt \to 0,$$

and hence

$$\int_{\mathcal{Q}\times[0,T]} v^2 \phi dxdt - 2\int_{\mathcal{Q}\times[0,T]} v v_h \phi dxdt + \int_{\mathcal{Q}\times[0,T]} v_h^2 \phi dxdt \to 0.$$

By the Cauchy Schwarz inequality, $v\phi \in L^1(\mathcal{Q}\times[0,T])$ and the second term converges to

$$\int_{\mathcal{Q}\times[0,T]} v^2 \phi dxdt.$$

So we have shown that $v_h^2 \to v^2$ in L^∞ weak star.

Last, using the same argument for $\phi = 1$, because $w_h^2 \to v^2$ in L^∞ weak star, we deduce that

$$\int_{\mathcal{Q}\times[0,T]} |w_h - v|^2 dxdt \to 0.$$

last, we get

$$\int_{\mathcal{Q}\times[0,T]} |w_h - v_h|^2 dxdt \to 0.$$

The conclusion comes from the fact that $\mathcal{Q}\times[0,T]$ is bounded so that $L^1(\mathcal{Q}\times[0,T]) \subset L^2(\mathcal{Q}\times[0,T])$

Second claim : We write

$$\nabla \tilde{v}_h^{|T} = \frac{1}{2|T|}\left(u_i \mathbf{n}_i^T + u_j \mathbf{n}_j^T + u_k \mathbf{n}_k^T\right) = \frac{1}{2|T|}\left((u_j - u_i)\mathbf{n}_j^T + (u_k - u_i)\mathbf{n}_k^T\right).$$

Then, we have

$$\begin{aligned}\Delta t \textstyle\sum_0^N \int_{\mathcal{Q}} ||\nabla \tilde{v}_h^n||^2 dxdt &= \Delta t \textstyle\sum_0^N \sum_{T, T\subset\mathcal{Q}} |T| \, \left|\left|(\nabla \tilde{v}_h^n)_{|T}\right|\right|^2 \\ &\leq \Delta t \textstyle\sum_0^N \sum_{T, T\subset\mathcal{Q}} \sum_{x_i, x_j \in T} \left|v_i^n - v_j^n\right|^2\end{aligned}$$

because the triangulation is regular (assumption H0). The same arguments as before enable to conclude. •

Remark 2.2 : One can also consider distributive schemes on meshes made of non triangular elements. In this case, all what we have said before is still true provided (a) that these element have a bounded number of edges, the bound beeing independant of h and (b) the following condition on the area of neighbor elements is true for any Ω bounded

$$\sum_{T_i \cap \Omega \neq \emptyset} \sum_{T_{i_l} \text{ neigbor of } T_i} | \, |T_i| - |T_{i_l}| \, | \to 0$$

when $h \to 0$. •

Lemma 2.2 *Let $\varphi \in C_0^1(R^2 \times R^+)$. With the same assumption as in the Theorem, we have*

$$\sum_i \Delta t \sum_i |C_i|(u_i^{n+1} - u_i^n)\varphi(x_i, t_n) + \int_{R^2 \times R^+} u\partial_t \varphi dx dt + \int_{R^2} u_0(x)\varphi(x, 0) \to 0$$

when $h \to 0$.

The proof is classical.

Lemma 2.3 *Let $\varphi \in C_0^1(R^2 \times R^+)$, $F \in (C^1(R^m))^{2m}$. We assume that F^h satisfies the assumption H2 (iii), and thate u_h satisfies those of theorem 1.1. Then, if*

$$\tilde{\varphi}^{n,T} = \frac{\varphi_{i_1}^n + \varphi_{i_2}^n + \varphi_{i_3}^n}{3},$$

(x_{i_1}, x_{i_2}, x_{i_3} are the vertices of T),

$$\Delta t \sum_{n,T} \tilde{\varphi}^{n,T} \sum_{x_i \in T} \phi_i^T(u_h^n) + \int_{R^2 \times R^+} F(u(x,t)) \;\; div \; \varphi(x,t) dx dt \to 0$$

when $h \to 0$.

Proof : Let $\varphi \in C_0^1(R^2 \times [0, +\infty[)$, Ω and T such that supp $(\varphi) \subset \Omega \times [0, T]$. We consider $T \in \mathcal{T}_h$ and

$$\tilde{\varphi}^{n,T} = \frac{\varphi_{i_1}^n + \varphi_{i_2}^n + \varphi_{i_3}^n}{3} = \pi_h \varphi(x_G, t_n)$$

where x_{i_1}, x_{i_2}, x_{i_3} are the vertices of T and x_G is its centroid. We have

$$\Delta t \textstyle\sum_{n,T} \tilde{\varphi}^{n,T} \sum_{i=1}^3 \phi_i^T(u_h^n) = \sum_{n,T} \int_{t^n}^{t^{n+1}} \int_T \tilde{\varphi}^{n,T} \;\; \text{div} \; F_T^h(u_h^n) dx dt$$

$$= \textstyle\sum_{n,T} \int_{t^n}^{t^{n+1}} \int_T \tilde{\varphi}^{n,T} \;\; \text{div} \; F_T^h(u_h^n) dx dt$$

$$+ \textstyle\sum_{n,T} \int_{t^n}^{t^{n+1}} \int_T (\pi_h \varphi(x_G, t^n) - \pi_h \varphi(x, t^n)) \;\; \text{div} \; F_T^h(u_h^n) dx dt.$$

From H2 *(iii)*, div $F_T^h(u_h^n)$ is bounded. The divergence formula can be applied,

$$\sum_T \int_T \pi_h\varphi(x,t^n) \ \text{div}\ F_T^h(u_h^n)dxdt = -\sum_T \int_T \nabla(\pi_h\varphi)\ .\ F_T^h(u_h^n)dx$$

$$+\sum_T \int_{\partial T} \pi_h\varphi(x,t^n) F_T^h(u_h^n)\ .\mathbf{n}dx\ .$$

Since $\pi_h\varphi$ is continuous and from H2 *(iii)*, the second term on the right hand side vanishes. Since φ is C^1 and the triangulation is regular, $\nabla(\pi_h\varphi)$ is uniformly bounded by C, independant of h. Thus we have

$$\Delta t\left|\sum_{n,T}\int_{t^n}^{t^{n+1}}\int_T \nabla(\pi_h\varphi)\ .\ F_T^h(u_h^n)dx - \sum_{n,T}\int_{t^n}^{t^{n+1}}\int_T \nabla\varphi\ .\ F(u)dx\right| \le$$

$$C\Delta t \sum_{n,T}\int_{T\times[t^n,t^{n+1}]} \left|\left|F^h(u_h)-F(u_h)\right|\right| dxdt$$

$$+\Delta t \sum_{n,T}\int_{T\times[t^n,t^{n+1}]} ||\nabla\pi_h\pi - \nabla\varphi||\ ||F(u_h)||\, dxdt$$

The first sum is less than $\left|\left|F^h(u_h)-F(u_h)\right|\right|_{L^1(\Omega\times[0,T])}$: it converges to 0 because $||u_h||_\infty$ is bounded uniformly in h. The second sum also because $L^1_{loc} \subset L^2_{loc}$.

Moreover, since u_h stays bounded and F is continuous, $F(u_h)$ stays bounded by a constant C. The second term of the righthand side sum is bounded by the L^1 norm of $\nabla\pi_h\varphi - \nabla\varphi$ which converges to 0 if the triangulation is uniform.

Last, by H2 *(ii)*, we have

$$\Delta t\sum_{n.T}\int_T |\pi_h\varphi^n(x_G)-\pi_h\varphi^n(x)|\ \left|\ \text{div}\ \ F_T^h(u_h^n(x))\right| dx \le$$

$$C\Delta t\sum_{n.T}\int_T \left|\frac{\pi_h\varphi^n(x_G)-\pi_h\varphi^n(x)}{h}\right|\left(\sum_{i,j=1,3}\left|u_j^n-u_i^n\right|\right)dx.$$

Then

$$\int_T\left|\frac{\pi_h\varphi^n(x_G)-\pi_h\varphi^n(x)}{h}\right|dx = \int_T\left|\nabla\left(\pi_h\varphi^n\right)\cdot\frac{x_G-x}{h}\right|dx \le Ch^2,$$

where C does not depend on h. We conclude by lemma 2.1.

Remark 2.4 : If div F_T^h is constant on T, then

$$\left|\left|\Delta t\sum_{n.T}\int_T (\pi_h\varphi^n(x_G)-\pi_h\varphi^n(x))\quad \text{div}\ \ F_T^h(u_h^n(x))dx\right|\right|$$

$$\le \Delta t\sum_{n,T}||\ \text{div}\ \ F_T^h(u_h^n)||\left|\int_T (\pi_h\varphi^n(x_G)-\pi_h\varphi^n(x))\,dx\right|$$

$$= \Delta t\sum_{n,T}||\ \text{div}\ \ F_T^h(u_h^n)||\left|\nabla_x\left(\pi_h\varphi^n\right)_{|T}\cdot\int_T (x_G-x)\,dx\right|$$

$$=0.$$

•

Proof of Theorem 1.1: We multiply (2) by $\varphi(x_i,t^n)|C_i|$, where φ is a test function in $C_0^1\left(R^2\times[0,+\infty[\right)$, such that supp $(\varphi)\subset\Omega\times[0,T]$ for Ω and T well chosen ($n\Delta t\le T$). The one sums on the indices $n=0,N$ and $i\in J$ such that $x_i\in\Omega$:

$$\sum_{i\in J}\sum_{n=0,N}|C_i|\left(u_i^{n+1}-u_i^n\right)\varphi(x_i,t^n)+\Delta t\sum_{i\in J}\sum_{n=0,N}\sum_{T;x_i\in T}\phi_i^T(u_h^n)\varphi(x_i,t^n)=0,$$

From lemma 2.2, we have

$$\sum_{i\in J}\sum_{n=0,N}|C_i|\left(u_i^{n+1}-u_i^n\right)\varphi(x_i,t^n)=-\int_{R^2\times R^+}u\partial_t\varphi-\int_{R^2}u_0\varphi(.,0)+o(1).$$

Then,

$$\begin{aligned}\sum_{i\in J}\sum_{T;x_i\in T}\phi_i^{T,n}\varphi_i^n&=\sum_{T\subset\Omega}\sum_{i=1,3}\phi_i^{T,n}\varphi_i^n\\&=\sum_{T\subset\Omega}\left(\tilde{\varphi}^{T,n}\sum_{i=1,3}\phi_i^{T,n}+\frac{1}{3}\sum_{i=1,3}\phi_i^{T,n}\left((\varphi_i^n-\varphi_j^n)+(\varphi_i^n-\varphi_k^n)\right)\right.\end{aligned}$$

where $\tilde{\varphi}^{T,n}=\frac{\varphi_1^n+\varphi_2^n+\varphi_3^n}{3}$. Lemma 2.3 gives

$$\Delta t\sum_{n=0,N}\sum_{T\subset\Omega}\tilde{\varphi}^{T,n}\sum_{i=1,3}\phi_i^{T,n}=-\int_{R^2\times R^+}F(u(x,t))\quad\text{div}\quad\varphi(x,t)dxdt+o(1).$$

Last, by (H1) and lemma 2.1:

$$\begin{aligned}\Delta t\left|\sum_{n,T}\phi_i^T(u_h^n)\left((\varphi_i^n-\varphi_j^n)+(\varphi_i^n-\varphi_k^n)\right)\right|&\le Ch\Delta t\sum_{n=0,N}\sum_{T\subset\Omega}|\phi_i^T(u_h^n)|\\&\le Ch^2\Delta t\sum_{n=0,N}\sum_{T\subset\Omega}\sum_{i,j=1,3}|u_j^n-u_i^n|-\end{aligned}$$

which ends the proof. •

3 Some examples

3.1 The finite volume schemes

Let us consider a finite volume scheme, with control volume the dual cells. The numerical flux is $\mathcal{F}$.

$$u_i^{n+1} = u_i^n - \frac{\delta t}{|C_i|} \sum_{j \text{ neighbor of } i} \mathcal{F}_{\mathbf{n}_{ij}}(u_i^n, u_j^n) \tag{4}$$

where $\mathbf{n}_{ij}$ is the normal vector between two cells C_i and C_j. Since $\sum_{j \text{ neigbor of } i} \mathbf{n}_{ij} = 0$, (4) can be rewritten as

$$\sum_{j \text{ neighbor of } i} \mathcal{F}_{\mathbf{n}_{ij}}(u_i^n, u_j^n) = \sum_{j \text{ neighbor of } i} \left(F_{\mathbf{n}_{ij}}(u_i^n, u_j^n) - F_{n_{ij}}(u_i^n)\right)$$

If we set

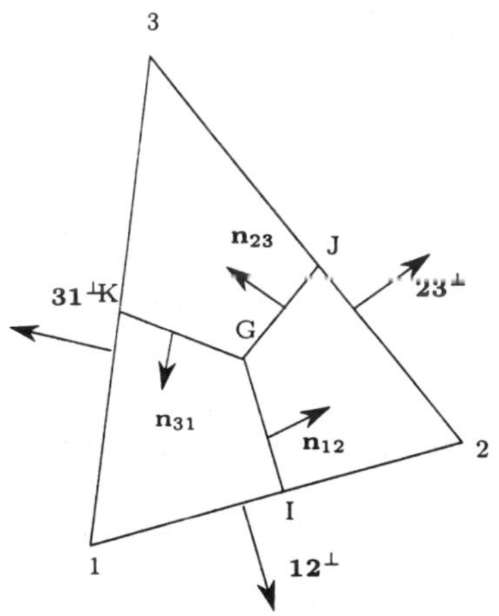

Fig. 1. Dual cell, normal vectors. The triangle is oriented. The nodes I, J, K are the midpoints of the edges [1, 2], [2, 3], [1, 3]

$$\phi_i^T = \mathcal{F}_{\mathbf{n}_{ij}}(u_i^n, u_j^n) + \mathcal{F}_{\mathbf{n}_{ik}}(u_i^n, u_k^n) - \frac{1}{2}\left(F_{\mathbf{n}_{ij}}(u_i^n) - F_{\mathbf{n}_{ik}}(u_i^n)\right),$$

since $\mathcal{F}_{\mathbf{n}_{ij}}(u_i^n, u_j^n) = -\mathcal{F}_{\mathbf{n}_{ji}}(u_j^n, u_i^n)$, we get

$$\sum_{i=1,3} \phi_i^T = \frac{1}{2}\sum_{i=1}^{3} F_{\mathbf{ij}^\perp}(u_i^n).$$

The approximation F^h is the $P1$ interpolation of F evaluated at the vertices of $\mathcal{T}_h$. If the flux is Lipschitz continuous, the assumption (H1) is true. The assumptions (H2)-*(i)* and (H2)-*(ii)* are true by construction.

Check of assumption (H2)-(iii) For $v \in X^h$, we write $||v||_{L^\infty(R^2)} \leq C$, and $F \in \left(C^1(R^m)\right)^{2m}$:

$$\begin{aligned} || \text{ div } F_T^h(v)|| &= \tfrac{1}{2|T|} \, ||(F(v_i) - F(v_k)) \cdot \mathbf{n}_i^T + (F(v_j) - F(v_k)) \cdot \mathbf{n}_j^T|| \\ &\leq \tfrac{C}{h} \sum_{i,j=1,3} ||v_i - v_j||, \end{aligned}$$

if the mesh satisfies (H0).

Check of assumption (H2)-(iv). Let $F^h(u) = (f^h(u), g^h(u))$. We show that for any sequence $(v_h)_h \in X^h$ that satisfies

$$\sup_h \sup_{x,y,t} |u_h(x,y,t)| \leq C \tag{5}$$

$$||u - u_h||_{(L^2_{loc}(R^2 \times R^+))^m} \, , \longrightarrow 0. \tag{6}$$

we have

$$||f^h(v_h) - f(v_h)||_{(L^1_{loc}(R^2 \times R^+))^m} \longrightarrow 0.$$

To show this, we consider the mass lumping operator $L_h : V^h \to X^h$. Let $\mathcal{Q}$ be an open bounded subset of $R^2 \times R^+$. Let $v_h^n \in X^h$ and $\tilde{v}_h^n = \pi_h v_h^n$, then $v_h^n = L_h \tilde{v}_h^n$, and we get

$$\begin{aligned} \int_{\mathcal{Q}} |f^h(v_h) - f(v_h)| \leq & \int_{\mathcal{Q}} |f^h(L_h \tilde{v}_h) - f^h(\tilde{v}_h)| \\ & + \int_{\mathcal{Q}} |f^h(\tilde{v}_h) - f(\tilde{v}_h)| + \int_{\mathcal{Q}} |f(\tilde{v}_h) - f(L_h \tilde{v}_h)| \, . \end{aligned}$$

The first term on the right vanishes. If $x \in T$, triangle which vertices are x_i, $i = 1, 3$, we get :

$$\pi_h \left[f(\tilde{v}_h) \right](x) = f\left(\tilde{v}_h(\mathbf{x})\right) + \sum_{i=1,3} \frac{df}{du} \left(\tilde{v}_h(\theta_i x + (1-\theta_i)x_i)\right) \nabla \tilde{v}_h \cdot (x_i - x) \, \phi_i(x),$$

with $\theta_i \in [0,1]$.

With any component $f_i \in C^1(R^m)$ of f, and from lemma 2.1, we have :

$$\left(\int_{\mathcal{Q}} \left| f_i^h(\tilde{v}_h) - f_i(\tilde{v}_h) \right|^2 \right)^{1/2} \leq Ch \, |\nabla \tilde{v}_h|_{0,\mathcal{Q}} \longrightarrow 0.$$

We also have the following error for the second term.

$$|\tilde{v}_h^n - L_h \tilde{v}_h^n|^2_{L^2(\Omega)} \leq \sum_{C_i \cap \Omega \neq \{\emptyset\}} h^2 |\nabla \tilde{v}_h^n|^2_{L^2(C_i)}.$$

In fact, for $x \in C_i$, we have $\tilde{v}_h^n(x) = L_h \tilde{v}_h^n(x) + \nabla \tilde{v}_h^n(\theta_i x + (1-\theta_i)x_i) \cdot (x - x_i)$, $\theta_i \in [0,1]$. The one integrates on the cells and sum up. Hence, from lemma 2.1 :

$$|\tilde{v}_h - L_h \tilde{v}_h|_{L^2(Q)} \longrightarrow 0.$$

Then we rewrite :

$$\int_Q |f(\tilde{v}_h) - f(L_h\tilde{v}_h)|^2 \, dx \leq \int_Q |f(\tilde{v}_h) - f(v)|^2 + \int_Q |f(v) - f(v_h)|^2 \, dx,$$

where v is the limit L^2 of v_h. The

$$|\tilde{v}_h - v|_{L^2} \leq |\tilde{v}_h - L_h\tilde{v}_h|_{L^2} + |v_h - v|_{L^2} \longrightarrow 0.$$

One can apply the dominated convergence theorem to get a subsequence $\tilde{v}_{h_k}$, $\int_Q |f(\tilde{v}_{h_k}) - f(v)|^2 \longrightarrow 0$. This ends the proof.

3.2 The system N scheme

This scheme has been described in [9]. It is first order and upwind, that is, given a triangle T, if all the eigenvalues of K_i are negative, then the associated residual vanishes. It is a generalisation of Roe's N scheme [3]. B. Perthame [7] has shown the convergence of the scalar N scheme in L^2.

We start from an hyperbolic system

$$\partial_t u + \partial_x f(W) + \partial_y g(W) = 0$$

which is linearised

$$\partial_t u + \bar{A}\partial_x u + \bar{B}\partial_y u = 0$$

The matrices $\bar{A}$ and $\bar{B}$ are chosen so that the linearised system is hyperbolic and so that

$$\int_{\partial T} (f,g)^T(u^h) \, . \, \mathbf{n}dl = \bar{A}\int_T \partial_x u^h dx + \bar{B}\int_T \partial_y u^h dx$$

where u^h is an interpolation of u that is described in details later.

The system N–scheme writes :

$$\phi_i^T(u) = -\sum_{j=1,3} K_i^+ N K_j^-(u_j - u_i).$$

The matrices K_i are $K_i = (\bar{A}, \bar{B})^T \, . \, \mathbf{n}_i$ where $\mathbf{n}_i$ is the inward normal vector of the edge facing the vertex $i \in \{1,2,3\}$ in T. It is the gradient of the P_1 basis function associated to this node, multiplied by $2|T|$. Since the linearised system is hyperbolique, K_i is diagonalisable in R. The matrices K_i^+ and K_i^- are the positive and negative part of K_i. The matrix N is

$$N = \left(K_1^- + K_2^- + K_3^-\right)^{-1}$$

if the sum is invertible. We show that for fluid mechanics, $K_1^- + K_2^- + K_3^-$ can be non invertible, but the N scheme is always defined. The N matrix is chosen so that

$$\sum_{i=1}^{3} \phi_i^T = \int_{\partial T} (f,g)^T(u^h) \cdot \mathbf{n} dl = \bar{A} \int_T \partial_x u^h dx + \bar{B} \int_T \partial_y u^h dx$$

In the case of fluid mechanics, one can consider the Roe-Struijs-Deconinck linearisation, by introducing the parameter vector $Z = \sqrt{\rho}(1, u, v, H)^T$. The flux and u are quadratic in Z,

$$u = \tfrac{1}{2} R(Z) Z,$$

$$F(u) = \tfrac{1}{2} S(Z) Z,$$

where $R(Z)$ and $S(Z)$ are linear in Z. Let be $Z^h = \pi_h Z$, we have

$$\int_T \text{ div } F(u(Z^h)) \, dx = \int_T \frac{\partial F}{\partial Z}_{|Z^h} \nabla Z^h \, dx = S(\overline{Z}) \nabla Z^h |T|$$

$$= S(\overline{Z}) R(\overline{Z})^{-1} R(\overline{Z}) \nabla Z^h |T| = \frac{\partial F}{\partial u}_{|\overline{Z}} \int_T \nabla u_{|Z^h} \, dx$$

$$= (\overline{A}, \overline{B}) \int_T \nabla u_{|Z^h} \, dx.$$

Then,

$$\int_T \nabla u(Z^h) dx = \frac{1}{2} R(\bar{Z}) \cdot \sum_{i=1,3} Z_i \mathbf{n}_i^T,$$

because Z is linear and u quadratic in Z. This gives

$$\phi^T = \sum_{i=1,3} \phi_i^T(u(Z^h)) = \sum_{i=1,3} K_i(\bar{Z}) Z_i,$$

avec

$$K_i = \frac{1}{2} S(\bar{Z}) \cdot \mathbf{n}_i^T.$$

We have $F^h(u) = F(u(\pi_h Z))$, this enable to get (H2) since $\pi_h Z$ is continuous on the edges. The residuals satisfy (H1)–(H2) because $||\mathbf{n}_i||$ is bounded by a term of the form Ch (if (H0) is true) and because the Jacobians $\bar{A}$ and $\bar{B}$ are continuous in $\pi_h Z$. Last, in [1], it is shown that this linearisation garanties the hyperbolicity of the linearised system.

3.3 The Low Diffusion Advection Scheme (LDA)

This scheme has been described in [9]. It is a generalisation of the scalar LDA scheme. The residual is

$$\phi_i^T = -K_i N \phi^T$$

where as for the N scheme, $N = (K_1^- + K_2^- + K_3^-)^{-1}$. The N matrix cannot be invertible, but the scheme is always defined as for the N scheme, for the same reasons.

The scheme is second order at steady state, and upwind.

3.4 The centered entropy conservative scheme

Here, we assume that (9) is symmetrisable. Here E is an entropy and G is the associated entropy flux. Using the entropy variables $v = \nabla E(u)^T$, it writes :

$$(\phi_i^S)^T(u) = \int_T \Lambda_i(x) \ \text{div } H(\pi_h v(x))dx.$$

where Λ_i is the P1–Lagrange basis function at node x_i and H is the flux written in term of the entropy variables $F(u) = H(v)$. We have :

$$\sum_{i=1,3} (\phi_i^S)^T(u) = \int_T \ \text{div } H(\pi_h u(x))dx,$$

which gives

$$F^h(u) = H(\pi_h v).$$

The properties of F^h can be proven with the same arguments as in §3.1.

The semi discrete scheme

$$|C_i|\frac{du_i}{dt} = - \sum_{T, x_i \in T} \phi_i^T$$

satisfies an entropy inequality. In fact, by left multiplication with v_i^T, we have

$$|C_i| v_i^T \cdot \frac{du_i}{dt} = - \sum_{T, x_i \in T} v_i^T \cdot (\phi_i^S)^T.$$

It is enough to show that $\Psi_i^T = v_i^T \cdot (\phi_i^S)^T$ is a residual, i.e. there exists an approximation of the entropy flux such that

$$\sum_{x_i \in T} \Psi_i^T = \int_T \ \text{div } G^h(v)dx.$$

But

$$\begin{aligned} \textstyle\sum_{x_i \in T} \Psi_i^T &= \textstyle\sum_{x_i \in T} \int_T v_i^T \ . \ \Lambda_i(x) \ \text{div } H(\pi_h v)dx \\ &= \textstyle\int_T \pi_h v^T \cdot \ \text{div } H(\pi_h v)dx = \int_T \ \text{div } G(\pi_h v)dx \end{aligned}$$

so that we have this property with $G^h = G(\pi_h v)$ (G is the entropy flux associated to F).

4 Consistancy with an entropy inequality

Let E be an entropy associated to G, an entropy flux. We say that the residual Φ_i^T satisfies an entropy inequality if there exists residuals Ψ_i^T and an approximation G^h of G which satisfies assumptions (H1), (H2) and the condition

$$\forall u \in X, E_u^T \cdot \sum_{i=1}^{3} \Phi_i^T \geq \sum_{i=1}^{3} \Psi_i^T. \tag{7}$$

We have the following result for the semi–discrete scheme

$$|C_i| \frac{du}{dt} + \sum_{T, x_i \in T} \Phi_i^T = 0$$

Theorem 4.1 *Under the assumptions of theorem 1.1 and the previous assumptions on the entropy residuals, the function u satisfies an entropy inequality for (E, G).*

The proof is identical to that of theorem 1.1. The assumption (H1) on Ψ_i^T can be replaced by a continuity assumption. An example is described in §3.4, at least for semi discrete schemes. We give additional results when the system is symmetrisable. The idea is to rewrite the system as

$$\partial_t u + \partial_x f(v) + \partial_y g(v) = 0$$

where $v = \nabla_u E(u)^T$.

4.1 The SUPG scheme

Let us consider the hyperbolic system

$$u_t + \text{ div H(v) } = 0$$

as above, v is the entropy variable associated to the entropy E. The matrix A_0 is the hessian of E with respect to the u variables

The SUPG scheme (semi-discrete version) read as follows. Let v^h be the P^1 interpolation of v, and φ^h any P^1 test function. We

$$\int_\Omega (\varphi^h \cdot \bar{A}_0 (v_h)_t + W_h^T \cdot \bar{A} \nabla v^h)\, dx = \sum_{i=1}^{n_t} \int_T (\tau \bar{A} \nabla \varphi^h) \cdot (\bar{A} \nabla v_h)\, dx + \text{ C.L.} \tag{8}$$

In (8), the matrix τ is symetric positive definite. We refer to [5] for more details. By omotting the boundary condition terms, (8) can be rewritten as

$$\mathcal{A}(v_h)_t + \Phi = 0$$

with $\Phi = (\phi_i)_{i=1,ns}$ et $\phi_i = \sum_{T, i \in T} \phi_i^T$ et

$$\phi_i^T = \int_T \Lambda_i \bar{A} \nabla v_h + \int_T (\tau_1 \bar{A} \nabla \Lambda_i) \cdot (\bar{A} \nabla V_h) dx.$$

Since the sum of the basis functions Λ_i is 1, the sum of the residuals on a triangle reduces to

$$\int_T \bar{A} \nabla v_h dx.$$

More over, by construction, we have

$$\sum_i v_h^T \cdot \phi_i^T \geq \sum_i v_h^T \cdot (\phi_i^S)^T$$

where $(\phi_i^S)^T$ is the residual for the entropy stable scheme defined in §3.4. This scheme dissipates entropy. The choice of τ is discussed in [5].

4.2 The Lax–Friedrichs scheme

The entropy stable scheme is

$$\Phi_i^T = \int_T \Lambda_i(x) \quad \text{div } H(\pi_h v) dx \tag{9}$$

where v is interpolated with piecewise linear functions, G stands for the flux in entropy variables.

The Lax Friedrichs scheme we consider here is

$$\Phi_i^T = \frac{1}{3} \sum_{i=1}^{3} \frac{1}{2} F_{n_i}(u_i) + \alpha \bar{A}_0 (\nabla V : \mathbf{n}_i) \tag{10}$$

where, if the component of V are V_l, $\nabla V : \mathbf{n}_i$ is the vector with component $\nabla V_l \cdot \mathbf{n}_i$. The matrix $\bar{A}_0$ is positive definite, it is the Hessain matrix A_0 evaluated at some averaged state.

The residual defined by (10) is conservative is $\sum_i \mathbf{n}_i = 0$. Several choices are possible, we consider here the vectors $\mathbf{n} = \overrightarrow{x_i x_j}$ where the x_is are the vertices of T. This choice simplifies the calculations.

One can rewrite (10) as

$$\phi_i^T = \sum_j C_{ij}(v_i - v_j)$$

where

$$C_{ij} = \frac{1}{2} A_{\mathbf{n}_i} + \alpha \beta_{ij} \bar{A}_0$$

with $\beta_{ij} = \nabla \Lambda_j \,.\, \mathbf{n}_i$ and $A_{n_i}^{\text{Roe}}$ is the Roe matrix in the direction $\mathbf{n}_i$ between the states u_i and u_j. When the edges are defined as above, on has $\beta_{ij} = -1$.

Similarly,

$$C_{ij}^S = \frac{1}{2|T|} \int_T \Lambda_i(x) A_{n_i}(\pi_h v) dx dy$$

To get $C_{ij} - C_{ij}^S \leq 0$, it is suffisant to have

$$\bar{A}_{n_i} - \frac{1}{|T|}\int_0^1 A_{n_i}(\xi V_i + (1-\xi)V_j)d\xi \leq -2\alpha \bar{A}_0 \beta_{ij}$$

i.e.

$$\bar{A}_0^{-1/2}\left(\bar{A}_{n_i} - \int_0^1 A_{n_i}(\xi V_i + (1-\xi)V_j)d\xi\right)\bar{A}_0^{-1/2} \leq -\alpha\beta_{ij}$$

and then

$$\rho\left(\bar{A}_0^{-1/2}\left[\bar{A}_{n_i} - \int_0^1 A_{n_i}(\xi V_i + (1-\xi)V_j)d\xi\right]\bar{A}_0^{-1/2}\right) \leq -\beta_{ij}\alpha$$

For any symmetric matrix, we have

$$\rho\left(A^{-1/2}BA^{-1/2}\right) \leq \rho(A^{-1})\,\rho(B)$$

so that a suffisant condition is

$$-\alpha\beta_{ij} \geq \max(\rho(A_{n_i}^{\text{Roe}}), \max_{(x,y)\in\in T}\rho(A_{n_i}(\pi_h v(x,y))))\;\rho(\bar{A}_0^{-1}) \tag{11}$$

i.e., when $\beta_{ij} = -1$,

$$\alpha \geq \max(\rho(A_{n_i}^{\text{Roe}}), \max_{(x,y)\in\in T}\rho(A_{n_i}(\pi_h v(x,y))))\;\rho(\bar{A}_0^{-1}) \tag{12}$$

Now, we have to provide tractable values for the spectral radii.

1. Case of $\bar{A}_0$. In 1D, $(q = u^2)$ $A_0 = C^T A C$ where C is

$$C = \begin{bmatrix} 1 & 0\,0 \\ q & 1\,0 \\ \frac{1}{2}q^2 & q\,1 \end{bmatrix}$$

(Galilean invariance) and

$$A = \begin{bmatrix} \frac{\mathcal{R}\gamma}{\rho(\gamma-1)} & 0 & -\frac{\mathcal{R}\gamma}{c^2\rho(\gamma-1)} \\ 0 & \frac{\mathcal{R}\gamma}{c^2\rho(\gamma-1)} & 0 \\ -\frac{\mathcal{R}\gamma}{c^2\rho(\gamma-1)} & 0 & \frac{\mathcal{R}\gamma^2}{\rho(\gamma-1)c^4} \end{bmatrix}$$

The eigenvalues of A are

$$\begin{aligned} \lambda_1 &= \frac{1}{2}\frac{\left(-\gamma^2 - \gamma c^4 + \sqrt{\gamma^2 c^8 - 2\gamma^3 c^4 + \gamma^4 + 4\gamma^2 c^4}\right)\mathcal{R}}{(-\gamma c^4 + c^4)\rho} \\ \lambda_2 &= \frac{1}{2}\frac{\left(-\gamma^2 - \gamma c^4 - \sqrt{\gamma^2 c^8 - 2\gamma^3 c^4 + \gamma^4 + 4\gamma^2 c^4}\right)\mathcal{R}}{(-\gamma c^4 + c^4)\rho} \\ \lambda_3 &= \frac{\gamma\mathcal{R}}{c^2\rho(\gamma-1)} \end{aligned} \tag{13}$$

From this, we deduce that if $\bar{A}_0$ is evaluated in an averaged state,

$$\rho(\bar{A}_0^{-1}) \leq \frac{\rho(C^{-1}(C^T)^{-1})}{\min(|\lambda_1|, |\lambda_2|, |\lambda_3|)}.$$

Then,

$$\begin{aligned} \rho(C^{-1}(C^T)^{-1})) &= \max(1, 1 + q^2 + \tfrac{q^4}{8} + \tfrac{|q|(q^2+4)}{8}\sqrt{q^2+8} \\ &\quad , 1 + q^2 + \tfrac{q^4}{8} - \tfrac{|q|(q^2+4)}{8}\sqrt{q^2+8}) \\ &= 1 + q^2 + \tfrac{q^4}{8} + \tfrac{|q|(q^2+4)}{8}\sqrt{q^2+8} \end{aligned} \tag{14}$$

In the two dimensional case, the eigenvalue λ_2 is double and q is replaced by $||\mathbf{u}||$ in (14).

2. Case of $\bar{A}_{n_i}$. We denote by A and B the fluxes defined in the conservative variables.

$$\bar{A}_{n_i} = (An_{x_i} + Bn_{y_i})A_0 = A_{n_i}A_0.$$

By mutiplying on the left and the right by $A_0^{-1/2}$, the symmetric matrix $A_0^{-1/2}\bar{A}_{n_i}A_0^{-1/2}$ is similar to A_{n_i}. From this we deduce that

$$\rho(A_0^{-1/2}\bar{A}_{n_i}A_0^{-1/2}) = \max(\mathbf{u}\cdot\mathbf{n}_i, \mathbf{u}\cdot\mathbf{n}_i + c, \mathbf{u}\cdot\mathbf{n}_i - c).$$

Since $\rho(A_0^{-1/2}\bar{A}_{n_i}A_0^{-1/2}) \leq \rho(\bar{A}_{n_i})\rho(A_0^{-1})$, we get

$$\rho(\bar{A}_{n_i})\rho(A_0^{-1}) \geq \max(\mathbf{u}\cdot\mathbf{n}_i, \mathbf{u}\cdot\mathbf{n}_i + c, \mathbf{u}\cdot\mathbf{n}_i - c). \tag{15}$$

Combining the inequalities (15), (14), (13) and (12), we get the following bound on α

$$\alpha \geq \max_{i=1,2,3}\max_{(x,y)\in T}\max((\mathbf{u}\cdot\mathbf{n}_i)(x,y), (\mathbf{u}\cdot\mathbf{n}_i + c)(x,y), (\mathbf{u}\cdot\mathbf{n}_i - c)(x,y)). \tag{16}$$

Without any surprise, we come back to a version of the classical lax Fridrich schemes, written in entropy variables.

5 Numerical examples

In this section, we very briefly describe some *preliminary* results obtained by the N scheme, a classical MUSCL scheme (interpolation on the conservative variables, van Leer van Albada limiter), the LDA scheme and a blending between the LDA and the N scheme to improve the accuracy. The last scheme write

$$\Phi_i = l\Phi_i^N + (Id - l)\Phi_i^{LDA}$$

where $l := l(\Phi_1^N, \Phi_2^N, \Phi_3^N)$ is a matrix that is built following the two principles :

1. Recover second order accuracy : if $\Phi = \sum_{i=1}^{3} \Phi_i^N = 0$, then $l = 0$,
2. a minimum principle on the specific entropy.

The conservation property is automatically recovered, whateve l. Details the construction can be found in [2]. The main idea of the construction is the following : all the numerical simulations tends to prove that the system N scheme satisfies a minimum principle on the entropy, which is the discrete analogue of a property proved by Tadmor [10]

$$s(x, t + \Delta t) \geq \min_{y, ||x-y|| \leq ||u||_\infty \Delta t} s(y, t)$$

where s is the physical entropy. The limiter matrix l is constructed in order to preserve this property.

The test case presented here is on a Naca 0012, $M_\infty = 0.85$ and 1 degree of incidence. We show the pressure coefficient c_p and the reduced entropy Σ,

$$c_p = \frac{p - p_\infty}{\frac{1}{2}\rho_\infty u_\infty^2}, \quad \Sigma = \frac{s}{s_\infty} - 1.$$

All the results are coonverged (L^∞ residual smaller that 10^{-6}). These results

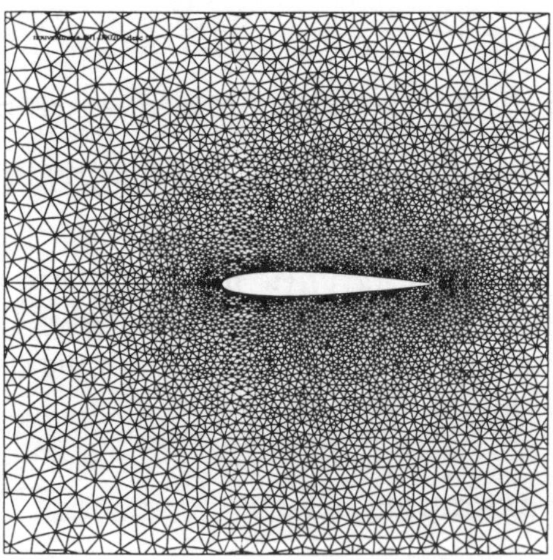

Fig. 2. Mesh

show that there is a hierarchy of solutions, from the N scheme to the blended scheme via the finite volume and the LDA schemes. First, the resolution of the shock by the finite volume scheme is rather poor, especially when it becomes weak, that is when the density of the mesh decreases in the present

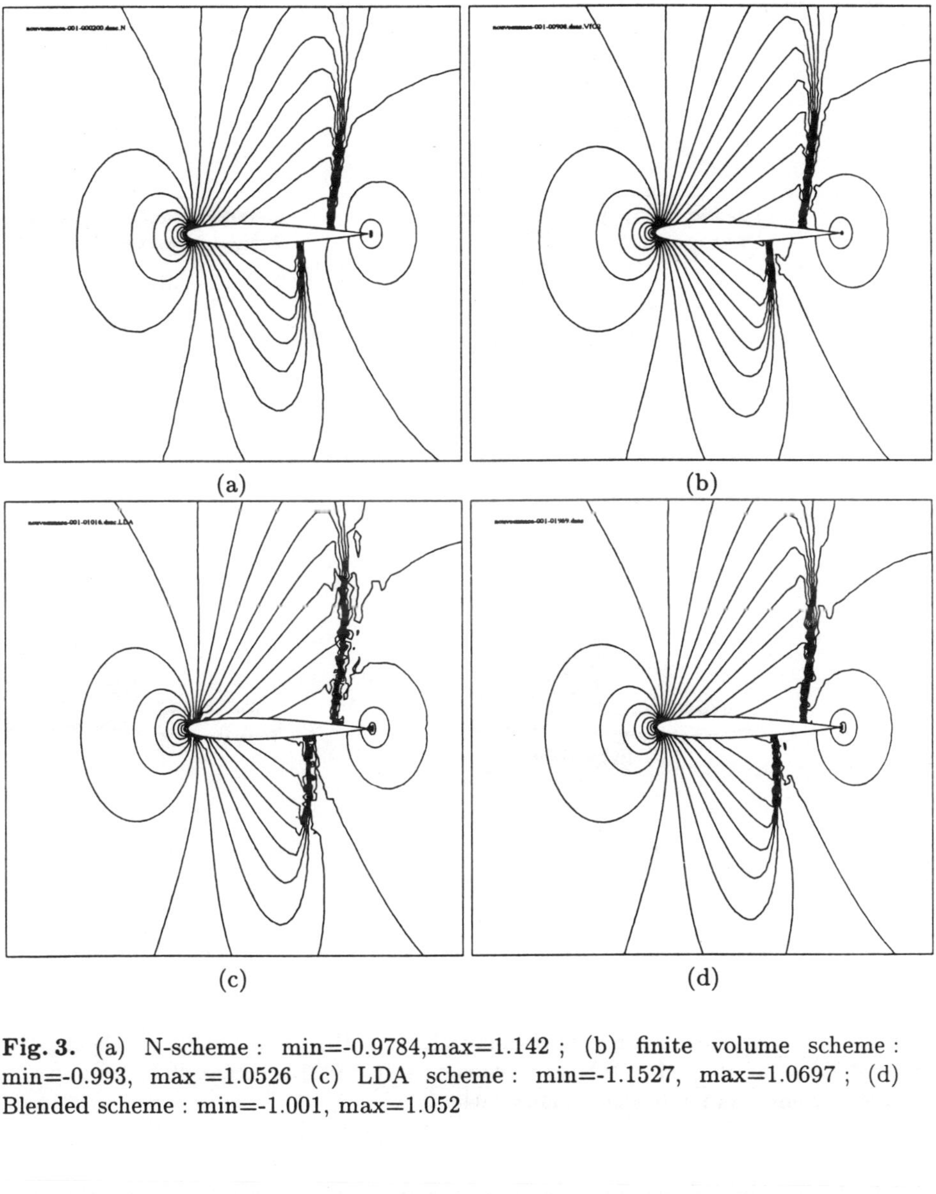

Fig. 3. (a) N-scheme : min=-0.9784,max=1.142 ; (b) finite volume scheme : min=-0.993, max =1.0526 (c) LDA scheme : min=-1.1527, max=1.0697 ; (d) Blended scheme : min=-1.001, max=1.052

cas (see upper and lower part of the shock). This is also true for the N scheme. The LDA scheme, as expected, is far from beeing monotone, see Figure 3 for example. However, the resolution of the slip line on the trailing edge is very sharp, compared to the N scheme and the finite volume one, see Figure 4 for example. This can also be seen from the Mach number, the density and velocity plots. The blended scheme is a good compromise between the LDA and N scheme. The entropy layer is diminished but Σ can become negative (but less that the finite volume scheme). There is no oscillation around the

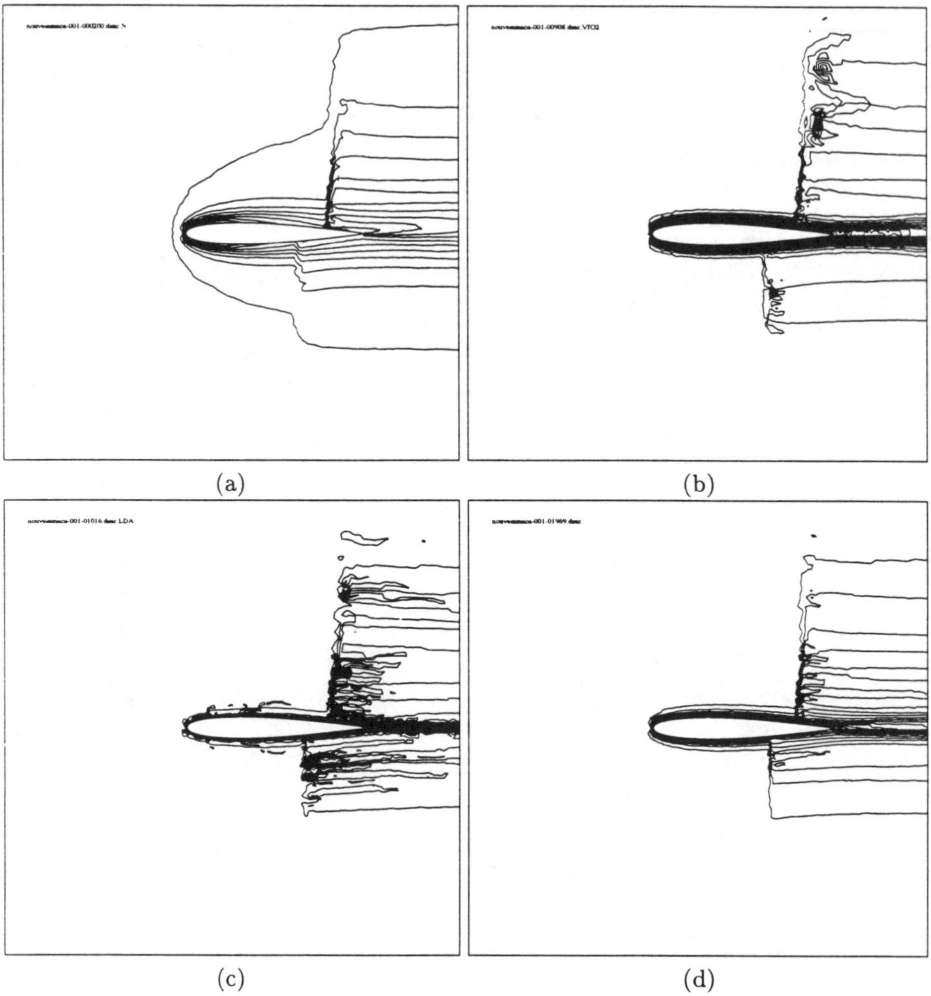

Fig. 4. (a) N-scheme : min=-0.9784,max=1.142 ; (b) finite volume scheme : min=-0.005, max=0.05414 ; (c) LDA scheme : min=-0.01239, max=0.08926 ; (d) Blended scheme : min=-0.00141, max=0.04861

shock. In [2], by adding additional constraints, it is possible to make the blended scheme satisfy an entropy inequality, as in §3.4.

6 Conclusion

We have given sufficient conditions on a distributive schemes and the numerical solution it gives that provide a Lax Wendroff like theorem. All the known distributive scheme satisfy these conditions, what is unclear are the conditions on the numerical solution. We also provide some preliminary results on a blended distributive scheme that seems to have a good on transonic and subsonic flows.

References

[1] R. Abgrall. Approximation of the multidimensional Riemann problem in Compressible Fluid Mechanics by a Roe type method. Technical Report 2343, INRIA, Septembre 1994.

[2] R. Abgrall. Construction of upwind, monotone, second order genuinely multidimensional schemes. Technical Report Mathématiques Appliquées de Bordeaux, work in progress.

[3] H. Deconinck, R. Struijs, G. Bourgeois, and P.L. Roe. Compact advection schemes on unstructured meshes. VKI Lecture Series 1993–04, Computational Fluid Dynamics, 1993.

[4] M. Geiben, D. Kröner, and M. Rokyta. A Lax-Wendroff type theorem for cell-centered finite volume schemes in 2-D. *Report No. 278, Sonderforschungsbereich 256, Rheinische Friedrich-Wilhelms-Universität, Bonn*, 1993.

[5] Th.J.R. Hughes and M. Mallet. A new finite element formulation for Computational Fluid Dynamics : IV a discontinuity–capturing operator for multidimensional advective–diffusive systems. *Computer Methods in Applied Mechanics and Engineering*, 58:329–336, 1986.

[6] P. Lax and B. Wendroff. Systems of conservation laws. *Comm. Pure Appl. Math.*, 13:381–394, 1960.

[7] B. Perthame. Convergence of n-schemes for linear advection equations. Technical Report 94036, Labôratoire d'Analyse Numérique, Paris VI, 1994.

[8] R. Struijs, H. Deconinck, and P. L. Roe. Fluctuation splitting schemes for the 2d Euler equations. *VKI LS 1991-01, Computational Fluid Dynamics*, 1991.

[9] E. van der Weide and H. Deconinck. Positive matrix distribution schemes for hyperbolic systems. In *Computational Fluid Dynamics '96*, pages 747–753. Wiley, 1996.

[10] E. Tadmor. A minimum entropy principle in the gas dynamics equations. Icase report 86–33, 1986.

A The system N scheme is well defined.

We show here that, for an hyperbolic symetrizable systeme, the system N scheme is well defined, that is the residual is always computable despites the occurence of an inverse matrix.

We consider the system

$$W_t + A_1 W_x + A_2 W_y = 0$$

where the matrices A_1 and A_2 are the Jacobian matrices of some fluxes evaluated at some average states. This system is assumed hyberbolic. For any vector $\mathbf{n} = (n_x, n_y)$, we define the matrix K_n by

$$K_n = A_1 n_x + A_2 n_y.$$

When $\mathbf{n}_i$ is indexed by i, we set K_i instead of $K_{\mathbf{n}_i}$. This matrix is diagonalizable, with real eigenvalues. The matrices $|K_n|$, K_n^+, K_n^- are the absolute values, the positive and negative part of K_n.

We consider the semi-decritized scheme

$$|C_i| \frac{dW_i}{dt} = - \sum_{i \text{ and } k \text{ neigbors}} M_{ki}(W_i - W_k) \tag{17}$$

where $|C_i|$ is the area of the dual cell at node M_i. We consider a triangle T which vertices are denoted by 1, 2, 3 to simplify notations. If the $\mathbf{n}_i$ are the inward normals to triangle T, the matrices M_{ki} are, for the system N scheme,

$$M_{ki} = -K_i^+ \left(\sum_{l=1,3} K_l^+ \right)^{-1} K_k^- \tag{18}$$

The problem is to see if, for any $x \in R^4$, the vectors $M_{ij}x$ are well defined. This reduces, in general, to see whether $\sum_{l=1,3} K_l^+$ is invertible.

Since the system is symetrisable, there exists symetric matrices A_0', A_1', A_2' , A_0' positive definite, such that telles que

$$A_1' = A_1 A_0' \ , \ A_2' = A_2 A_0'.$$

By introducing $V = A_0' W$, equation (17) rewrites

$$A_0' V_t + A_1' V_x + A_2' V_y = 0. \tag{19}$$

By setting $K_n' = A_1' n_x + A_2' n_y$, it is clear that $K_n' = K A_0'$. Since $A_0' > 0$, by left and right multiplication by $(A_0)^{-1/2}$ we see that

$$K_n'^+ = K^+ A_0' \ , \ K_n'^- = K^- A_0'$$

$$|K_n'| = |K| A_0'$$

Last $\sum_i \mathbf{n}_i = 0$ implies $\sum_i K_i = \sum_i K_i' = 0$.

Since A_1' and A_2' are symetric, the K_i' are also symetric, as well as their positive and negative part and their absolute value. Hence,

$$\sum_i K_i'^+ \geq 0$$
$$\sum_i K_i'^- \leq 0$$
$$\sum_i |K_i'| \geq 0.$$

This shows that $\sum_i |K_i|$, $\sum_i K_i^+$, $-\sum_i K_i^-$ have positive eigenvalues (by left multiplication by $A_0^{1/2}$ and right multiplication by $A_0^{-1/2}$).

If one of the matrices K_i^+ has a system of strickly positive eigenvalues, $\sum_i K_i^+$ has only stricly positive eigenvalues. In this case, $\sum_i K_i^+$ is invertible.

Assume now there exists x such that $\sum_i K_i^+ x = 0$ By setting $y = A_0^{-1}x$, we have $\sum_i K_i^{'+} y = 0$. Hence,

$$0 = (\sum K_i^+ x|x) = \sum_i (K_i^{'+} y|y).$$

If there exists i for which $(K_i^+ y|y) > 0$, y cannot be in the null space of $\sum K_i^{'+}$ unless $y = 0$ and then $x = 0$.

Assume that $K_i^{'+} y = 0$ for $i = 1, 2, 3$, i.e. $K_i x = 0$, with $x \neq 0$. Coming back to the definition of K_i, since two among the vectors $\mathbf{n}_1$, $\mathbf{n}_2$ et $\mathbf{n}_3$ are linearly independant, we have $Ax = Bx = 0$: The matrices A and B have a common eigenvector with the eigenvalue 0.

The eigenvectors of A are

$$R_A^+ = \begin{pmatrix} 1 \\ u+a \\ v \\ H+ua \end{pmatrix} \quad R_A^- = \begin{pmatrix} 1 \\ u-a \\ v \\ H-ua \end{pmatrix} \quad R_A^0 = \begin{pmatrix} 1 \\ u \\ v \\ (u^2+v^2)/2 \end{pmatrix} \quad R_A^t = \begin{pmatrix} 0 \\ 0 \\ 1 \\ v \end{pmatrix}$$

associated to the eigenvalues $u + a$, $u - a$, u et u. The eigenvectors of B are

$$R_B^+ = \begin{pmatrix} 1 \\ u \\ v+a \\ H+ua \end{pmatrix} \quad R_B^- = \begin{pmatrix} 1 \\ u \\ v-a \\ H-ua \end{pmatrix} \quad R_B^0 = \begin{pmatrix} 1 \\ u \\ v \\ (u^2+v^2)/2 \end{pmatrix} \quad R_B^t = \begin{pmatrix} 0 \\ -1 \\ 0 \\ u \end{pmatrix}$$

associated to the eigenvalues $v + a$, $v - a$, v, v.

The only solution to the problem is then $u = v = 0$ and $x = \lambda R_A^0 = \lambda R_B^0$ (stagnation point) unless $a = 0$ which corresponds to vacum.

We show now that even in the case $u = v = 0$, $M_{ij}x$ can be given a meaning. More precisely, we show there exists a decomposition of the state space R^4, $R^4 = RR_A^0 \bigoplus H$ where H contains the eigenvectors of A_1 and A_2 that are different of R_A^0.

Lemma A.1 *If A and B have a common eigenvector, and if there exists $A_0 > 0$ that symetrizes A and B, then, there exists a vector space H which can be explicitly computes such that the other eigenvectors of A and B belongs to H.*

Proof. The matrices AA_0 and BA_0 are symetric hence $A' = A_0^{-1/2} A A_0^{1/2}$ and $B' = A_0^{-1/2} B A_0^{1/2}$ are also symetric (left and right muliplication by $A_0^{-1/2}$). They are similar to A and B.

Let $\{r_k\}_{k=1,n}$ and $\{r_k^{'}\}_{k=1,n}$ be systems of eigenvectors of A and B respectively. We assume that $r_1 = r_1^{'}$.Then $\{A_0^{-1/2} r_k\}_{k=1,n}$ and $\{A_0^{-1/2} r_k^{'}\}_{k=1,n}$ are systems of eigenvectors of symetric matrices. They are orthogonal.Then

$$R^n = RA_0^{-1/2}r_1 \bigoplus (RA_0^{-1/2}r_1)^{\perp}.$$

Clearly, $A_0^{-1/2}r_k' \in (RA_0^{-1/2}r_1)^{\perp}$ and $A_0^{-1/2}r_k \in (RA_0^{-1/2}r_1)^{\perp}$ for $k > 1$. By setting

$$H = A_0^{1/2}(RA_0^{-1/2}r_1)^{\perp},$$

we get the result.

The for any $x \in R^4$, we can write

$$x = \lambda(x)R_A^0 + x^{\perp}, x^{\perp} \in H$$

where $\lambda(x) = A_0 x \cdot x$. We have

$$M_{ij}x = \left(\frac{(\mathbf{u}|\mathbf{n_i})^+ \, (\mathbf{u}|\mathbf{n_j})^-}{\sum_{i=1,3}(\mathbf{u}|\mathbf{n_i})^+} \right) \lambda(x)R_0 + M_{ij}x^{\perp}.$$

When $\mathbf{u} \longrightarrow 0$, the first term tends to 0 because

$$\left| \frac{(\mathbf{u}|\mathbf{n_i})^+ \, (\mathbf{u}|\mathbf{n_j})^-}{\sum_{i=1,3}(\mathbf{u}|\mathbf{n_i})^+} \right| \leq |(\mathbf{u}|\mathbf{n_j})^-| \longrightarrow 0$$

and $M_{ij}x^{\perp}$ converges to a finite limit because no eigenvalue vanishes and $x^{\perp}$ also converges to a finite limit.

Kinetic schemes for solving Saint-Venant equations on unstructured grids

M.O. Bristeau and B. Perthame

INRIA-Rocquencourt, Projet M3N
BP 105
78153 Le Chesnay Cedex, France
E-mail: Marie-Odile.Bristeau@inria.fr, Benoit.Perthame@ens.fr

Key Words: SAINT-VENANT SYSTEM , SHALLOW WATER SYSTEM, FINITE VOLUME METHOD, KINETIC SCHEMES, UNSTRUCTURED GRID.

Dedicated to Ph. Roe on the occasion of his 60th birthday

1 Introduction

We consider the Saint-Venant system, also called shallow-water equations, which is a simple and usual model to describe, for instance, the flows in rivers or coastal areas. This system can be derived from the three dimensional incompressible Euler system with a free surface. Using several assumptions (hydrostatic pressure, vertical homogeneity of horizontal velocities, small height of water...), it allows to describe the flow at time $t \geq 0$, and at the point $x \in \mathbb{R}^2$, through the height of water $h(t,x) \geq 0$, and its velocity $u(t,x) \in \mathbb{R}^2$, by the hyperbolic system

$$\frac{\partial h}{\partial t} + \mathrm{div}(hu) = 0, \tag{1}$$

$$\frac{\partial hu}{\partial t} + \mathrm{div}(hu \otimes u) + \nabla(\frac{g}{2}h^2) + gh\nabla Z = 0, \tag{2}$$

where g denotes the gravity intensity and $Z(x)$ the bottom height, and therefore $h+Z$ is the level of water surface. Usually, several other terms are added to this system in order to take into account frictions on the bottom and the surface and other physical features. Here we neglect these additional terms and we take $Z = 0$ in order to simplify the presentation.

A classical approach for solving hyperbolic systems on general triangular grids consists in using finite volume schemes where the control volume is the mediane based dual cell as it was introduced several years ago for aerodynamics problems (see [2] for instance). In this paper we describe a possible resolution of the system (1), (2) using this approach together with a kinetic solver. Let us recall that finite volume methods require to compute fluxes at the control volumes interfaces, and the overall stability of the method requires some upwinding in the interpolation of the fluxes, this is usually called the 'building block' or 'approximate Riemann solver'. Among those solvers, a very good and popular compromise between stability and accuracy is the so called Roe solver (see [9], [4] for instance). Among other solvers, the kinetic solvers are simpler and more stable, especially they are well known to be able to treat vacuum ($h = 0$ here, it corresponds to dry soils, then the system losses hyperbolicity) and satisfy precise in-cell entropy inequalities. We refer to [7] for a survey of the theoretical properties of these schemes, [8] for the actual implementation in two dimensions. Their main drawback is the lack of accuracy on steady contact discontinuities (they are computed exactly by Roe solver) which makes them unable to compute boundary layers or shear flows in classical fluid dynamics problems. For Saint-Venant equations this problem does not exist however, since the only possible discontinuities are shocks, and therefore the kinetic schemes might be a better compromise between accuracy, stability and efficiency. In this paper we present the method and first numerical tests in this direction. However, we would like to point out that other well known difficulties still have to be treated, which have been pointed out by several authors (e.g. to keep equilibrium states [5], to use larger time steps [6], to modify existing schemes in order to allow h to vanish [3]).

Another advantage of kinetic schemes is that it allows to derive a scheme for the advection of temperature $T(t, x)$ (or a concentration of pollutant) which is both monotone in T and conservative in hT. This additional, and decoupled equation is written in a conservative form as

$$\frac{\partial hT}{\partial t} + \operatorname{div}(huT) = 0, \tag{3}$$

but 'monotinicity' is better seen on the developed form

$$\frac{\partial T}{\partial t} + u \cdot \nabla T = 0. \tag{4}$$

After presenting the scheme for Saint-Venant equations in Section 2, the advection of temperature will be treated in Section 3. In Section 4 we present some first numerical results.

2 Kinetic scheme for St-Venant equations

We now present the finite volume method on a general triangular grid. We give some notations in a first subsection and the kinetic solver is presented in a second subsection.

2.1 Notations and general presentation

We use a triangulation of $\mathbb{R}^2$ which vertices are denoted X_i. The control volume (cell) of center X_i is denoted C_i and its boundary is formed by the edges E_α joining the centers of mass of the triangles around X_i. We use the following notations : $|C_i| = area(C_i)$, $l_\alpha = length(E_\alpha)$, n^i_α is the outward unit normal to E_α and, generically $j = j(i, \alpha)$ denotes the label of the neighbour cell to C_i along E_α, n^j_α its outward unit normal (and thus $n^i_\alpha = -n^j_\alpha$).

To present the general finite volume method, we consider a system of first order conservation laws

$$\frac{\partial U}{\partial t} + \mathrm{div} F(U) = 0. \tag{5}$$

For Saint-Venant system (1),(2) with $Z = 0$, we have: $U = (h, hu)^t$ and $F(U) = (hu, hu \otimes u + \frac{g}{2}h^2 Id)^t$. Then, the finite volume scheme writes

$$|C_i|U_i^{n+1} - |C_i|U_i^n + \Delta t \sum_\alpha l_\alpha F(U_i, U_j, n^i_\alpha) = 0, \tag{6}$$

where $F(U_i, U_j, n^i_\alpha)$ denotes an interpolation of the normal component of the flux $F(U) \cdot n^i_\alpha$ along the edge E_α. This interpolation is usually performed using a one dimensional solver since locally the problem looks like a planar discontinuity.

2.2 Kinetic solver

In the local basis associated with the edge E_α, we set $x = (x_1, x_2)$, $u = (u_1, u_2)$ where 1 refers to the outward normal direction, and 2 to the tangential direction. Then, the normal flux is simply given by $F(U) \cdot n^i_\alpha = \left(hu_1, h(u_1)^2 + \frac{g}{2}h^2, hu_1u_2\right)^t$. We obtain a consistent scheme in choosing the kinetic flux splitting to interpolate the normal fluxes along E_α

$$F(U_i, U_j, n^i_\alpha) = F_+(U_i) \cdot n^i_\alpha - F_-(U_j) \cdot n^i_\alpha \tag{7}$$

where the flux decomposition is given, in the local basis, by the formulas (close to that of the Van Leer scheme)

$$F_+(U) \cdot n = \begin{pmatrix} h\, c \left((Ma_+)^2 - (Ma_-)^2\right)/2 \\ h\, c^2 \left((Ma_+)^3 - (Ma_-)^3\right)/3 \\ h\, cu_2 \left((Ma_+)^2 - (Ma_-)^2\right)/2 \end{pmatrix} \tag{8}$$

where $c = \sqrt{3gh/2}$ is a number close to the celerity $\sqrt{gh}$ which determines the eigenvalues of the Saint-Venant system, and

$$Ma_+ = Max(0, \frac{u_1}{c} + 1), \quad Ma_- = Max(0, \frac{u_1}{c} - 1).$$

The part F_- of the numerical flux is deduced by symmetry, or by the consistancy relation $F_+(U)\cdot n - F_-(U)\cdot n = F(U)\cdot n$.

To explain the origin of these formulas we recall the kinetic interpretation of the system (1), (2) with $Z=0$. In one dimension, it is formally obtained integrating in $\xi \in \mathbb{R}$ the kinetic equation on the microscopic density $f(t,x,\xi)$

$$\frac{\partial f}{\partial t} + \xi \frac{\partial f}{\partial x} = 0, \tag{9}$$

with -here χ denotes the indicator function-,

$$f(\xi) = \frac{h}{2c}\chi(|\xi - u| \leq c). \tag{10}$$

The formulas (8) are just obtained with an upwind finite volume discretisation of the equation (9) and can be written also

$$\begin{array}{rl} F_+^h = & \int_{\xi\geq 0} \xi f(\xi) d\xi, \\ F_+^{hu_1} = & \int_{\xi\geq 0} \xi^2 f(\xi) d\xi, \\ F_+^{hu_2} = & u_2 F_+^h. \end{array} \tag{11}$$

3 Kinetic scheme for advection of temperature

Before we explain the monotonicity and conservation properties of the kinetic scheme for the advection of temperature, we need a better derivation of the kinetic scheme.

3.1 More on kinetic schemes

From a theoretical point of view, it is better to give a direct kinetic interpretation of the two dimensional Saint-Venant equations. We just take $\xi \in \mathbb{R}^2$, $c = \sqrt{gh}$ and define the equilibrium state in ξ as

$$f(\xi) = \frac{h}{\pi c^2}\chi(|\xi - u| \leq c). \tag{12}$$

Then, again to solve (1), (2) with $Z = 0$ is equivalent to solve, with such an f,

$$\frac{\partial f}{\partial t} + \xi \cdot \nabla_x f = 0, \tag{13}$$

(but of course this equilibrium is not preserved by the evolution and the scheme consists in a projection to the equilibrium at each time step). This amounts to notice that the equilibrium (12) satisfies

$$\begin{array}{l} h = \int_{\mathbb{R}^2} f(\xi) d\xi, \\ hu = \int_{\mathbb{R}^2} \xi f(\xi) d\xi, \\ hu \otimes u + \frac{g}{2} h^2 I_d = \int_{\mathbb{R}^2} \xi \otimes \xi f(\xi) d\xi. \end{array} \tag{14}$$

Although this approach leads to a numerical scheme which is more expensive than (8), and does not exhibit better properties in practice, it is useful to get a direct and simple proof of these theoretical properties.

We may discretize (13) in (t, x) with an upwind finite volume scheme, keeping ξ continuous. We also choose at time t^n an equilibrium state (12) for $f(\xi)$ and therefore we define

$$\chi_i^n(\xi) = \frac{h}{\pi c}\chi(|\xi - u| \le c). \tag{15}$$

This yields

$$|C_i|f_i^{n+1}(\xi) - |C_i|\chi_i^n(\xi) + \Delta t \sum_\alpha l_\alpha \big[(\xi \cdot n_\alpha^i)_+ \chi_i^n(\xi) - (\xi \cdot n_\alpha^i)_- \chi_j^n(\xi)\big] = 0. \tag{16}$$

Integrating this difference equation in ξ against the weights 1 and ξ yields the finite volume discretisation (6) of the Saint-Venant system with a flux splitting of the type(7). The half-fluxes are now given by

$$F_+^h(U_i).n_\alpha^i = \int_{\mathbb{R}^2} (\xi \cdot n_\alpha^i)_+ \chi_i^n(\xi) d\xi,$$

$$F_+^{hu}(U_i).n_\alpha^i = \int_{\mathbb{R}^2} (\xi \cdot n_\alpha^i)_+ \xi \chi_i^n(\xi) d\xi.$$

These formulas are genuinely two dimensional variants of the simpler formulas (8), (11).

This scheme is *stable* under the CFL condition

$$\Delta t \left(|u_i^n| + \sqrt{g h_i^n}\right) \sum_{\alpha \text{ around } X_i} l_\alpha \le |C_i|, \qquad \forall i. \tag{17}$$

Indeed, with this condition one can readily see that, for all ξ, $f_i^{n+1}(\xi)$ is a convex combination of the $\chi_j^n(\xi)$. In particular we deduce for instance that $f_i^{n+1}(\xi)$ is nonnegative and therefore h_i^n is also nonnegative.

3.2 Advection of temperature

We now consider the equation (3) for the advection of temperature. We discretize it in the conservative form as

$$|C_i|h_i^{n+1}T_i^{n+1} - |C_i|h_i^n T_i^n + \Delta t \sum_\alpha l_\alpha \big[T_i^n F_+^h(U_i) \cdot n_\alpha^i - T_j^n F_-^h(U_j) \cdot n_\alpha^i\big] = 0. \tag{18}$$

We have the following properties

Lemma 1 *Under the CFL condition (17), the conservative scheme (18) preserves positivity of T_i^n, and satisfies*

$$T_i^{n+1} \leq \max_j T_j^n.$$

Proof. The proof relies on the kinetic interpretation of this scheme. It is the ξ integral of the scheme (see (16))

$$|C_i| f_i^{n+1}(\xi) T_i^{n+1} - |C_i| \chi_i^n(\xi) T_i^n + \Delta t \sum_\alpha l_\alpha \big[(\xi \cdot n_\alpha^i)_+ \chi_i^n(\xi) T_i^n$$

$$- (\xi \cdot n_\alpha^i)_- \chi_j^n(\xi) T_j^n \big] = 0.$$

To prove the maximum principle for instance, we just write, still because $f_i^{n+1}(\xi)$ is a convex combination of the $\chi_j^n(\xi)$,

$$|C_i| f_i^{n+1}(\xi) T_i^{n+1} \leq$$

$$\max_j T_j^n \left[\chi_i^n(\xi) \big(|C_i| - \Delta t \sum_\alpha l_\alpha (\xi \cdot n_\alpha^i)_+ \big) + \Delta t \sum_\alpha l_\alpha (\xi \cdot n_\alpha^i)_- \chi_j^n(\xi) \right]$$

$$= |C_i| \; f_i^{n+1}(\xi) \; \max_j T_j^n.$$

This gives the maximum principle.

4 Numerical Results

The first order scheme defined in the previous sections is extended to a "formally" second order one using a MUSCL like extension (see [10]) with a Van Albada type limitor. The following numerical results have been obtained with this second order scheme. For the time integration we use a fourth stage Runge-Kutta scheme. We have tested the method on two dam-break problems.

The first test case is equivalent to a one dimensional problem for which the analytical solution can be computed. The problem consists of a rectangular channel with a discontinuous initial condition which represents the water level upstream and downstream of a dam. The length of the domain is 1. and the width is 0.1, the boundary discretization step is 0.01. We use the unstructured mesh shown in Fig: 1, it contains 1348 nodes and 2474 triangles. We assume that the velocity is zero everywhere and the waterheight $h = 1.$ for $x \leq 0.5$ and $h = 0.$ (dry soil) for $x > 0.5$. We also assume that the temperature of water is constant. The following solution has been computed at time $t = 0.05$, in Fig: 2 (resp. Fig: 3, Fig: 4) we compare the computed waterheight (resp. x-component of velocity, temperature) plotted with a dotted line with the exact ones plotted with a continuous line. For the temperature, the difficulty is only to compute the discontinuity corresponding to the limit of the water on the dry soil.

The second test case concerns a partial failure of a dam in a 200 x 200 m basin as depicted in Fig: 5. This problem has been computed by many authors (see e.g. [1]). The initial waterheight is $h_1 = 10.m$ on the left hand side and $h_1 = 5.m$ on the right of an idealized dam. Water is released into the downstream side through a breach 75. m wide. The discretization step of the boundaries is 2.5 m, the mesh contains about 8100 nodes and 15800 triangles. In Fig: 6 we show the watersurface elevation at time $t = 7.2s$. The level lines are plotted in Fig: 7 and the velocity field in Fig: 8. On this fine mesh, the vertices on each side of the breach are accurately computed.

Acknowledgements The authors would like to thank J.M. Hervouet and B. Mohammadi for helpful comments and suggestions.

References

[1] ALCRUDO F. AND GARCIA-NAVARRO P., A High-Resolution Godunov-type Scheme in Finite Volumes for the 2D Shallow-water Equations, *Int. J. for Numerical Methods in Fluids*, Vol 16, 489-505 (1993)

[2] ANGRAND F., DERVIEUX A., BOULARD V., PERIAUX J. AND VIJAYA-SUNDARAM G., Transonic Euler simulation by means of finite element explicit schemes. In *AIAA-83*, (1984).

[3] BUFFARD T., GALLOUET T. AND HERARD J.-M., Un schéma simple pour les équations de Saint-Venant. *Internal report EDF-DER-HE-41/97/022/A* (1997).

[4] GODLEWSKI E. AND RAVIART P.-A., *Numerical approximations of hyperbolic systems of conservation laws*, *Applied Mathematical Sciences* 118, Springer-Verlag, New York, (1996).

[5] GREENBERG J.M. AND LEROUX A.-Y., A well balanced scheme for the numerical processing of source terms in hyperbolic equations. *SIAM J. Num. Anal.* 33, 1-16 (1996).

[6] HOLDAHL R., HOLDEN H. AND LIE K.-A., Unconditionally stable splitting methods for the shallow water equations. *Preprint NUST Trondheim* (1998).

[7] PERTHAME B. An introduction to kinetic schemes for gas dynamics. In *An introduction to recent developments in theory and numerics for conservation laws. L.N. in Computational Sc. and Eng.*, *5*, D. Kroner, M. Ohlberger and C. Rohde editors. Springer (1998).

[8] PERTHAME B. AND QIU Y. A variant of Van Leer's method for multidimensional systems of conservation laws. *J. Comp. Phys.* 112(2), 370-381 (1994).

[9] ROE P. L., Approximate Riemann solvers, parameter vectors and difference schemes. *J. Comp. Phys.* 43, 357-372 (1981).

[10] VAN LEER B., Towards the Ultimate Conservative Difference Schemes. V. A Second Order Sequel to the Godunov's Method. *J. Comp. Phys.*, 32 (1979)

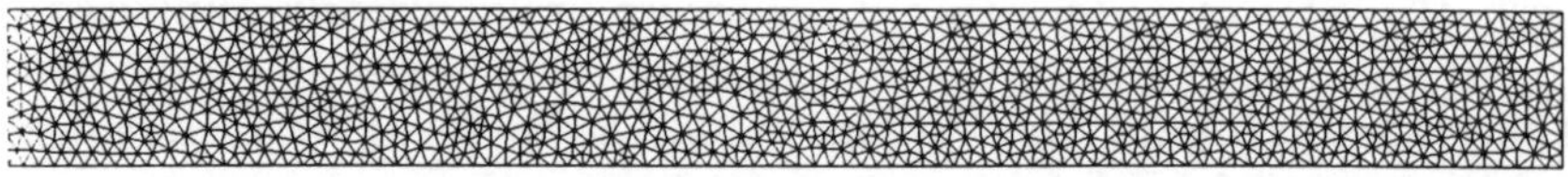

Figure 1: Mesh of the rectangular channel

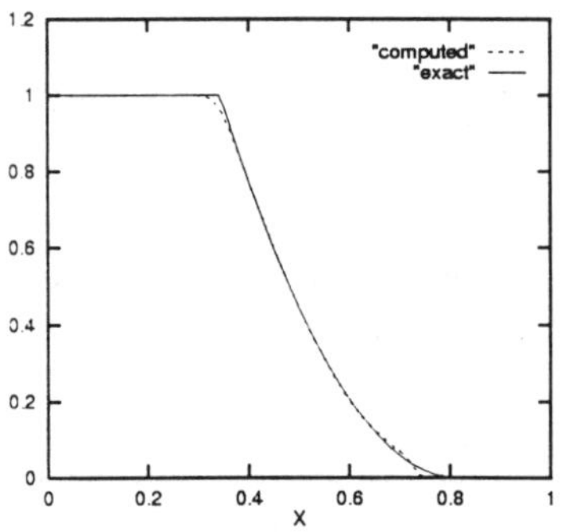

Figure 2: Waterheight at $t = 0.05$.

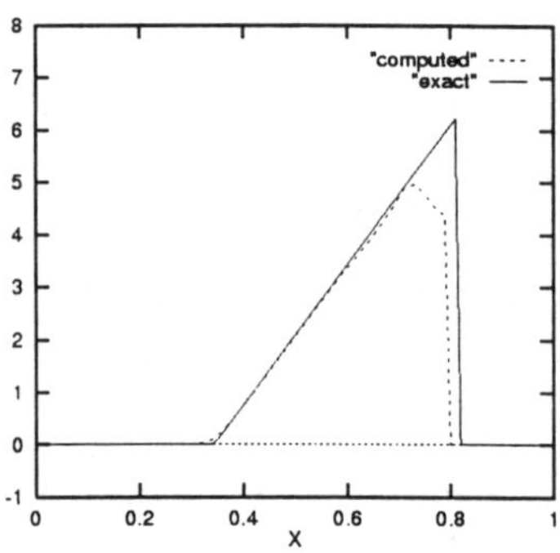

Figure 3: Velocity at $t = 0.05$.

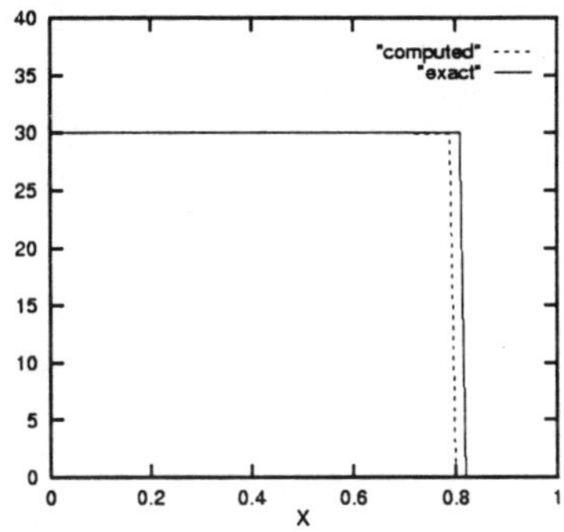

Figure 4: Temperature at $t = 0.05$.

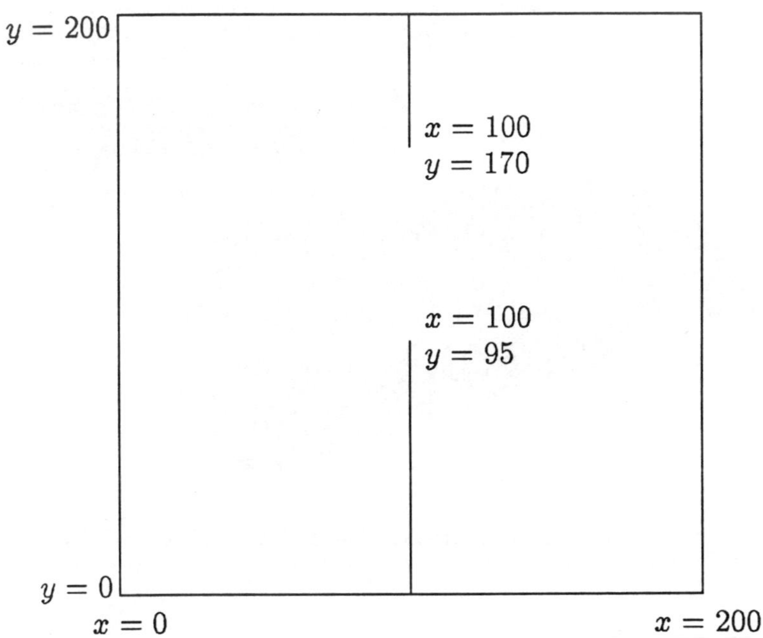

Figure 5: Geometry of the basin with dam breach.

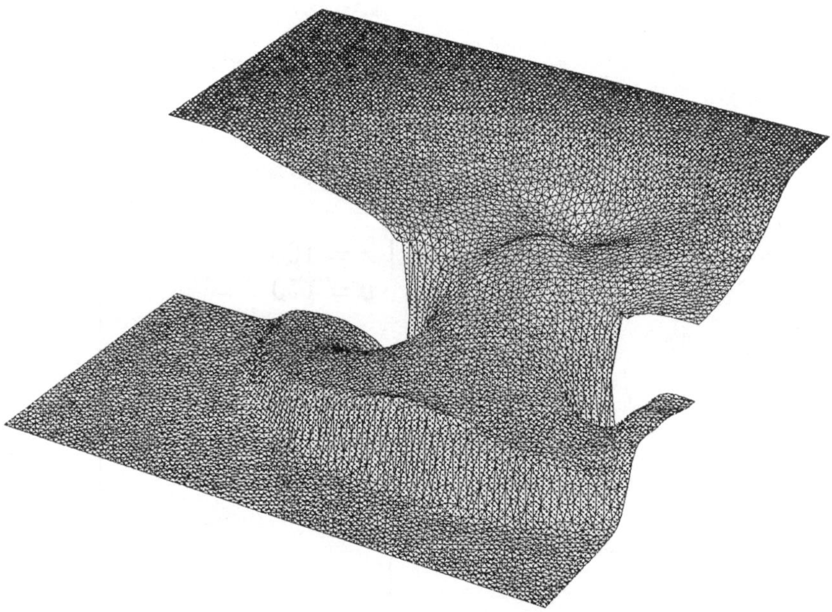

Figure 6: Partial dam break: Water elevation at $t = 7.2s$.

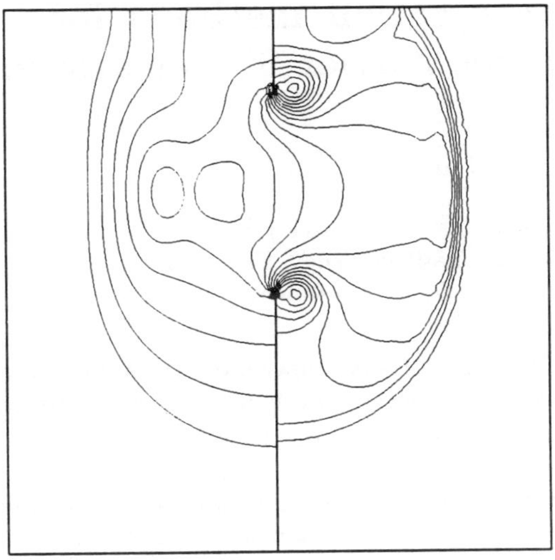

Figure 7: Partial dam break: Waterheight contours at $t = 7.2s$.

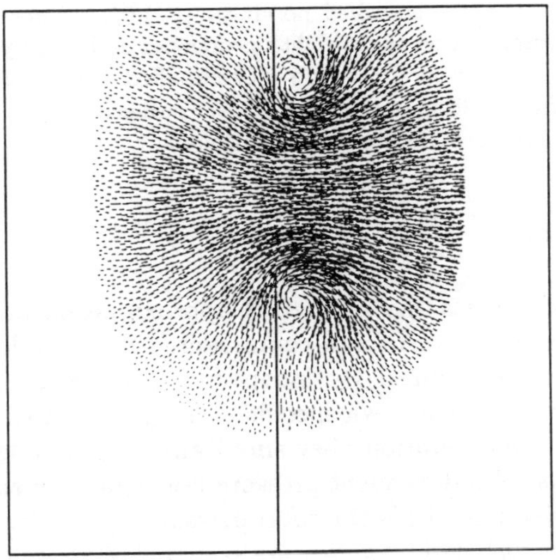

Figure 8: Partial dam break: Velocity field at $t = 7.2s$.

Nonlinear projection methods for multi-entropies Navier-Stokes systems

Christophe BERTHON and Frédéric COQUEL

LAN-CNRS, tour 55 65, 5 Etage 4, place jussieu, 75252 Paris Cedex 05, FRANCE.
ONERA, BP 72, 92322 Châtillon Cedex, FRANCE.
email: *coquel@ann.jussieu.fr*, *berthon@onera.fr*

Key Words: NAVIER-STOKES EQUATIONS, TURBULENCE MODELS, NON CONSERVATIVE PRODUCTS, TRAVELLING WAVE SOLUTIONS, GODUNOV TYPE METHODS, ROE LINEARIZATION, NONLINEAR PROJECTION.

Abstract. This paper is devoted to the numerical approximation of the compressible Navier-Stokes equations with several independent temperatures. Several models derived in plasma physics or in turbulence typically enter the proposed framework. The striking novelty over the usual Navier-Stokes equations stems from the impossibility to recast equivalently the present system in full conservation form. Classical finite volume methods ar shown to grossly fail in the capture of travelling wave solutions that are of primary interest in the present work. We propose a systematic and effective procedure, the so-called nonlinear projection operator, for correcting the errors while preserving all the stability properties satisfied by suitable Godunov type methods. This nonlinear procedure is exemplified when coupled with an exact Roe solver and the resulting method is shown numerically to yield approximate solutions in close agreement with exact solutions.

1 Introduction

The present work treats the numerical approximation of the solutions of the Navier-Stokes equations for a compressible fluid modelled by N, $N \geq 2$, independent temperatures. By independent, we mean that each of the temperatures comes with its own specific entropy so that smooth solutions of the system under consideration obey simultaneously N independent entropy balance equations. N independent pressure laws can be then defined and the sum of all the pressures yields the total pressure.

Despites that such fluid models are seen below to exhibit several close relationships with the usual Navier-Stokes system, the fundamental discrepancy stays in the lack of an admissible change of variables that recasts the governing equations in full conservation form. Indeed, the three conservation laws ruling density, momentum and total energy must be supplemented by $(N-1)$ of the entropy balance equations that govern the N temperatures. However and without restrictive modelling assumptions to be put on the viscosities and the conductivities, none of the N entropy balance equations boils down to a conservation law. These equations generally involve non conservative products that account for dissipative phenomena : namely the entropy dissipation rates. The governing PDE's system can be thus given the following abstract form :

$$\partial_t \mathbf{v} + \partial_x f(\mathbf{v}) = \mathcal{E}(\mathbf{v}, \partial_{xx}^2 \mathbf{v}), \quad \mathbf{v} \in I\!R^{N+2}, \quad x \in I\!R,\ t > 0, \tag{1}$$

where the right hand side involves diffusive terms in conservation form as well as the entropy dissipation rates in non conservation form. Here smooth solutions of (1) satisfy without condition the following additional and non trivial balance equation (namely the Nth entropy balance equation which has not been addressed yet) :

$$\partial_t U(\mathbf{v}) + \partial_x F(\mathbf{v}) = \mathcal{E}_N(\mathbf{v}, \partial_{xx}^2 \mathbf{v}), \tag{2}$$

where the right hand side is directly inferred from the dissipative phenomena modelled in (1).

Such systems can be understood as a natural extension of the classical Navier-Stokes equations, i.e. equipped with a single entropy balance equation, in that they actually occur in several distinct physical settings. They arise for instance in plasma physics where the electrons temperature must be distinguished from the mixture temperature of the other (heavy) species. They can be also recognized under that form within the frame of the so-called "two transport" equations models for turbulent compressible flows where the averaged thermodynamic temperature must be distinguished from the specific turbulent kinetic energy k (or for short turbulent temperature). All these models are addressed below with the assumption of a large Reynolds number.

The numerical capture of the viscous shock layers coming with (1), is of primary importance in the present work. Since the Reynolds numbers of interest are large, these layers display the character of a shock wave in that they differ from their end states only in a small interval of rapid transition. Hence for mesh refinements of practical interest, the associated discrete profiles stay largely under resolved. Our purpose is actually to correctly capture the two end states, $\mathbf{v}_L$ and $\mathbf{v}_R$, of a given shock layer together with its relevant speed of propagation σ whitout resolving sharply the viscous layer itself.

It is quite well-known that such an issue does not raise special difficulties within the standard frame of the Navier-Stokes equations, e.g. in conservation form. Indeed, the triples $(\sigma; \mathbf{v}_L, \mathbf{v}_R)$ are solutions of the classical Rankine-Hugoniot relations and finite volume methods in conservation form readily ensure their suitable capture. At the very core of this success, is the independence property of (say) the right end state $\mathbf{v}_R = \mathbf{v}_R(\sigma; \mathbf{v}_L)$ with respect to the precise modelling of the diffusive phenomena (see for instance Gilbarg [9]). Put in other words, the definition of $\mathbf{v}_R(\sigma; \mathbf{v}_L)$ stays completely free from the resulting entropy dissipation rate and in this sense, its numerical capture stays also free from the rate of numerical dissipation coming with a given scheme in conservation form.

The situation turns out to be completely different in the setting of the extended Navier-Stokes equations (1). Its non conservation form makes this time the triple $(\sigma; \mathbf{v}_L, \mathbf{v}_R)$ to heavily depend on the precise shape of the diffusive tensor. In deep contrast with the setting of systems in conservation form, such a dependence stays at the basis of recent works devoted to hyperbolic systems involving non conservative products (see Lefloch [14], Dal Maso-LeFloch-Murat [7], Raviart-Sainsaulieu [16]) and it will be numerically illustrated below (cf. Figures 2). When rephrased in the setting of numerical methods, this dependence implies in turn that the numerical viscous form of the method must fit the exact diffusive tensor in (1). Such a requirement has received a theoretical foundation in a work by Hou-LeFloch [11] and is actually the starting point of some numerical methods as deviced for instance by Karni [12]. In this work, fitness is measured via a modified equations approach according to which unsuited troncature errors are forced to cancel up to a given order but such a correction turns out to be unsufficient (see [11] for proofs and numerical illustrations and also [5] for an enlightening presentation). In this work, we propose another approach based on the analysis of the discrete dissipation rates of a given Lax entropy. As pointed out below, it turns out that the end states of under resolved viscous shock layers of (1) require for their correct capture to prescribe explicitely the Lax entropy dissipation rate $\mathcal{E}_N(\mathbf{v}, \partial^2_{xx}\mathbf{v})$ in (2). This requirement will be explained below but notice from now on that it asks the numerical methods to satisfy in turn an imposed rate of entropy dissipation. This non standard issue is precisely the main motivation of the present work.

In order to illustrate the negative consequence of a failure in the above requirement, let us consider the numerical results displayed in figure 1. These are obtained using a standard and consistent finite volume method which is briefly derived in the forthcoming numerical section. This problem enters the

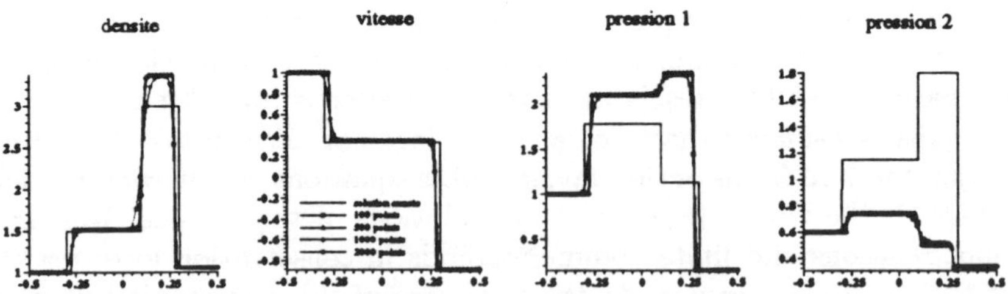

Fig. 1. The first figure

present setting for extended Navier-Stokes equations with two independent specific entropies when choosing two strictly positive viscosities and two identically zero conductivities. The discrete solutions stay virtually non sensitive with respect to the mesh refinements (from 100 points up to 2000 points) and clearly seem to coincide with a given function. However such a "limit" function exhibits large errors with respect to the exact solution plotted in solid lines. Let us underline that the above result does not contradict the celebrated Lax-Wendroff convergence theorem because the numerical method under consideration cannot be put in conservation form since the system (1) to be tackled cannot.

For the sake of simplicity in the notations, we focuse ourselves in the sequel to the case of two independent specific entropies. Governing PDE's will be furthermore addressed using only one space variable. Let us underline that most of the results we state below extend in a straightforward way to higher space dimensions, taking advantage of the rotational invariance of the equations. In the same way, the numerical method we propose will be seen to easily extend to additional independent entropies.

The format of the present paper is as follows. The first section describes the convective-diffusive system under consideration with a special emphasis put on the analysis of its travelling wave solutions (i.e. viscous shock layers). A close characterization of the triples $(\sigma; \mathbf{v}_L, \mathbf{v}_R)$ will be proposed on the basis of generalized jump relations. These relations are given some convenient forms that in turn will be shown to encode the small scale effects taking place within viscous shock layers. The second section is devoted to the numerical approximation of the travelling wave solutions for mesh refinements of practical interest. Such solutions are thus by definition under resolved. We first underline that despite relevant Godunov type methods actually enjoy several stability properties, their corresponding numerical rate of entropy dissipation stays always smaller than the required one in (2) and as a result, the discrepancies with the exact solutions can only unendly amplify with time. This error analysis will then suggest the introduction of a nonlinear projection step that enforce the validity of the generalized jump conditions at the discrete level. Such a nonlinear projection may be understood as a systematic way to correct standard Godunov type methods in order to achieve a much better agreement between exact and discrete solutions. Furthermore, approximate solutions performed thanks to the nonlinear method satisfy the required positivity preserving properties in addition to several (nonlinear) stability requirements. Several numerical results are displayed, intending to illustrate the benefits of the proposed nonlinear projection. Most of the proofs of the forthcoming statements are omitted and we refer the reader to the companion papers [1], [2] and [3] for the details.

2 Mathematical model

We consider a gas with density ρ and velocity u, which is modelled by two independent pressure laws p and p_T, associated with two constant adiabatic exponents $\gamma > 1$ and $\gamma_T > 1$. The system of PDE's that governs such a fluid model writes :

$$\begin{cases} \partial_t \rho + \partial_x \rho u = 0, & x \in I\!R,\ t > 0, \\ \partial_t \rho u + \partial_x (\rho u^2 + p + p_T) = \partial_x \left((\mu + \mu_T) \partial_x u \right), \\ \partial_t p + \partial_x p\, u + (\gamma - 1) p \partial_x u = (\gamma - 1) \left(\mu \left(\partial_x u \right)^2 + \partial_x (\kappa \partial_x T) \right), \\ \partial_t p_T + \partial_x p_T\, u + (\gamma_T - 1) p_T \partial_x u = (\gamma_T - 1) \left(\mu_T \left(\partial_x u \right)^2 + \partial_x (\kappa_T \partial_x T_T) \right), \end{cases} \tag{3}$$

where the involved temperatures respectively read $T = p/\rho$ and $T_T = p_T/\rho$. This convective-diffusive system can be understood as an extension of the standard Navier-Stokes equations when considering an additional PDE for governing an additional pressure. Depending on the closure relations for the viscosities μ, μ_T and the thermal conductivities κ and κ_T, several distinct physical models enter the present framework. Let us quote for instance plasma models (see [6] for instance) but also turbulence models (see [1] and below). In this section, all these transport coefficients are assumed to be fixed positive constants for the sake of simplificity in the discussion. The reader is referred to section 3 for the case of varying coefficients.

Similarly to the classical Navier-Stokes equations, the smooth solutions of system (3) obey additional governing equations as we now state :

Lemma 2.1 *Smooth solutions of (3) satisfy the following conservation law:*

$$\partial_t \rho E + \partial_x (\rho E + p + p_T) u = \partial_x ((\mu + \mu_T) u \partial_x u) + \partial_x (\kappa \partial_x T) + \partial_x (\kappa_T \partial_x T_T), \tag{4}$$

where the total energy ρE is defined by :

$$\rho E = \frac{(\rho u)^2}{2\rho} + \frac{p}{\gamma - 1} + \frac{p_T}{\gamma_T - 1}. \tag{5}$$

Smooth solutions satisfy in addition the following balance equations :

$$\partial_t \rho \log(s) + \partial_x \rho \log(s) u = \frac{\gamma - 1}{T} \left(\mu (\partial_x u)^2 + \partial_x (\kappa \partial_x T) \right), \tag{6}$$

$$\partial_t \rho \log(s_T) + \partial_x \rho \log(s_T) u = \frac{\gamma_T - 1}{T_T} \left(\mu_T (\partial_x u)^2 + \partial_x (\kappa_T \partial_x T_T) \right), \tag{7}$$

where the specific entropies are respectively given by

$$s := \frac{p}{\rho^\gamma}, \quad s_T := \frac{p_T}{\rho^{\gamma_T}}. \tag{8}$$

Consequently, smooth solutions of (3) obey :

$$\mu_\tau \frac{T}{\gamma - 1}\left(\partial_t \rho \log(s) + \partial_x \rho \log(s) u\right) - \mu \frac{T_\tau}{\gamma_\tau - 1}\left(\partial_t \rho \log(s_\tau) + \partial_x \rho \log(s_\tau) u\right)$$
$$= \partial_x(\mu_\tau \kappa \partial_x T - \mu \kappa_\tau \partial_x T_\tau), \qquad (9)$$

where the right hand side follows under the assumption of two constant viscosities μ and μ_τ.

The three balance equations (4), (6) and (7) can be proved to be the only non trivial additional equations for smooth solutions (up to some standard nonlinear tranforms in s and s_τ). As a consequence and besides several close relationships with the usual Navier-Stokes system (see Berthon [1] for a deeper insight), the very discrepancy stays in the lack of four non trivial conservation laws. Indeed and without restrictive modelling assumptions (see below), none of the equations (6), (7) and (9) boils down to a conservation law. As a consequence, (3) cannot be recast, generally speaking, in full conservation form.

Our purpose here is to highlight after the pioneering works by LeFloch [14], Raviart-Sainsaulieu [16] and Sainsaulieu [18] that the non conservation form met by (3) makes the end states of viscous shock layers to intrinsically depend on the closure relations for the transport coefficients μ, μ_τ and κ, κ_τ. Under the assumption of two constant viscosities, such a dependence more precisely occurs in term of the ratio of μ and μ_τ. In order to assess this issue, let us focuse our attention on the non standard balance equation (9). Its straightforward derivation reflects a cancellation property. Namely the entropy balance equations (6) and (7) are not independent but actually evolve proportionally to the ratio of the two viscosities μ and μ_τ. Indeed and at least formally, (9) yields once integrated with respect to the space variable :

$$\frac{\mu_\tau}{\mu + \mu_\tau}\left\{\int_{\mathbb{R}} \frac{T}{\gamma - 1}(\partial_t \rho \log(s) + \partial_x \rho \log(s) u) dx\right\} -$$
$$\frac{\mu}{\mu + \mu_\tau}\left\{\int_{\mathbb{R}} \frac{T_\tau}{\gamma_\tau - 1}(\partial_t \rho \log(s_\tau) + \partial_x \rho \log(s_\tau) u) dx\right\} = 0, \qquad (10)$$

so that the evolution in time of the two entropies must be kept in balance according to the ratio of the two viscosities. Let us underline that by contrast with (6) and (7) where entropy dissipation rates are actually independently imposed, the weighted equation (9) exhibits a compared rate of both the entropy dissipations.

To further assess this claim, let us adopt temporarily some restrictive modelling assumptions on both the viscosities and conductivities coefficients. Noticing that (9) reexpresses equivalently as :

$$\mu_\tau \frac{\rho^{\gamma-1}}{\gamma - 1}\left(\partial_t \rho s + \partial_x \rho s u\right) - \mu \frac{\rho^{\gamma_\tau - 1}}{\gamma_\tau - 1}\left(\partial_t \rho s_\tau + \partial_x \rho s_\tau u\right) =$$
$$\partial_x(\mu_\tau \kappa \partial_x T - \mu \kappa_\tau \partial_x T_\tau). \qquad (11)$$

the following result easily follows :

Lemma 2.2 *(i) Assume that μ and μ_τ are two given positive contants and $\kappa = \kappa_\tau = 0$. If moreover $\gamma = \gamma_\tau$, then (11) reduces to the following conservation law :*

$$\partial_t \rho\,(\mu_\tau s - \mu s_\tau) + \partial_x \rho\,(\mu_\tau s - \mu s_\tau)\, u = 0. \tag{12}$$

(ii) Let μ_0 and $\mu_{\tau 0}$ denote two positive constants and assume that $\mu = \mu_0 T$, $\mu_\tau = \mu_{\tau 0} T_\tau$ with $\kappa = \kappa_\tau = 0$. Then (9) coincides with the following conservation law :

$$\partial_t \rho \left(\frac{\mu_{\tau 0}}{\gamma - 1} \log(s) - \frac{\mu_0}{\gamma_\tau - 1} \log(s_\tau) \right) + \partial_x \rho \left(\frac{\mu_{\tau 0}}{\gamma - 1} \log(s) - \frac{\mu_0}{\gamma_\tau - 1} \log(s_\tau) \right) u = 0. \tag{13}$$

These restrictive assumptions therefore allow for additional non trivial conservation laws that are encoded in the non standard balance equation (9) or its equivalent form (11). When considering their associated Rankine-Hugoniot condition, they clearly indicate that the end states of the viscous shock layers under consideration actually depend on the ratio of both viscosities. The above restrictive assumptions will no longer be adopted in the sequel. The dependence we have just pointed out is numerically illustrated in figure 2 in a more general setting in which the system (3) does not admit an equivalent full conservation form. For a given left end state $\mathbf{v}_L$ and a given velocity σ, the

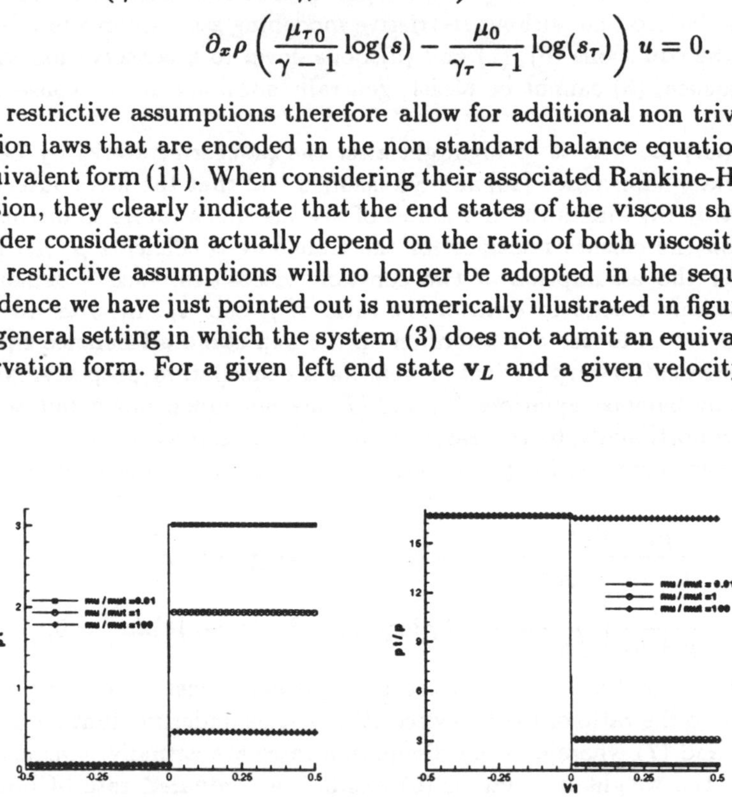

Fig. 2. Right end state $\mathbf{v}_R$ as a function of μ/μ_τ for a given $(\sigma, \mathbf{v}_L)$.

required right end states $\mathbf{v}_R$ are defined when solving numerically the nonlinear EDO's system governing travelling wave solutions (see next section) for various ratios of the viscosities. Furthermore and in a sense described below, (9) or (11) continue to play a major role in a general setting since they encode a generalized jump condition which turns out to play a central role for our numerical purpose.

2.1 Travelling wave solutions and jump relations

In this section, we derive generalized jump relations that are needed to characterize the triple $(\sigma; \mathbf{v}_L, \mathbf{v}_R)$ associated with a given viscous shock layer. In that aim, let us first recall that a travelling wave solution of (3) is a particular solution in the form $\mathbf{v}(x,t) = \widehat{\mathbf{v}}(x - \sigma t)$ with

$$\lim_{\xi \to -\infty} \widehat{\mathbf{v}}(\xi) = \mathbf{v}_L, \quad \lim_{\xi \to +\infty} \widehat{\mathbf{v}}(\xi) = \mathbf{v}_R, \quad \xi = x - \sigma t, \tag{14}$$

where the triple $(\sigma; \mathbf{v}_L, \mathbf{v}_R)$ is prescribed. In addition to (14), it thus satisfies the following nonlinear EDO's system :

$$-\sigma d_\xi \widehat{\mathbf{v}} + d_\xi f(\widehat{\mathbf{v}}) = \mathcal{E}(\widehat{\mathbf{v}}, d^2_{\xi\xi}\widehat{\mathbf{v}}), \quad \xi \in I\!R, \tag{15}$$

where we have used the abstract form (1) of system (3). Neglecting the conductivity coefficients κ and κ_τ in (3), Berthon-Coquel [2] have proved global existence and uniqueness of smooth travelling wave solutions of (3) for general (nonlinear) viscosity functions.

Our purpose is to highlight, after LeFloch [14], Raviart-Sainsaulieu [16], that the end states of integral curve solution of (14), (15) does not depend on the amplitude of the diffusive tensor modelled in (15) but just on its shape. Put in other words, the end states as well as the velocity σ do not depend on the Reynolds number under consideration. Indeed let us introduce, after [14] and [16], the function

$$\mathbf{v}_\delta(x,t) = \widehat{\mathbf{v}}\left(\frac{x - \sigma t}{\delta}\right),$$

where $\delta > 0$ denotes a positive rescaling parameter. For all $\delta > 0$, $\mathbf{v}_\delta(x,t)$ turns out to be a travelling wave solution of (3) but for the viscosities μ^δ, ${\mu_\tau}^\delta$ and the conductivities κ^δ, κ_τ^δ defined by

$$\mu^\delta = \delta\mu, \quad {\mu_\tau}^\delta = \delta\mu_\tau, \quad \kappa^\delta = \delta\kappa, \quad \kappa_\tau^\delta = \delta\kappa_\tau.$$

The function $\mathbf{v}_\delta(x,t)$ is obviously associated with the same triple $(\sigma, \mathbf{v}_L, \mathbf{v}_R)$ for all values of the rescaling parameter $\delta > 0$: the expected independence property with respect to the Reynolds number thus follows. As a consequence, the following results hold true (see [3] for a proof and extensions to a more general setting):

Theorem 2.3 *Assume that μ, μ_τ, κ and κ_τ denote positive constants. Then the triple $(\sigma, \mathbf{v}_L, \mathbf{v}_R)$ associated with the resulting dissipative tensor necessarily obeys the jump relations :*

$$-\sigma[\rho] + [\rho u] = 0, \tag{16.i}$$
$$-\sigma[\rho u] + [\rho u^2 + p + p_\tau] = 0, \tag{16.ii}$$
$$-\sigma[\rho E] + [(\rho E + p + p_\tau)u] = 0, \tag{16.iii}$$

and necessarily satisfies the two entropy inequalities :

$$-\sigma[\rho s] + [\rho s u] > 0, \tag{17.i}$$
$$-\sigma[\rho s_\tau] + [\rho s_\tau u] > 0. \tag{17.ii}$$

In order to specify (17), let us consider for all $\delta > 0$, $\mathbf{v}_\delta$ a rescaled travelling wave for the same triple $(\sigma; \mathbf{v}_L, \mathbf{v}_R)$. Let us then define with clear notations:

$$\int_{\mathbb{R}} \frac{\widehat{\rho_\delta}^{\gamma-1}}{\gamma - 1} \left(\partial_t(\widehat{\rho s})_\delta + \partial_x(\widehat{\rho s u})_\delta\right)(\xi)\, d\xi = \zeta, \tag{18}$$
$$\int_{\mathbb{R}} \frac{\widehat{\rho_\delta}^{\gamma_\tau-1}}{\gamma_\tau - 1} \left(\partial_t(\widehat{\rho s_\tau})_\delta + \partial_x(\widehat{\rho s_\tau u})_\delta\right)(\xi)\, d\xi = \zeta_\tau. \tag{19}$$

Then the total masses ζ and ζ_τ of the two entropy inequalities are bounded and do not depend on the rescaling parameter $\delta > 0$ but only depend on the ratio of the viscosities through the following identity:

$$\mu_\tau \zeta(\sigma; \mathbf{v}_L, \mathbf{v}_R) - \mu \zeta_\tau(\sigma; \mathbf{v}_L, \mathbf{v}_R) = 0. \tag{20}$$

Rephrasing the above result, the triple $(\sigma; \mathbf{v}_L, \mathbf{v}_R)$ is entirely characterized by the classical jump relations (16.i), (16.ii), (16.iii) and the generalized jump condition (20). It can be seen that when reexpressed for a travelling wave solution, the averaged balance equation (10) exactly coincides with the generalized jump relation (20). Despite that (20) cannot be given generally speaking a close form of expression, it seems convenient to reexpress it equivalently when introducing two suitable averages as follows :

$$\overline{\frac{\rho^{\gamma-1}}{\gamma-1}}\Big\{-\sigma[\rho s] + [\rho s u]\Big\} = \zeta(\sigma; \mathbf{v}_L, \mathbf{v}_R), \tag{21.i}$$
$$\overline{\frac{\rho^{\gamma_\tau-1}}{\gamma_\tau-1}}\Big\{-\sigma[\rho s_\tau] + [\rho s_\tau u]\Big\} = \zeta_\tau(\sigma; \mathbf{v}_L, \mathbf{v}_R), \tag{21.ii}$$

so that (20) now reads :

$$\frac{\mu_\tau/\mu}{1 + \mu_\tau/\mu} \overline{\frac{\rho^{\gamma-1}}{\gamma - 1}}\Big\{-\sigma[\rho s] + [\rho s u]\Big\}$$
$$-\frac{\mu/\mu_\tau}{1 + \mu/\mu_\tau} \overline{\frac{\rho^{\gamma_\tau-1}}{\gamma_\tau - 1}}\Big\{-\sigma[\rho s_\tau] + [\rho s_\tau u]\Big\} = 0. \tag{22}$$

Remark 2.4 In view of the jump relations (16) and (22), one of the two thermodynamic entropies, either ρs or ρs_τ, must be obviously understood as a nonlinear function of the four remaining independent variables $(\rho, \rho u, \rho E, .)$.

Notice after [14] and [16], that the family of travelling wave solutions $\{\hat{\mathbf{v}}\}_{\delta>0}$, for a given triple $(\sigma; \mathbf{v}_L, \mathbf{v}_R)$, can be seen to converge as δ goes to zero in the L^1_{loc} strong topology to the following step function :

$$\mathbf{v}_0(x,t) = \begin{cases} \mathbf{v}_L, & x < \sigma t, \\ \mathbf{v}_R, & x > \sigma t. \end{cases} \tag{23}$$

Let us underline that this so-called shock-solution satisfies by construction the (generalized) jump conditions (16), (22). Numerically speaking, under resolved viscous shock layers are nothing else but an approximation, relevant or not, of the underlying shock solution (23).

The relation (22) clearly suggests that the system (3) admits two equivalent limit systems in full conservation form when the ratio of viscosities μ/μ_τ or μ_τ/μ goes to zero. Since the jump relations are explicitly known in the two limit cases, the following statement allows for asymptotic expansions of the generalized jump condition (20) respectively in $\dfrac{\mu_\tau}{\mu}$ and $\dfrac{\mu}{\mu_\tau}$. More precisely we have (see [3] for a proof and detailled expansions):

Theorem 2.5 *Let us assume that* $\kappa = \kappa_\tau = 0$.

When the ratio $\dfrac{\mu_\tau}{\mu}$ *goes to zero with* μ *set at a fixed positive constant, travelling wave solutions of the system (3) converge uniformly over* $\mathbb{R}$ *to travelling wave solutions of the following system of conservation laws:*

$$\begin{cases} \partial_t \rho + \partial_x \rho u = 0, \qquad x \in \mathbb{R},\ t > 0, \\ \partial_t \rho u + \partial_x(\rho u^2 + p + p_\tau) = \partial_x(\mu \partial_x u), \\ \partial_t \rho E + \partial_x(\rho E + p + p_\tau)u = \partial_x(\mu u \partial_x u), \\ \partial_t \rho s_\tau + \partial_x \rho s_\tau u = 0, \\ \partial_t \rho s + \partial_x \rho s u \leq 0. \end{cases} \tag{24}$$

Conversely, if $\dfrac{\mu}{\mu_\tau}$ *goes to zero with* μ_τ *set at a fixed positive constant, travelling wave solutions of the system (3) converge uniformly over* $\mathbb{R}$ *to travelling wave solutions of the following system of conservation laws:*

$$\begin{cases} \partial_t \rho + \partial_x \rho u = 0, \qquad x \in \mathbb{R},\ t > 0, \\ \partial_t \rho u + \partial_x(\rho u^2 + p + p_\tau) = \partial_x(\mu_\tau \partial_x u), \\ \partial_t \rho E + \partial_x(\rho E + p + p_\tau)u = \partial_x(\mu_\tau u \partial_x u), \\ \partial_t \rho s + \partial_x \rho s u = 0, \\ \partial_t \rho s_\tau + \partial_x \rho s_\tau u \leq 0. \end{cases} \tag{25}$$

2.2 Equivalent formulations and Convexity properties

Here the extended Navier-Stokes system (3) is given two equivalent formulations for smooth solutions that allow for some convexity properties of importance in the forthcoming numerical derivations. Indeed several stability properties will be inherited from convexity. In this way, the equivalent formulations involve nonlinear versions of the specific entropies s_τ and s_τ, namely $S = g(s)$ and $S_\tau = h(s_\tau)$, so that (say) the entropy ρS, is strictly convex when understood as a function of the independent variables $(\rho, \rho u, \rho E, \rho S_\tau)$. To enforce the convexity property, both the functions g and h are assumed to obey (see Berthon [1] and Berthon-Coquel [4] for a proof):

f is a strictly decreasing and strictly convex $C^2(\mathbb{R}_+, \mathbb{R}_+)$ function such that for $\alpha = \max(\gamma, \gamma_\tau)$

$$f''(x) > \frac{\alpha - 1}{\alpha}(-f'(x)), \quad \text{for all } x > 0. \tag{26}$$

Notice that the above requirements exclude the "natural" choice $g = h = Id$.

Equipped with these relevant nonlinear transforms, the first equivalent system for smooth solutions we consider is obtained when choosing (for instance) $(\rho, \rho u, \rho E, \rho S_\tau)$ as independent variables and thus writes :

$$\begin{cases} \partial_t \rho + \partial_x \rho u = 0, \qquad x \in \mathbb{R},\ t > 0, \\ \partial_t \rho u + \partial_x(\rho u^2 + p + p_\tau) = \partial_x\left((\mu + \mu_\tau)\partial_x u\right), \\ \partial_t \rho E + \partial_x(\rho E + p + p_\tau)u = \partial_x\left((\mu + \mu_\tau)u\partial_x u\right) + \partial_x(\kappa \partial_x T) + \partial_x(\kappa_\tau \partial_x T_\tau), \\ \partial_t \rho S_\tau + \partial_x \rho S_\tau u = \dfrac{\gamma_\tau - 1}{\rho^{\gamma_\tau - 1}} h'(h^{-1}(S_\tau))\left(\mu_\tau(\partial_x u)^2 + \partial_x(\kappa_\tau \partial_x T_\tau)\right). \end{cases} \tag{27}$$

Here the pair $(\rho S, \rho S u)$ must be understood, under the assumption (26), as a Lax entropy pair but that must satisfy without additional conditions for smooth solutions the following imposed rate of entropy dissipation :

$$\partial_t \rho S + \partial_x \rho S u = \frac{\gamma - 1}{\rho^\gamma} g'(g(S))\left(\mu(\partial_x u)^2 + \partial_x(\kappa \partial_x T)\right). \tag{28}$$

We next propose a second equivalent system in which a straightforward extension of the balance equation (11) is explicitely involved. Let us indeed consider :

$$\partial_t \rho + \partial_x \rho u = 0, \qquad x \in \mathbb{R},\ t > 0, \tag{29.i}$$

$$\partial_t \rho u + \partial_x(\rho u^2 + p + p_\tau) = \partial_x\left((\mu + \mu_\tau)\partial_x u\right), \tag{29.ii}$$

$$\partial_t \rho E + \partial_x(\rho E + p + p_\tau)u = \partial_x\left((\mu + \mu_\tau)u\partial_x u\right) + \partial_x(\kappa \partial_x T) + \partial_x(\kappa_\tau \partial_x T_\tau), \tag{29.iii}$$

$$\begin{aligned} &\mu\left(\frac{\rho^{\gamma_\tau - 1}}{\gamma_\tau - 1}\frac{1}{h'(h^{-1}(S_\tau))}\left(\partial_t \rho S_\tau + \partial_x \rho S_\tau u\right)\right) - \\ &\qquad \mu_\tau\left(\frac{\rho^{\gamma - 1}}{\gamma - 1}\frac{1}{g'(g^{-1}(S))}\left(\partial_t \rho S + \partial_x \rho S u\right)\right) = \\ &\qquad\qquad \partial_x(\mu_\tau \kappa \partial_x T - \mu \kappa_\tau \partial_x T_\tau). \end{aligned} \tag{29.iv}$$

Since the above system involves five partial derivatives in time for only four independent variables, one of the two Lax entropies, either ρS or ρS_τ must be understood as a nonlinear function of the four remaining independent variables $(\rho, \rho u, \rho E, .)$. Let us underline that (29.iv) is actually equivalent to the non standard balance equation (11) and is thus intrinsically associated with the jump relation (20) (see Berthon-Coquel [4] for the details).

3 On a physical example

In this section, our purpose is to briefly exemplify the proposed framework on the basis of a crude model for turbulent compressible flows. Such a model is derived when assuming a constant turbulent mixing lenght. Let us underline that more sophisticated models, the so-called "two transport" equations models, also enter the present framework but the required material falls beyond the scope of the present paper. The reader is referred to the companion works [1], [4] for the details.

Under the assumption of a large Reynolds number, the PDE's system we are interested in is usually written under the following (unsuited) form :

$$\partial_t \rho + \partial_x \rho u = 0, \qquad x \in \mathbb{R},\ t > 0, \tag{30.i}$$

$$\partial_t \rho u + \partial_x(\rho u^2 + p + \frac{2}{3}\rho k) = \partial_x\left((\mu + \mu_T)\partial_x u\right), \tag{30.ii}$$

$$\partial_t \rho E + \partial_x(\rho E + p + \frac{2}{3}\rho k)u = \partial_x\left((\mu + \mu_T)u\partial_x u\right) + \partial_x(\kappa \partial_x T) + \partial_x(\kappa_T \partial_x T_T), \tag{30.iii}$$

$$\partial_t \rho k + \partial_x \rho k\, u = \left\{ -\frac{2}{3}\rho k \partial_x u + \mu_T\,(\partial_x u)^2 \right\} + \partial_x(\kappa_T \partial_x T_T) - \rho\epsilon, \tag{30.iv}$$

where ρk denotes the so-called turbulent kinetic energy. Here the laminar viscosity $\mu := \mu(T)$ obeys the standard Sutherland law while the turbulent viscosity μ_T is modelled by :

$$\mu_T = C_\mu L \rho \sqrt{k}, \tag{31}$$

where L, the so-called turbulent mixing lenght, is assumed to be a fixed positive real number and C_μ denotes some positive constant of the model. At last the relaxation term $\rho\epsilon$ in (30) follows from the definition:

$$\epsilon = \frac{k^{3/2}}{L}. \tag{32}$$

Introducing the following convenient notation :

$$p_T = (\gamma_T - 1)\rho k, \quad \gamma_T = \frac{5}{3}. \tag{33}$$

the system (30) readily finds the following form :

$$\partial_t \rho + \partial_x \rho u = 0, \qquad x \in \mathbb{R},\ t > 0,$$
$$\partial_t \rho u + \partial_x(\rho u^2 + p + p_T) = \partial_x\left((\mu + \mu_T)\partial_x u\right),$$
$$\partial_t \rho E + \partial_x(\rho E + p + p_T)u = \partial_x\left((\mu + \mu_T)u\partial_x u\right) + \partial_x(\kappa \partial_x T) + \partial_x(\kappa_T \partial_x T_T),$$
$$\partial_t p_T + \partial_x p_T\, u + (\gamma_T - 1)p_T \partial_x u = (\gamma_T - 1)\left\{\mu_T\,(\partial_x u) + \partial_x(\kappa_T \partial_x T_T) - \rho\epsilon\right\},$$

when redistributing suitably the products in non conservation form in (30). The above system is nothing else but the system (3) up to some relaxation term. Let us underline that both the viscosities functions nonlinearly depends on the unknown and that the non standard balance equation (11) extends to :

$$\mu_r\Big\{\frac{\rho^{\gamma-1}}{\gamma-1}(\partial_t\rho s+\partial_x\rho s u)-\partial_x(\kappa\partial_x T)\Big\}-$$
$$\mu\Big\{\frac{\rho^{\gamma_r-1}}{\gamma_r-1}(\partial_t\rho s_r+\partial_x\rho s_r u)-\partial_x(\kappa_r\partial_x T_r)\Big\}=0. \qquad (35)$$

A straightforward extension of the generalized jump condition (20) thus easily follows when suitably redifining both (18) and (19) according to (35).

4 Godunov methods with nonlinear projections

This section is devoted to the numerical approximation of the smooth solutions of the convective-diffusive system (3) with a special emphasis put on the satisfaction at the discrete level of the generalized jump condition (22). As pointed out in the previous sections, several equivalent forms for smooth solutions exist and by construction they are all associated with the same generalized jump conditions.

As proved below, such an equivalence principle, valid for exact solutions, turns out to be lost in general at the discrete level within the frame of classical finite volume methods. Indeed such methods when derived for one equivalent form and another systematically produce grossly different approximate solutions that exhibit large errors with the exact solution. As pointed out below, errors find essentially their roots in the numerical approximation of the first order extracted system. Roughly speaking, if suitable finite difference formulae for approximating the second order operator can be actually deviced so that they yield no further error, on the contrary the errors induced by classical approximate Riemann solvers cannot be avoided. Referred as to L^2 projection errors in the sequel, these are indeed shown below to result from the averaging procedure of neighbouring approximate Riemann solutions over each computational cell. Essentially due to the Jensen inequality, linear averagings induce a too large numerical entropy dissipation rate in comparison with the one to be prescribed. This in turn forbids the right hand sides in (6) and (7) to be kept in balance at the discrete level according to (10). This can only result at the discrete level in a violation of the required generalized jump condition (29.iv).

To enforce the validity of (29.iv), we introduce below a nonlinear projection procedure. Involved as an additional and last step to any given standard finite volume method, this procedure provides a systematic and effective correction in redistributing the L^2 projection errors between the two numerical entropy dissipation rates in (6) and (7) in order to keep in balance their evolution in time according to (29.iv). As a benefit, discrete solutions obtained by nonlinear projection actually achieve a much better agreement with exact solutions.

This nonlinear correction procedure is proved in addition to preserve all the stability properties met by the underlying classical Approximate Riemann solver, namely and for the relevant Riemann solvers : positivy properties, a full set of discrete entropy inequalities and a maximum principle for both the specific entropies.

The numerical method we propose relies on a splitting of operators. It is made of three distinct steps. The first two steps coincide with a standard finite volume method for the system (27) and makes use of a given approximate Riemann solver. The third step details the nonlinear projection operator. We then conclude in presenting the required discretization of the initial data.

For the sake of concisness in the forthcoming developments, we do not address the discretization of the involved Fourier laws. We thus let $\kappa = \kappa_\tau = 0$ and we refer the reader to the companion paper [4] for the required discrete formulae.

4.1 Godunov type methods and L^2 projections

Let Δt and Δx respectively denote the time and space increments, chosen to be constant without restriction. The numerical approximate solution, $\mathbf{v}_{\Delta x} : \mathbb{R} \times \mathbb{R}_+ \to \Omega$, is as usual supposed to be piecewise constant and we set using classical notations :

$$\mathbf{v}_{\Delta x}(x,t) = \mathbf{v}_i^n, \quad (x,t) \in (x_{i-1/2}, x_{i+1/2}) \times (t^n, t^{n+1}), \quad i \in Z, n \in N. \tag{36}$$

First step : Extracted first order system ($t^n \to t^{n+1,=}$)
Assume that the discrete solution $\mathbf{v}_h(x,t^n)$ is known at the time level t^n. In order to evolve it in time, we propose to solve as a first step the following Cauchy problem :

$$\begin{cases} \partial_t \rho + \partial_x \rho u = 0, \qquad x \in \mathbb{R},\ t > 0, \\ \partial_t \rho u + \partial_x (\rho u^2 + p + p_\tau) = 0, \\ \partial_t \rho E + \partial_x (\rho E + p + p_\tau) u = 0, \\ \partial_t \rho S_\tau + \partial_x \rho S_\tau u = 0, \end{cases} \tag{37}$$

when prescribing the initial data to $\mathbf{v}_h(x,t^n)$. Weak solutions of the above hyperbolic systems are asked to satisfy the following Lax entropy inequality

$$\partial_t \rho S + \partial_x \rho S u \leq 0, \tag{38}$$

to rule out unphysical solutions. For convenience in the discussion, the problem (37), (38) is solved exactly. Then under the following CFL like condition :

$$\frac{\Delta t}{\Delta x}\max|\lambda_i(\mathbf{v})| \leq \frac{1}{2}, \tag{39}$$

the solution is classically made of neighbouring and non interacting elementary Riemann solutions. This solution is then classically averaged over each cell. Let denote by $g : \Omega \times \Omega \to \mathbb{R}^4$ the associated Lipschitz continuous numerical flux function. Setting $g^n_{i+1/2} = g(\mathbf{v}^n_i, \mathbf{v}^n_{i+1})$, the updated solution then reads :

$$\mathbf{v}^{n+1,=}_i = \mathbf{v}^n_i - \frac{\Delta t}{\Delta x}\{g^n_{i+1/2} - g^n_{i-1/2}\}, \quad i \in Z. \tag{40}$$

The following result easily follows using standard arguments (see for instance Godlewski-Raviart [10]) :

Lemma 4.1
Under the CFL condition (39), the unknown ρS_τ updates according to the following identity :

$$(\rho S_\tau)^{n+1,=}_i - (\rho S_\tau)^n_i + \frac{\Delta t}{\Delta x}\Big\{\{\rho S_\tau u\}^n_{i+1/2} - \{\rho S_\tau u\}^n_{i-1/2}\Big\} = 0, \tag{41}$$

while the Lax entropy pair $(\rho S, \rho S u)$ obeys the following discreie entropy inequality :

$$\{\rho S\}(\mathbf{v}^{n+1,=}_i) - (\rho S)^n_i + \frac{\Delta t}{\Delta x}\Big\{\{\rho S u\}^n_{i+1/2} - \{\rho S u\}^n_{i-1/2}\Big\} = \mathcal{E}^n_i \leq 0 \tag{42}$$

Wr refer the reader to the companion paper [4] for a precise definition of the numerical flux functions in (41) and (42). Let us underline that the numerical rate of entropy dissipation in (42) is generally strictly negative. Indeed the L^2 projection of the exact solution of (37), (38) cannot in general preserve ρS since by the well-known Jensen inequality, we have :

$$\{\rho S\}(\mathbf{v}^{n+1,=}_i) \leq \frac{1}{\Delta x}\int_{x_{i-1/2}}^{x_{i+1/2}} \{\rho S\}\,(\rho, \rho u, \rho E, \rho S_\tau)\,(x, t^{n+1,=})dx. \tag{43}$$

By contrast, the specific entropy S_τ is simply advected by the flow according to system (37) and is thus preserved at the L^2 projection step. This discrepancy in the rates of entropy dissipation, strictly negative versus zero, results in a failure for satisfying the expected balance equation (11) at the discrete level. Notice that this negative result cannot be bypassed when using instead of the exact Godunov scheme a relevant entropy satisfying approximate Riemann solver (see [10] for examples).

Second step : Diffusive operator ($t^{n+1,=} \to t^{n+1,-}$)
The discrete solution $\mathbf{v}_h(x, t^{n+1,=})$ is next evolved in time to the date $t^{n+1,-}$ when solving with the initial data $\mathbf{v}_h(x, t^{n+1,=})$:

$$\begin{aligned}
\partial_t \rho &= 0, && (44.\text{i})\\
\partial_t \rho u &= \partial_x((\mu + \mu_\tau)\partial_x u), && (44.\text{ii})\\
\partial_t \rho E &= \partial_x((\mu + \mu_\tau)u\partial_x u), && (44.\text{iii})\\
\partial_t \rho S_\tau &= \frac{\gamma_\tau - 1}{\rho^{\gamma_\tau - 1}} h'(h^{-1}(S_\tau))\mu_\tau(\partial_x u)^2. && (44.\text{iv})
\end{aligned}$$

In that aim, we suggest to adopt the following implicite finite difference scheme :

$$\begin{aligned}
\rho_i^{n+1,-} &= \rho_i^{n+1,=}, && (45.\text{i})\\
(\rho u)_i^{n+1,-} &= (\rho u)_i^{n+1,=} + \Delta t \overline{\partial_x((\mu + \mu_\tau)\partial_x u_i}^{n+1,-}, && (45.\text{ii})\\
(\rho E)_i^{n+1,-} &= (\rho E)_i^{n+1,=} + \Delta t \overline{\partial_x((\mu + \mu_\tau)u\partial_x u)_i}^{n+1,-}, && (45.\text{iii})\\
(\rho S_\tau)_i^{n+1,-} &= (\rho S_\tau)_i^{n+1,=} + \Delta t \overline{\frac{\gamma_\tau - 1}{\rho^{\gamma_\tau - 1}} h'(h^{-1}(S_\tau))\mu_\tau(\partial_x u)^2}_i^{n+1,-}, && (45.\text{iv})
\end{aligned}$$

where we have set :

$$\begin{aligned}
&\overline{\partial_x((\mu + \mu_\tau)\partial_x u_i}^{n+1,-} = \frac{\mu + \mu_\tau}{\Delta x^2}\left(M^n u_{i+1} - 2M^n u_i + M^n u_{i-1}\right), && (46.\text{i})\\
&\overline{\partial_x(\mu + \mu_\tau)u\partial_x u_i}^{n+1,-} = \frac{\mu + \mu_\tau}{2\Delta x^2}\left((M^n u_{i+1})^2 - 2(M^n u_i)^2 + (M^n u_{i-1})^2\right), && (46.\text{ii})\\
&\overline{\frac{\gamma_\tau - 1}{\rho^{\gamma_\tau - 1}} h'(h^{-1}(S_\tau))\mu_\tau(\partial_x u)^2}_i^{n+1,-} = \frac{\gamma_\tau - 1}{(\rho_i^{n+1,-})^{\gamma_\tau - 1}} \overline{h'(h^{-1}(S_\tau))}_i^{n+1,-} \times\\
&\qquad \frac{\mu_\tau}{2\Delta x^2}\left((M^n u_{i+1} - M^n u_i)^2 + (M^n u_i - M^n u_{i-1})^2\right), && (46.\text{iii})
\end{aligned}$$

together with

$$\overline{h'(h^{-1}(S_\tau))}_i^{n+1,-} = \begin{cases} \dfrac{S_{\tau i}^{n+1,-} - S_{\tau i}^{n+1,=}}{h^{-1}(S_{\tau i}^{n+1,-}) - h^{-1}(S_{\tau i}^{n+1,=})}, & \text{if } S_{\tau i}^{n+1,-} \neq S_{\tau i}^{n+1,=} \\ h'(h^{-1}(S_{\tau i}^{n+1,-})), & \text{otherwise.} \end{cases} \tag{47}$$

In (46), M^n denotes a time averaging operator given by :

$$M^n X = \frac{X^{n+1,-} + X^{n+1,=}}{2} \tag{48}$$

Let us underline that the first two finite difference operators, (46.i), (46.ii), preserve by construction the conservation property to be satisfied by the unknowns ρu and ρE while as expected, the last operator always achieves the signe of $h'(h^{-1}(S_\tau))$.

Besides and as pointed out below, the benefit of these formulae is twofold.

In the one hand, since the density is kept constant during this second step, the implicit equations (45.ii) stay completely decoupled from the others, namely (45.iii) and (45.iv).

Therefore solving (45) just amounts in practice to invert a positive definite symmetric matrix for the unknown $M^n u$. $(\rho E)_i^{n+1,-}$ can be then evaluated. Turning considering the last unknown $(\rho S_r)_i^{n+1,-} = h({s_r}_i^{n+1,-})$, it turns out that by construction ${s_r}_i^{n+1,-}$ explicitely reads :

$$
\begin{aligned}
{s_r}_i^{n+1,-} &= {s_r}_i^{n+1,=} + \\
&\frac{\gamma_r - 1}{(\rho_i^{n+1,-})^{\gamma_r}} \frac{\mu_r \Delta t}{2\Delta x^2} \left((M^n u_{i+1} - M^n u_i)^2 + (M^n u_i - M^n u_{i-1})^2 \right). \quad (49)
\end{aligned}
$$

In the second hand, straightforward calculations yield from the above formulae the following identity :

$$
\begin{aligned}
\{s\}(\mathbf{v}_i^{n+1,-}) &= s_i^{n+1,=} + \\
&\frac{\gamma_r - 1}{(\rho_i^{n+1,-})^{\gamma_r}} \frac{\mu_r \Delta t}{2\Delta x^2} \left((M^n u_{i+1} - M^n u_i)^2 + (M^n u_i - M^n u_{i-1})^2 \right). \quad (50)
\end{aligned}
$$

Hence during this second step, the proposed finite difference operators allow for preserving at the discrete level the expected balance between the two rates of entropy dissipation. Indeed, we easily get from (49) and (50) :

$$
\begin{aligned}
&\mu_r \frac{(\rho_i^{n+1,-})^{\gamma-1}}{\gamma - 1} \left\{ (\rho s)_i^{n+1,-} - (\rho s)_i^{n+1,=} \right\} \\
&-\mu \frac{(\rho_i^{n+1,-})^{\gamma_r-1}}{\gamma_r - 1} \left\{ (\rho s_r)_i^{n+1,-} - (\rho s_r)_i^{n+1,=} \right\} = 0. \quad (51)
\end{aligned}
$$

Let us underline that other finite difference formulae are actually possible but up to our knowledge, such formulae systematically produce (strictly) negative errors in the discrete entropy balance equation for ρS.

Summary of the first two steps : Classical L^2 projection methods. The above two steps yield a standard finite volume method for approximating the solutions of (27). Such a method will be referred in the sequel as to a classical L^2 projection method. The properties of interest are stated below :

Lemma 4.2
Under the CFL condition (39), the variable ρS_r satisfies :

$$
\begin{aligned}
(\rho S_r)_i^{n+1,-} - (\rho S_r)_i^n + \frac{\Delta t}{\Delta x} \left\{ \{\rho S_r u\}_{i+1/2}^n - \{\rho S_r u\}_{i-1/2}^n \right\} &= \\
\Delta t \overline{\frac{\gamma_r - 1}{\rho^{\gamma_r - 1}} h'(h^{-1}(S_r)) \mu_r (\partial_x u)^2}_i^{n+1,-} &\leq 0, \quad (52)
\end{aligned}
$$

while simultaneously, the Lax entropy pair $(\rho S, \rho Su)$ satisfies the discrete entropy inequality :

$$\{\rho S\}(\mathbf{v}_i^{n+1,-}) - (\rho S)_i^n + \frac{\Delta t}{\Delta x}\Big\{\{\rho S u\}_{i+1/2}^n - \{\rho S u\}_{i-1/2}^n\Big\} =$$

$$\mathcal{E}_{c_i}^n + \Delta t \overline{\frac{\gamma-1}{\rho^{\gamma-1}} g'(g^{-1}(S))\mu(\partial_x u)^2}_i^{n+1,-} \leq 0, \qquad (53)$$

where the last term obeys a definition similar to (46.iii).

Besides, the classical L^2 projection method can be shown to satisfy several desirable stability properties : namely it is positivity preserving and obeys discrete maximum principles for both specific entropies S and S_r (see Tadmor [19] for the setting of 3×3 Euler equations). Nevertheless, (52) and (53) are easily seen to yield the following discrete form for (11) :

$$\mu_r \Big\{\overline{\frac{\gamma-1}{\rho^{\gamma-1}} g'(g^{-1}(S))}\Big\}^{-1} \Big\{\{\rho S\}(\mathbf{v}_i^{n+1,-}) - (\rho S)_i^n + \frac{\Delta t}{\Delta x}\Delta\{\rho S u\}_{i+1/2}^n\Big\} -$$

$$\mu \Big\{\overline{\frac{\gamma_r-1}{\rho^{\gamma_r-1}} h'(h^{-1}(S_r))}\Big\}^{-1} \Big\{(\rho S_r)_i^{n+1,-} - (\rho S_r)_i^n + \frac{\Delta t}{\Delta x}\Delta\{\rho S_r u\}_{i+1/2}^n\Big\} =$$

$$\mu_r \Big\{\overline{\frac{\gamma-1}{\rho^{\gamma-1}} g'(g^{-1}(S))}\Big\}^{-1} \times \mathcal{E}_{c_i}^n \neq 0. \qquad (54)$$

This strongly suggests that the classical L^2 projection method can only fail in satisfying the generalized jump condition (22). The reader is referred to [1] for a rigourous proof and to the numerical results below for an illustration of the negative consequences of such a failure.

4.2 Nonlinear projection methods.

According to the discrete balance equation (54), standard finite volume methods induce a too large numerical rate of entropy dissipation for ρS in comparison with that of ρS_r that in turn precludes the satisfaction of (22). Here, we propose to add as an additional step to classical L^2 projection methods a nonlinear procedure, the so-called nonlinear projection step, which purpose is precisely to correct the former errors. Indeed, the aim of the third step we propose is to redistribute the errors between the two rates of entropy dissipation in order to keep them in balance according to (22). Let us underline that the numerical procedure derived below inherits by construction all the desirable stability properties satisfied by relevant approximate Riemann solvers.

Third step : Nonlinear projection ($t^{n+1,-} \to t^{n+1}$)

In order to preserve the required conservation properties, let us define :

$$\rho_i^{n+1} = \rho_i^{n+1,-}, \quad (\rho u)_i^{n+1} = (\rho u)_i^{n+1,-}, \quad (\rho E)_i^{n+1} = (\rho E)_i^{n+1,-}. \quad (55)$$

Then to enforce the validity of the generalized jump condition at the discrete level (20), we propose to seek for $(\rho S_r)_i^{n+1}$ as a solution of :

$$\begin{aligned}\mu_r \overline{\frac{\rho^{\gamma-1}}{\gamma-1}\frac{1}{g'(g^{-1}(S))}}_i^{n+1,-} &\times \Big(\{\rho S\}(\rho_i^{n+1,-}, (\rho u)_i^{n+1,-}, (\rho E)_i^{n+1,-}, (\rho S_r)_i^{n+1}) \\ &-(\rho S)_i^n + \tfrac{\Delta t}{\Delta x}\Delta\{\rho S u\}_{i+1/2}^n\Big) - \quad (56)\\ \mu \overline{\frac{\rho^{\gamma_r-1}}{\gamma_r-1}\frac{1}{h'(h^{-1}(S_r))}}_i^{n+1,-} &\left((\rho S_r)_i^{n+1} - (\rho S_r)_i^n + \frac{\Delta t}{\Delta x}\Delta\{\rho S_r u\}_{i+1/2}^n\right) = 0.\end{aligned}$$

The above nonlinear problem in the unknown $(\rho S_r)_i^{n+1}$ can be shown to admit a unique nonnegative solution as soon as the approximate Riemann solver involved in the first step (4.1) obeys discrete entropy inequalities for the Lax pair $(\rho S, \rho S u)$. This in turn uniquely defines $(\rho S)_i^{n+1}$ according to :

$$(\rho S)_i^{n+1} = \{\rho S\}\left(\rho_i^{n+1}, (\rho u)_i^{n+1}, (\rho E)_i^{n+1}, (\rho S_r)_i^{n+1}\right).$$

Let us furthermore underline that when understanding $\{\rho S_r\}$ as a function of the unknowns $\mathbf{w}_i^{n+1} = (\rho_i^{n+1}, (\rho u)_i^{n+1}, (\rho E)_i^{n+1}, (\rho S)_i^{n+1})$, $(\rho S)_i^{n+1}$ can be easily seen to be the unique solution of the symmetric version of (56) :

$$\begin{aligned}\mu_r \overline{\frac{\rho^{\gamma-1}}{\gamma-1}\frac{1}{g'(g^{-1}(S))}}_i^{n+1,-} &\left((\rho S)_i^{n+1} - (\rho S)_i^n + \frac{\Delta t}{\Delta x}\Delta\{\rho S u\}_{i+1/2}^n\right) - \\ \mu \overline{\frac{\rho^{\gamma_r-1}}{\gamma_r-1}\frac{1}{h'(h^{-1}(S_r))}}_i^{n+1,-} &\left(\{\rho S_r\}(\mathbf{w}_i^{n+1}) - (\rho S_r)_i^n + \frac{\Delta t}{\Delta x}\Delta\{\rho S_r u\}_{i+1/2}^n\right) = 0.\end{aligned} \quad (57)$$

In other words, the two entropies $(\rho S)_i^{n+1}$ and $(\rho S_r)_i^{n+1}$ now play a symmetric role. Numerical methods based on (56) are referred in the sequel as to nonlinear L^2 projection methods. The nonlinear projection step (56) allows to prove in addition the following stability results :

Theorem 4.3
Under the CFL restriction (39), the following discrete entropy inequalities are satisfied :

$$\begin{aligned}&\{\rho\phi(S)\}\left(\rho_i^{n+1}, (\rho u)_i^{n+1}, (\rho E)_i^{n+1}, (\rho S_r)_i^{n+1}\right) - \\ &\qquad (\rho\phi(S))_i^n + \frac{\Delta t}{\Delta x}\Delta_i\{\rho\phi(S)u\}_{i+1/2}^n \le 0, \quad (58.\text{i})\\ &\{\rho\psi(S_r)\}\left(\rho_i^{n+1}, (\rho u)_i^{n+1}, (\rho E)_i^{n+1}, (\rho S_r)_i^{n+1}\right) - \\ &\qquad (\rho\psi(S_r))_i^n + \frac{\Delta t}{\Delta x}\Delta_i\{\rho\psi(S_r)u\}_{i+1/2}^n \le 0, \quad (58.\text{ii})\end{aligned}$$

for all strictly decreasing and convex functions ϕ and ψ. The following maximum principles for the specific entropies S and S_τ are met :

$$S_i^{n+1} \leq \max(S_{i-1}^n, S_i^n, S_{i+1}^n), \tag{59.i}$$

$$S_{\tau i}^{n+1} \leq \max(S_{\tau i-1}^n, S_{\tau i}^n, S_{\tau i+1}^n). \tag{59.ii}$$

Since the two functions g and h are strictly decreasing under the assumption (26), both the pressures p_i^{n+1} and $p_{\tau i}^{n+1}$ stay positive as soon as the density ρ^{n+1} is positive.

The proof of the above statement is detailled in [1]. Let us conclude the present paragraph in highlighting that the nonlinear projection step (63) is by construction consistent with the two limit cases in full conservation form we have put forward in Theorem 2.5. Indeed, sending formally (for instance) the viscosity μ_τ to zero, makes (63) degenerate into the following expected relations (see (24)):

$$\begin{aligned}(\rho S_\tau)_i^{n+1} &= (\rho S_\tau)_i^n - \frac{\Delta t}{\Delta x}\Delta_i\{\rho S_\tau u\}_{i+1/2}^n, \\ (\rho\phi(S)_i^{n+1} &- (\rho\phi(S)_i^n + \frac{\Delta t}{\Delta x}\Delta_i\{\rho\phi(S)u\}_{i+1/2}^n \leq 0.\end{aligned} \tag{60}$$

Similarly and since in view of (57), the nonlinear projection does not break the symmetry in the roles played by ρS and ρS_τ, we get from (63) when sending formally to zero the viscosity μ, the expected relations (see (25)) :

$$\begin{aligned}(\rho S)_i^{n+1} &= (\rho S)_i^n - \frac{\Delta t}{\Delta x}\Delta_i\{\rho S u\}_{i+1/2}^n, \\ (\rho\psi(S_\tau)_i^{n+1} &- (\rho\psi(S_\tau)_i^n + \frac{\Delta t}{\Delta x}\Delta_i\{\rho\psi(S_\tau)u\}_{i+1/2}^n \leq 0.\end{aligned} \tag{61}$$

In the sequel, the nonlinear projection operator (56) is coupled with an exact Roe linearization for the system (3) as proposed in [6].

4.3 Discretization of the initial data

For the sake of completness, we conclude the description of the method we propose in detailling the projection onto piecewise constant functions of the initial data $(\rho_0(x), (\rho u)_0(x), p_0(x), p_{\tau 0}(x))$ for system (3).

L^2 projection step
Let us first define from the initial data the following three real valued functions $\{\rho E\}_0(x)$, $\{\rho S\}_0(x)$ and $\{\rho S_\tau\}_0(x)$. Let us then introduce with clear notations the following five L^2 projections :

$$\rho_i^{0,-} = \frac{1}{\Delta x}\int_{x_{i-1/2}}^{x_{i+1/2}} \rho_0(x)\, dx, \qquad (\rho u)_i^{0,-} = \frac{1}{\Delta x}\int_{x_{i-1/2}}^{x_{i+1/2}} (\rho u)_0(x)\, dx,$$

with similar definitions for $(\rho E)_i^{0,-}$, $(\rho S)_i^{0,-}$ and $(\rho S_\tau)_i^{0,-}$.

The averages $\rho_i^{0,-}$, $(\rho u)_i^{0,-}$ and $(\rho E)_i^{0,-}$ being fixed, it is obviously impossible to simultaneously prescribe both $(\rho S)_i^{0,-}$ and $(\rho S_\tau)_i^{0,-}$ without introducing some ambiguity in the definition of the required pressures $p_i^{0,-}$ and $p_{\tau i}^{0,-}$. It indeed suffices to note that the convexity of the mapping $\{\rho S\}$ implies by the Jensen inequality the following (generally strict) inequality :

$$\{\rho S\}\left(\rho_i^{0,-}, (\rho u)_i^{0,-}, (\rho E)_i^{0,-}, (\rho S_\tau)_i^{0,-}\right) \leq \frac{1}{\Delta x}\int_{x_{i-1/2}}^{x_{i+1/2}} (\rho S)_0(x)dx = (\rho S)_i^{0,-}.$$

The next and last step aims at restoring uniqueness in the definition of both the pressures.

Nonlinear projection step In order to preserve the required conservation of density, momentum and total energy, let us set for all $i \in Z$:

$$\rho_i^0 = \rho_i^{0,-}, \quad (\rho u)_i^0 = (\rho u)_i^{0,-}, \quad (\rho E)_i^0 = (\rho E)_i^{0,-}. \tag{62}$$

Then let us redefine $(\rho S_\tau)_i^0$ as the solution of the following nonlinear equation :

$$\mu_\tau \overline{\frac{(\rho)^{\gamma-1}}{\gamma-1}\frac{1}{g'(g^{-1}(S))}}_i^{0,-}\left(\{\rho S\}\left(\rho_i^0, (\rho u)_i^0, (\rho E)_i^0, (\rho S_\tau)_i^0\right) - (\rho S)_i^{0,-}\right) - \mu \overline{\frac{(\rho)^{\gamma_\tau-1}}{\gamma_\tau-1}\frac{1}{h'(h^{-1}(S_\tau))}}_i^{0,-}\left((\rho S_\tau)_i^0 - (\rho S_\tau)_i^{0,-}\right) = 0, \tag{63}$$

where the involved two averages are given by the consistent definitions :

$$\overline{\frac{(\rho)^{\gamma-1}}{\gamma-1}\frac{1}{g'(g^{-1}(S))}}_i^{0,-} = \frac{(\rho_i^{0,-})^{\gamma-1}}{\gamma-1}\frac{1}{g'(g^{-1}(S_i^{0,-}))}, \tag{64}$$

together with a similar definition for $\overline{\frac{(\rho)^{\gamma_\tau-1}}{\gamma_\tau-1}\frac{1}{h'(h^{-1}(S_\tau))}}_i^{0,-}$. The present second step clearly corresponds to the nonlinear projection we have described in the above section. Let us conclude with the following statement :

Theorem 4.4
The nonlinear projection step (63) uniquely determines the pressures p_i^0, $p_{\tau i}^0$ and preserves their positivity as soon as the initial density $\rho_0(x)$ is positive.

4.4 The usual approach for solving (3)

For the sake of comparison, we end the present section when briefly recalling the most usual (if not systematic) approach for approximating the solutions of systems in non conservation form like (3). According to this approach, all the non conservative products are rejected to the right hand side of the governing equations and are treated as "source terms". When applied to (27), the first step (4.1) is therefore concerned with the following first order extracted system in conservation form :

$$\begin{cases} \partial_t \rho + \partial_x \rho u = 0, \qquad x \in I\!R,\ t > 0, \\ \partial_t \rho u + \partial_x(\rho u^2 + p + p_\tau) = 0, \\ \partial_t \rho E + \partial_x(\rho E + p + p_\tau)u = 0, \\ \partial_t p_\tau + \partial_x p_\tau u = 0, \end{cases} \tag{65}$$

while in a next step, the "source term" $-(\gamma_\tau - 1)p_\tau \partial_x u$ is to be given some *ad hoc* finite difference approximation. We refer for instance the reader to [15] and [13] concerning the details. The (severe) drawbacks in the resulting numerical schemes are illustrated below. Let us furthermore underline after Forestier-Herard-Louis [8] that exact Riemann solutions of (65) all preserve the positivity of the total pressure (far away from vacuum) but do not necessarily keep non negative the partial pressure p. By contrast, both the pressures p and p_τ in (3) can be shown to stay positive (again far away from vacuum). Such schemes are referred in the sequel as to classical methods. In what follows, Riemann solutions of (65) are approximated using one of the most widely used method : namely the Roe scheme [17].

5 Numerical results

The ability of the three discussed schemes in the capture of viscous shock layers for (3) is evaluated when testing their sensitivity in the prediction of the end states with respect to the mesh refinement. The initial data of the Cauchy problems to be solved are made of two constant states, the discontinuity being located at $x = 0$. The associated exact solutions thus correspond to smooth regularizations of Riemann solutions for the underlying first order system extracted from (3). They are thus made up generally speaking of the juxtaposition of travelling wave solutions and "rarefaction" solutions. Discrete solutions are systematically compared with the exact solutions obtained when integrating the EDO's system for governing travelling wave solutions.

All the calculations described below have been performed according to the following strategy. An exact Roe type linearization for system (37) yields an approximate Riemann solver to solve the first step (4.1) (see [6], [1] for the detailled formulae). Successive grids refinements, ranging from 100 to 2000 cells, are considered. The CFL number is fixed at the constant value 0.5.

Four test cases, labelled from A to D, are addressed. Problems A, B and C are directly motivated by the three distinct regimes that underly the flow model under consideration and that are dictated by the amplitude of the viscosities ratio μ_τ/μ. Namely, they involve a viscosities ratio which is successively small, close and large with respect to unity. Here, both the viscosities μ and μ_τ are assumed to be fixed positive constants. The reader is referred to Berthon [1] for the setting of varying viscosities. The problem D treats the case of non zero heat conductivities. Problem D treats the full model (3) when considering non zero heat conductivities.

In all the benchmarks discussed below, the Reynolds number is set at the constant value $Rey = 10^5$. For the benchmark D, the Prandtl numbers are respectively set to $Pr = 0.72$ and $Pr_\tau = 0.9$. The associated initial data are defined in table 1. Test case D admits the same initial data as problem B.

Test	γ	γ_τ	μ_τ/μ	ρ	u	p	p_τ
A	1.4	1.6	0.01	1	1	1	0.6
				1.05518	-0.88895	0.15031	0.33869
B, D	1.4	1.6	1	1	1	1	0.6
				1.92678	-1.25451	2.74247	2.09561
C	1.4	1.6	100	1	1	1	0.6
				1.01595	-0.93415	0.24142	0.15528

Table 1.

Test cases A, B and C. The Theorem (2.5) indicates that the benchmarks A and C are both asymptotically close to two limit systems in conservation form. Namely, case A makes the entropy ρS_τ to be asymptotically driven by a conservation law while ρS inherits such a property for problem C. By contrast, the solution of problem B stays far away from these two limit situations. All the figures assess that the usual numerical strategy (4.4) grossly fails to properly restore the correct end states in the three regimes. Turning considering the L^2 projection method, the discrete solutions agree with the exact ones only in case A as expected since ρS_τ is close to be a conservative variable. Such a property non longer holds for problems B and C and consequently large errors occur. These two schemes furthermore suffer from a dramatic sensitivity with respect to mesh refinements for problem C in that discrete solutions do not seem to converge to a given limit function even for the finest proposed grids. By contrast and concerning benchmarks A and B, the discrete solutions stay non sensitive with respect to the mesh refinement but the "limit" function does not coincide with the expected exact solution.

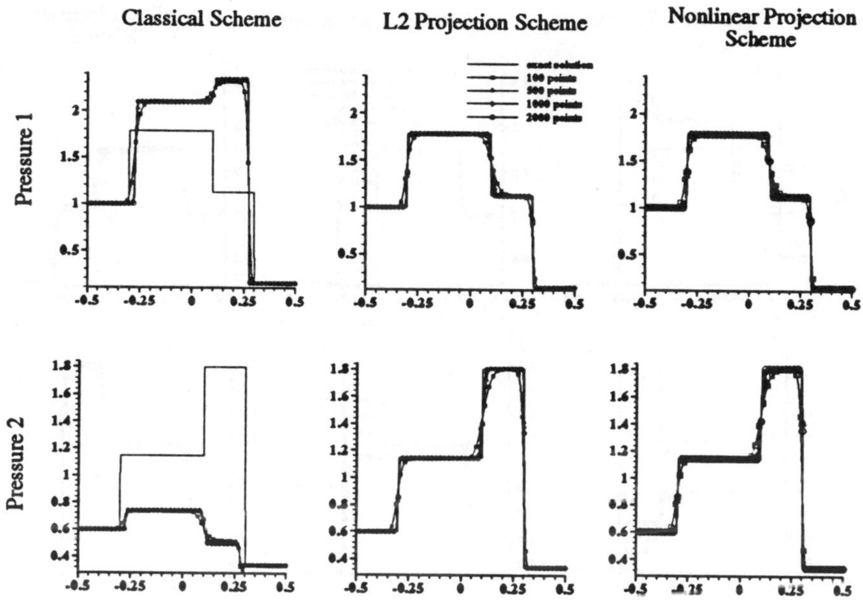

Fig. 3. Problem A : $\mu_\tau/\mu << 1$

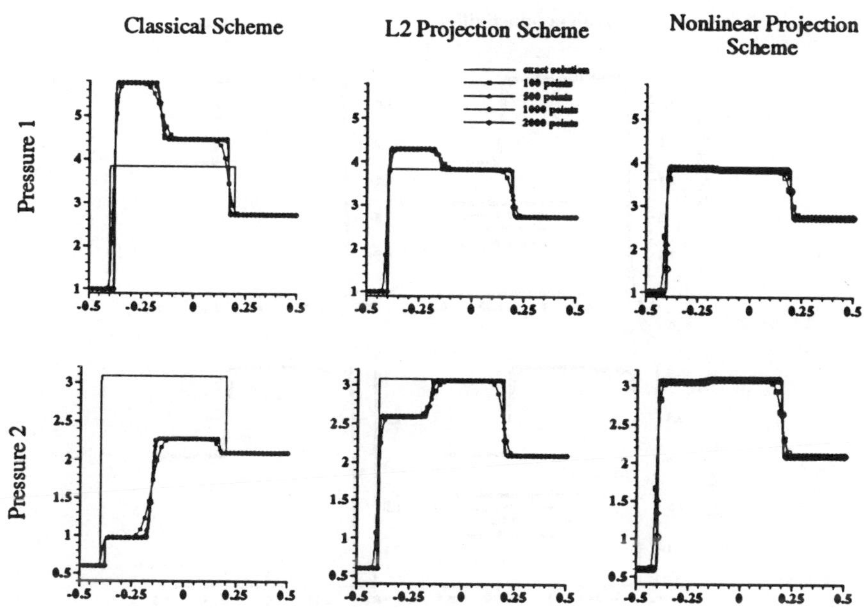

Fig. 4. Problem B : $\mu_\tau/\mu = 1$

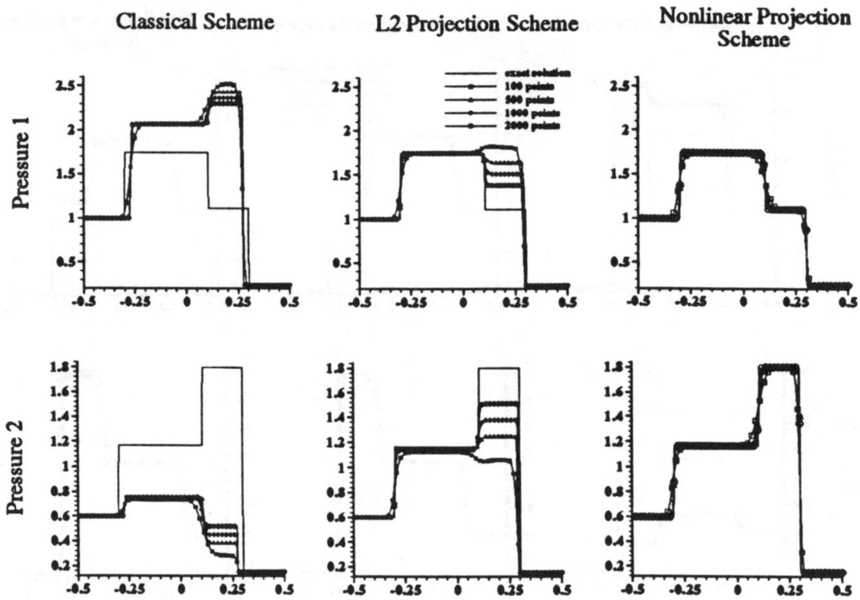

Fig. 5. Problem C : $\mu_\tau/\mu >> 1$

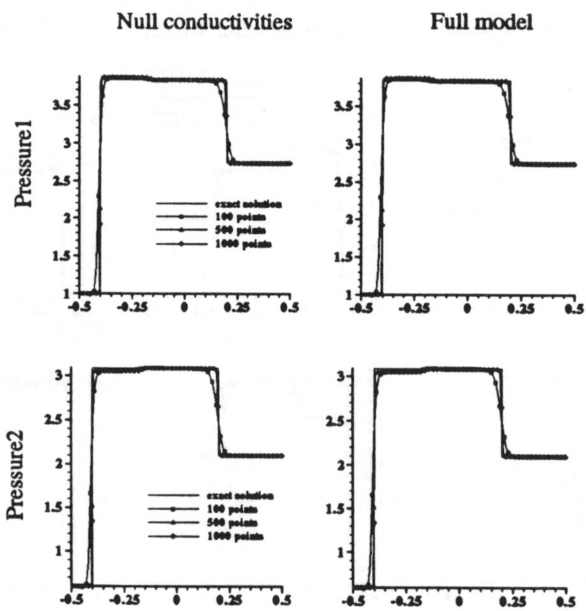

Fig. 6. Problem D : $\mu_\tau/\mu = 1$

Turning considering the nonlinear L^2 projection method, it produces approximate solutions that achieve a fairly good (if not excellent) agreement with the exact solutions while staying almost non sensitive with Δx in the three investigated regimes.

Test case D. The present problem studies the sensitivity of the nonlinear projection method with respect to heat conductivities. The choice of the initial data (namely the one of problem B) makes the solution to stay far away from two fully conservative limit cases. In this sense, the problem to be solved is the more difficult. The numerical results displayed in figure 6 clearly indicates that the discrete solutions again stay non sensitive with respect to mesh refinements in the presence of Fourier laws. This illustrates the benefits of the nonlinear projection method we have proposed.

References

[1] C. BERTHON, Contributions to the numerical analysis of the compressible Navier-Stokes equations with two specific entropies. Applications to turbulent compressible flows. *PhD dissertation* (in French) University Paris VI, january 1999.

[2] C. BERTHON AND F. COQUEL, A non conservative system modelling compressible turbulent flows. Part 1 : existence of travelling wave solutions, *work in preparation*, see also Proceedings of the 7th International Conference on Hyperbolic Problems, Zurich 1998.

[3] C. BERTHON AND F. COQUEL, A non conservative system modelling compressible turbulent flows. Part 2 : approximate generalized jump conditions, *work in preparation.*

[4] C. BERTHON AND F. COQUEL, Nonlinear projection methods for systems in non conservation form. *work in preparation.*

[5] J. F. COLOMBEAU, A. Y. LEROUX, A. NOUSSAIR AND B. PERROT, Microscopic profiles of shock waves and ambiguities in multiplications of distributions, *SIAM, J. of Numer. Anal.*, 26, No 4, 871–883 1989.

[6] F. COQUEL AND C. MARMIGNON, A Roe-type linearization for the Euler equations for weakly ionized multi-component and multi-temperature gas, Proceedings of the AIAA 12th CFD Conference, San Diego (USA) 1995.

[7] G. DAL MASO, P. LEFLOCH AND F. MURAT, Definition and weak stability of a non conservative product, *J. Math. Pures Appl.*, 74, 483–548 1995.

[8] A. FORESTIER, J.M. HERARD AND X. LOUIS, A Godunov type solver to compute turbulent compressible flows, *C.R.A.S. Paris*, 324, Série I, No 8, pp. 919-926 1997.

[9] D. GILBARG, The existence and limit behavior of the one-dimensional shock layer, *Amer. J. Math.*, 73, 256–274 1951.

[10] E. GODLEWSKI AND P.A. RAVIART, **Hyperbolic systems of conservations laws**, Applied Mathematical Sciences, Vol 118, Springer 1996.

[11] T. Y. HOU AND P. G. LEFLOCH, Why nonconservative schemes converge to wrong solutions : error analysis, *Math. of Comp.*, Vol 62, No 206, 497–530 1994.

[12] S. KARNI, Viscous shock profiles and primitive formulations, *SIAM J. Numer. Anal.*, 29, No 6, 1592–1609 1992.

[13] B. LARROUTUROU AND C. OLIVIER, On the numerical appproximation of the K-eps turbulence model for two dimensional compressible flows, *INRIA report*, No 1526 1991.

[14] P.G. LEFLOCH, Entropy weak solutions to nonlinear hyperbolic systems under non conservation form. *Comm. Part. Diff. Equa.* 13, No 6, 669–727 (1988).

[15] B. MOHAMMADI AND O. PIRONNEAU, Analysis of the K-Epsilon Turbulence Model, Research in Applied Mathematics, Masson Eds 1994.

[16] P. A. RAVIART AND L. SAINSAULIEU, A nonconservative hyperbolic system modelling spray dynamics. Part 1. Solution of the Riemann problem, *Math. Models Methods in App. Sci.*, 5, No 3, 297–333 1995.

[17] P.L. ROE, Approximate Riemann solvers, parameter vectors and difference schemes, *J. Comp. Phys.*, 43, 357–372 1981.

[18] L. SAINSAULIEU, Travelling waves solutions of convection-diffusion systems whose convection terms are weakly nonconservative, *SIAM J. Appl. Math.*, 55, No 6, 1552–1576 1995.

[19] E. TADMOR, A minimum entropy principle in the gas dynamics equations, *Appl. Numer. Math.*, No 2, 211–219 1986.

A hybrid fluctuation splitting scheme for two-dimensional compressible steady flows

P. De Palma*, G. Pascazio**, M. Napolitano**

* Dipartimento di Ingegneria Meccanica, Università di Roma "Tor Vergata", via di Tor Vergata 110, 00130 Roma, ITALY
** Istituto di Macchine ed Energetica, Politecnico di Bari, Via Re David 200, 70125 Bari, ITALY
email: *pietro@imedepa.poliba.it pascazio@poliba.it napolita@poliba.it*

Key Words: FLUCTUATION SPLITTING, MULTIDIMENSIONAL UPWINDING, TRANSONIC FLOWS

Abstract. This paper provides a genuinely multidimensional upwind scheme which is robust and (second-order) accurate at all flow regimes of interest for aerodynamic and turbomachinery applications. Firstly, a review of the state of the art of multidimensional upwind schemes for scalar advection equations is provided; then, the case of the Euler system is considered. For such a case, it is shown that, within the framework of fluctuation splitting residual-distribution schemes (or, equivalently, of linear finite elements), one can obtain: i) a linear second-order-accurate matrix scheme which is fully adequate for solving subsonic flows; ii) a nonlinear scalar scheme which is optimal for supersonic shockless flows and also resolves monotonically shocks separating two supersonic-flow regions; iii) a linear first-order-accurate matrix scheme which resolves strong shocks without spurious oscillations. A fully satisfactory methodology is therefore proposed by combining such three methods within a hybrid approach, which selects the most suitable scheme for each single element in the discrete domain, depending on the local flow conditions. The methodology is finally extended to the case of the (Reynolds averaged) Navier–Stokes equations using a standard finite element discretization of the viscous terms. Results are provided which demonstrate the accuracy and robustness of the proposed approach for several test cases involving inviscid, laminar and turbulent flows.

1 Introduction

In recent years it has become clear that a major advance in the accuracy of CFD codes for compressible viscous flows can be achieved by bringing multidimensional upwinding to the level of maturity of one-dimensional high resolution schemes. In a first effort to achieve such a goal, the authors have developed a very efficient local-adaptive multigrid method (Catalano 97b) based on a simple-wave decomposition of the Euler equations (Roe 86b). Such a decomposition leads to the solution of an equivalent set of scalar advection equations by means of accurate upwind Fluctuation Splitting (FS) discretizations (Struijs 91) and very efficient time integration schemes based on optimal smoothers and multigrid (Deconinck 93b, Catalano 97b). Unfortunately, such a methodology, although robust and very efficient, is limited to first-order accuracy and is thus inadequate for viscous flow calculations (Paillère 94, Catalano 96). Therefore, a characteristic decomposition of the Euler equations has been considered to develop a more accurate genuinely multidimensional methodology for compressible flows. However, the coupling terms in the compatibility equations render the application of standard FS schemes less appropriate than in the case of pure scalar advection equations and the design of optimal smoothers impossible. In fact, both stability and accuracy problems may arise from a straightforward extension of such schemes to scalar equations with source terms. Therefore, a new class of matrix FS schemes has been introduced, by generalizing the procedure proposed for scalar advection equations to the discretization of systems with non commuting Jacobians (van der Weide 96, Catalano 97a). On one hand, two linear matrix schemes have been successfully designed by extending the definition of the corresponding scalar FS schemes: the matrix N scheme, which captures shocks monotonically, and the matrix LDA scheme, which is second-order-accurate. On the other hand, two nonlinear schemes combining the two merits above have been proposed (van der Weide 96, Catalano 97a), which cannot be considered satisfactory with respect to robustness.

At present, the design of an accurate and robust nonlinear residual-distribution scheme for systems has eluded the efforts of many researchers. And, indeed, developing one such scheme appears a formidable task from an analytical viewpoint because no Total Variation Diminishing (TVD) scheme can exist for hyperbolic systems in more than one space dimension, energy being the only functional known to be bounded (Friedrichs 54, Liu 96). In a recent work by the authors (Catalano 97d), this difficulty has been circumvented by introducing the nonlinearity into the matrix FS formulation in a new way; namely, by extending the concept of the nonlinear advection velocity to the Euler system, after recasting it into a set of two advection equations plus an acoustic sub-system (hyperbolic-elliptic splitting) (Mesaros 95, Paillère 95). The resulting method is optimal for supersonic flow conditions and works reasonably well for subsonic smooth flows. However, it cannot resolve shocks monotonically and, being based upon the van Leer preconditioning (van Leer 91), becomes ill conditioned as the local Mach number, M, vanishes. Therefore, the search for a single FS scheme being accurate and robust in the range $0 \leq M \leq 2$ still goes on. Nevertheless, in addition to the above nonlinear FS scheme one has: the matrix LDA scheme of (van der Weide 96), which is so far unsurpassed for solving (subsonic) flows without discontinuities; and the matrix N scheme, which can resolves strong shocks monotonically.

Therefore, a new idea is proposed and pursued in the present work: namely, to develop a hybrid approach which, among the three aforementioned schemes, selects the most appropriate one at each computational cell, according to the local flow conditions.

In the remainder of this paper, the FS methodology is described at first for the simple case of a linear scalar advection equation and compared with a similar approach proposed by Sidilkover (94). The case of the Euler system is then considered. After a brief review of multidimensional upwind schemes for systems, the linear matrix LDA scheme (which is second-order-accurate) and the N scheme (which is monotone but only first-order-accurate), as well as the nonlinear scheme are described in detail. The new hybrid approach is then presented and applied to a series of well documented inviscid flows, including the very severe test cases of supersonic and hypersonic flows past a circular cylinder. An accurate and robust FS scheme being available at last, its extension to the Navier–Stokes and Reynolds averaged Navier–Stokes equations is provided, using a standard finite element discretization of the viscous terms. Laminar supersonic flow past a NACA 0012 airfoil is computed at first to test the viscous methodology without any interference by turbulence modeling effects. Then, the very severe transonic turbulent flow through the VKI-LS59 cascade is used to fully demonstrate the accuracy and robustness of the proposed hybrid FS scheme. Finally, a few concluding remarks are provided together with an indication of the future directions of this research.

2 Scalar advection equation

Consider the scalar advection equation with constant velocity $\boldsymbol{\lambda} = a\,\boldsymbol{i} + b\,\boldsymbol{j}$:

$$\frac{\partial w}{\partial t} = -\left(a\frac{\partial w}{\partial x} + b\frac{\partial w}{\partial y}\right), \tag{1}$$

to be discretized in space by compact multidimensional residual-distribution schemes. The scheme has to satisfy two requirements: 1) positivity, in order to avoid spurious oscillations in the numerical solution; 2) second-order accuracy. In the present framework of residual-distribution schemes on triangles, positivity can be effectively studied by means of the convection matrix $\mathcal{C}_T$ (Mizukami 85). Discretizing equation (1) by an Euler explicit time integration scheme, one has:

$$w_i^{n+1} - w_i^n = -\frac{\Delta t}{S_i}\sum_T (\mathcal{C}_T \mathcal{W}_T)_i, \tag{2}$$

where $\mathcal{W}_T = [w_1, w_2, w_3]^T$ is the vector containing the three unknowns at the vertices of each triangle contributing to node i. Obviously, the scheme satisfies the local Positivity (P) property if, provided a CFL condition is satisfied, the matrix $\mathcal{C}_T$ is of nonnegative type, namely, the diagonal terms are nonnegative and the off-diagonal terms are nonpositive. Furthermore, the notion of second-order accuracy is replaced with that of Linearity Preservation (LP): a scheme satisfies the LP property, provided it is able to preserve an initial linear solution of the steady problem (Struijs 91). An LP scheme is second-order-accurate for an homogeneous scalar advection equation (Paillère 94, Catalano 96). A generalization of Godunov's theorem shows that a linear scheme cannot satisfy both properties P and LP (Struijs 91).

Two different residual-distribution approaches will be considered here: the one originally introduced by Roe (Roe 87), known as Fluctuation Splitting (FS), and the one proposed by Sidilkover (Sidilkover 94).

In the FS method, if the variable w is assumed to vary bilinearly over each triangle, the fluctuation, namely, the flux balance over each cell,

$$\phi_T = -\int_T \left(a\frac{\partial w}{\partial x} + b\frac{\partial w}{\partial y} \right) \mathrm{d}S, \tag{3}$$

can be evaluated as:

$$\phi_T = -\sum_{j=1}^{3} k_j w_j, \qquad k_j = \frac{1}{2}\boldsymbol{\lambda} \cdot \boldsymbol{n}_j l_j, \tag{4}$$

l_j being the length of the side j, opposite to node j, and $\boldsymbol{n}_j$ denoting the corresponding inward normal unit vector (figure 1). With reference to fig-

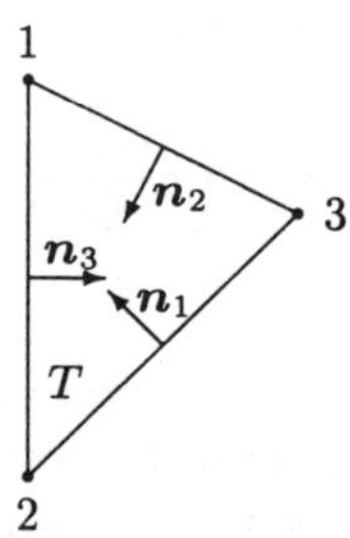

Fig. 1. Basic triangular cell.

ure 2, linearly interpolated values of w at the inflow and outflow points can be evaluated as:

$$w_{\text{in}} = \frac{\sum_{j=1}^{3} k_j^- w_j}{\sum_{j=1}^{3} k_j^-}, \qquad w_{\text{out}} = \frac{\sum_{j=1}^{3} k_j^+ w_j}{\sum_{j=1}^{3} k_j^+}, \tag{5}$$

where $k_j^+ = \max(0, k_j)$, $k_j^- = \min(0, k_j)$. Then, the fluctuation can be written in compact form as (Paillère 95):

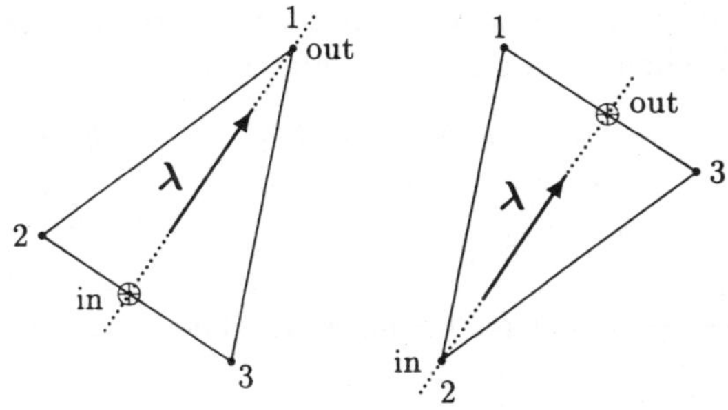

a. One-inflow side. b. Two-inflow sides.

Fig. 2. Definition of inflow and outflow points.

$$\phi_T = -\sum_{j=1}^{3} k_j w_j = -\sum_{j=1}^{3} k_j^+ (w_{\text{out}} - w_{\text{in}}). \tag{6}$$

Equation (6) shows that the fluctuation is zero when the same values of w occur at the inflow and outflow points or, equivalently, when w is constant along streamlines.

Multidimensional upwind schemes are obtained by assigning a fraction $\beta_{T,j}$ of the fluctuation to each downstream vertex j of the triangle, defined by the condition $k_j \geq 0$, with $\sum_{j=1}^{3} \beta_{T,j} = 1, \forall T$, for conservation. Two configurations are possible with either one or two downstream nodes, as shown in figure 2. For the configuration 2a, all of the fluctuation must be sent to the single downstream node, thus satisfying both properties P and LP. For the nontrivial configuration 2b, the various FS schemes choose different distribution coefficients $\beta_{T,j}$ trying again to satisfy the same properties.

The residual at node i is finally computed as:

$$\left(\frac{\partial w}{\partial t}\right)_i = \frac{1}{S_i} \sum_T \beta_{T,i} \phi_T, \tag{7}$$

where the sum is extented to all triangles sharing node i.

Two linear schemes have been successfully tested over a wide range of flow problems. The N scheme, which is the optimal first-order-accurate positive scheme, is provided by the following distribution coefficients:

$$\phi_{T,j}^N = \beta_{T,j}^N \phi_T = -k_j^+ (w_j - w_{\text{in}}). \tag{8}$$

The LDA scheme, which is designed so as to satisfy the property LP, is obtained as:

$$\phi_{T,j}^{LDA} = \beta_{T,j}^{LDA} \phi_T = k_j^+ \left(\sum_{j=1}^{3} k_j^+\right)^{-1} \phi_T = -k_j^+ (w_{\text{out}} - w_{\text{in}}), \tag{9}$$

where the fluctuation has been replaced by its expression, equation (6). For the case of the N scheme the convection matrix $\mathcal{C}_T$,

$$\mathcal{C}_T = S^+ \begin{bmatrix} -k_1^+(k_2^- + k_3^-) & k_1^+ k_2^- & k_1^+ k_3^- \\ k_2^+ k_1^- & -k_2^+(k_1^- + k_3^-) & k_2^+ k_3^- \\ k_3^+ + k_1^- & k_3^+ k_2^- & -k_3^+(k_1^- + k_2^-) \end{bmatrix}, \tag{10}$$

with $S^+ = \dfrac{1}{\sum_{j=1}^{3} k_j^+}$, is always nonnegative. The convection matrix of the LDA scheme,

$$\mathcal{C}_T = S^+ \begin{bmatrix} k_1^+ k_1 & k_1^+ k_2 & k_1^+ k_3 \\ k_2^+ k_1 & k_2^+ k_2 & k_2^+ k_3 \\ k_3^+ k_1 & k_3^+ k_2 & k_3^+ k_3 \end{bmatrix}, \tag{11}$$

can be verified to be nonnegative only in the one-target configuration, as anticipated. In order to design a scheme satisfying both properties P and LP, it is necessary to make the distribution nonlinear. As proposed in (Mizukami 85, Roe 90), a nonlinear advection velocity can be used in equation (1) without changing the steady state residual:

$$\boldsymbol{\lambda}^\omega = \boldsymbol{\lambda} + \omega\, \boldsymbol{s}, \qquad \boldsymbol{s} = \frac{-w_y \boldsymbol{i} + w_x \boldsymbol{j}}{\mid \boldsymbol{\nabla} w \mid}. \tag{12}$$

In such a way, a degree of freedom, namely, the auxiliary speed ω, is intro-

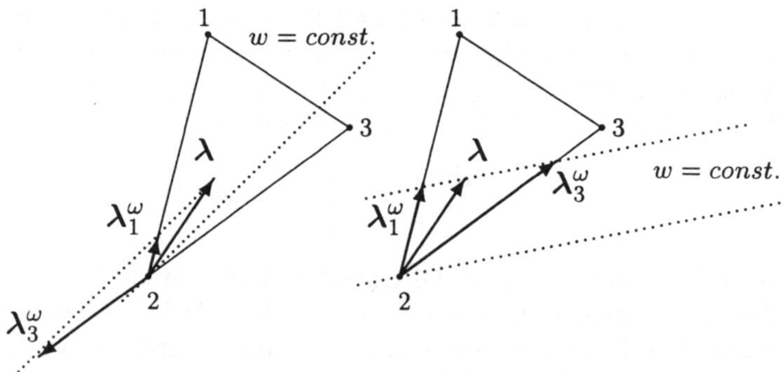

Fig. 3. Nonlinear scheme configurations.

duced to determine the appropriate advection velocity for the sought upwind scheme. The strategy used to evaluate ω makes use of the *level line*, namely, the line of constant w. Such a strategy is briefly recalled here, for completeness. In the two-target configuration (with respect to $\boldsymbol{\lambda}$), see figure 2b, two situations may occur: i) the level line passing through the upwind node 2 cuts the outflow edge, i.e. $(w_3 - w_2)(w_1 - w_2) < 0$ (figure 3a); ii) the level line passing through the upwind node 2 does not cut the outflow edge, i.e. $(w_3 - w_2)(w_1 - w_2) > 0$ (figure 3b). In both cases one looks for two values of $\boldsymbol{\lambda}^\omega$ being aligned with the two sides 2-1 and 2-3, respectively, $\boldsymbol{\lambda}_1^\omega$ and $\boldsymbol{\lambda}_3^\omega$. In the first case only one of such $\boldsymbol{\lambda}^\omega$ is downwind so that the scheme is one-target. In particular, if $(w_1 - w_3)\,\phi_T < 0$, see figure 3a, the entire fluctuation is sent to node 1 and ω is computed by requiring that $k_3^\omega = 0$, namely:

$$\omega = -\frac{\boldsymbol{\lambda} \cdot \boldsymbol{n}_3}{\boldsymbol{s} \cdot \boldsymbol{n}_3}; \tag{13}$$

for $(w_1 - w_3)\,\phi_T > 0$, the entire fluctuation is sent to node 3 and ω is computed by requiring that $k_1^\omega = 0$, namely:

$$\omega = -\frac{\boldsymbol{\lambda} \cdot \boldsymbol{n}_1}{\boldsymbol{s} \cdot \boldsymbol{n}_1}. \tag{14}$$

In the second situation, see figure 3b, both $\boldsymbol{\lambda}^\omega$ are downwind so that the scheme must be two-target and any value of ω which guarantees that β_1 and β_3 are both nonnegative provides a positive distribution. The different non-linear schemes proposed so far (Mizukami 85, Roe 90, Struijs 91) only differ in the linear ($\omega = 0$) upwind distribution employed when such a situation occurs. In particular, the SUPG scheme is used in (Mizukami 85), whereas the PSI scheme proposed in (Struijs 91) smoothly blends into the N scheme as soon as both $\boldsymbol{\lambda}^\omega$ become downwind, thus significantly improving the robustness of the overall method. In the latter case, the convection matrix associated with the scheme is obtained from equation (10) employing k_j^ω in place of k_j and remains nonnegative. This is particularly easy to verify by rewriting the PSI scheme using an argument due to Sidilkover and Roe (Sidilkover 95), that a positive and linearity preserving scheme can be obtained by applying a limiter function to the distribution coefficients of a linear positive scheme. In fact, using the N scheme contributions to the downwind nodes 1 and 3, and defining the ratio

$$r = -\frac{\Phi_{T,1}^N}{\Phi_{T,3}^N},$$

the PSI scheme distribution can be written as:

$$\begin{aligned} \Phi_{T,1}^{PSI} &= \Phi_{T,1}^N \left(1 - \tfrac{\Psi(r)}{r}\right), \\ \Phi_{T,3}^{PSI} &= \Phi_{T,3}^N \left(1 - \Psi(r)\right), \end{aligned} \tag{15}$$

where Ψ is the minmod function.

Let us now consider the second residual-distributions scheme, proposed by Sidilkover (Sidilkover 94). A non-orthogonal (ξ, η) coordinate system aligned with two edges of the triangle, e. g., the vectors $\boldsymbol{e}_1$ and $\boldsymbol{e}_2$ in figure 4, is introduced, so that equation (1) can be rewritten as:

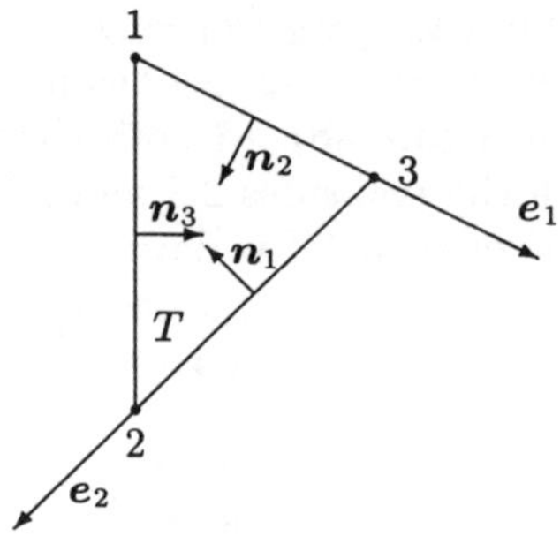

Fig. 4. Local coordinate system.

$$\frac{\partial w}{\partial t} = -\left(\alpha \frac{\partial w}{\partial \xi} + \beta \frac{\partial w}{\partial \eta}\right), \tag{16}$$

where

$$\alpha = \frac{\boldsymbol{\lambda} \cdot (l_2 \boldsymbol{n}_2)}{\theta}, \quad \beta = \frac{\boldsymbol{\lambda} \cdot (l_1 \boldsymbol{n}_1)}{\theta}, \quad \theta = \boldsymbol{e}_1 \cdot (l_2 \boldsymbol{n}_2) \equiv \boldsymbol{e}_2 \cdot (l_1 \boldsymbol{n}_1).$$

The fluctuation is then evaluated as the sum of *one-dimensional* contributions along the directions of the local coordinate system:

$$\phi_T = \phi_T^{\xi} + \phi_T^{\eta}, \tag{17}$$

where

$$\phi_T^{\xi} = -S_T \frac{\alpha(w_3 - w_1)}{l_1}, \quad \phi_T^{\eta} = -S_T \frac{\beta(w_2 - w_3)}{l_2}, \tag{18}$$

S_T being the triangle area. The distribution procedure is carried out according to standard *one-dimensional* upwinding along the three sides of the triangle:

$$\begin{aligned}
S_1 w_1^{n+1} &= S_1 w_1^n + \frac{\Delta t}{2}(\phi_T^{\xi} - \mathrm{sign}(\alpha)\phi_T^{\xi}), \\
S_2 w_2^{n+1} &= S_2 w_2^n + \frac{\Delta t}{2}(\phi_T^{\eta} + \mathrm{sign}(\beta)\phi_T^{\eta}), \\
S_3 w_3^{n+1} &= S_3 w_3^n + \frac{\Delta t}{2}[(\phi_T^{\xi} + \mathrm{sign}(\alpha)\phi_T^{\xi}) + (\phi_T^{\eta} - \mathrm{sign}(\beta)\phi_T^{\eta})].
\end{aligned} \tag{19}$$

Such a scheme is positive but only first-order-accurate. Introducing the quantity

$$Q = -\frac{\phi_T^{\xi}}{\phi_T^{\eta}},$$

and a limiter function $\Psi(Q)$, a nonlinear scheme is obtained by replacing the linear residual contributions ϕ_T^{ξ} and ϕ_T^{η} in equation (19) with the following nonlinear ones:

$$\phi_T^{\xi^*} = \phi_T^{\xi}\left(1 - \frac{\Psi(Q)}{Q}\right), \quad \phi_T^{\eta^*} = \phi_T^{\eta}(1 - \Psi(Q)). \tag{20}$$

Such a nonlinear scheme is conservative since

$$\phi_T = \phi_T^{\xi^*} + \phi_T^{\eta^*} = \phi_T^{\xi} + \phi_T^{\eta};$$

moreover, it is positive and linearity preserving provided that

$$0 \leq \Psi(Q) \leq 1, \quad 0 \leq \frac{\Psi(Q)}{Q} \leq 1, \tag{21}$$

and

$$\Psi(1) = 1, \tag{22}$$

respectively. Higher-order-accurate steady-state solutions without oscillations have been obtained by the author (Sidilkover 94). However, both the linear and nonlinear schemes are not genuinely multidimensional upwind as long as the choice of the local coordinate directions, aligned with two sides of each triangle, is independent of the advection velocity $\boldsymbol{\lambda}$.

3 Euler equations

3.1 Characteristic formulations

The Euler equations are written in quasi-linear form as:

$$\frac{\partial U}{\partial t} = -\left(A\frac{\partial U}{\partial x} + B\frac{\partial U}{\partial y}\right), \tag{23}$$

where U is the vector of the conservative variables and A and B are the Jacobian matrices.

Taking a linear combination of equations (23) one obtains an equivalent system of four compatibility equations with three degrees of freedom, namely, the unit vectors $\boldsymbol{m}^k, k = 2, 3, 4$,

$$\frac{\partial V}{\partial t} + \boldsymbol{\Lambda} \cdot \boldsymbol{\nabla} V + S = 0. \tag{24}$$

In equations (24), $\boldsymbol{\Lambda}$ is the diagonal matrix containing the bicharacteristic vectors $\boldsymbol{\lambda}^{1,2} = \boldsymbol{u}$, $\boldsymbol{\lambda}^{3,4} = \boldsymbol{u} + c\boldsymbol{m}^{3,4}$ ($\boldsymbol{u}$ and c being the velocity vector and the speed of sound, respectively), V is the vector of the characteristic variables and S is the vector of the coupling terms:

$$\partial V = R^{-1}\partial U = \begin{pmatrix} \partial\rho - \partial p/c^2 \\ \boldsymbol{s}^2 \cdot \partial\boldsymbol{u} \\ \boldsymbol{m}^3 \cdot \partial\boldsymbol{u} + \partial p/\rho c \\ \boldsymbol{m}^4 \cdot \partial\boldsymbol{u} + \partial p/\rho c \end{pmatrix}, \qquad S = \begin{pmatrix} 0 \\ \boldsymbol{s}^2 \cdot \boldsymbol{\nabla} p/\rho \\ c\boldsymbol{s}^3 \cdot (\boldsymbol{s}^3 \cdot \boldsymbol{\nabla})\boldsymbol{u} \\ c\boldsymbol{s}^4 \cdot (\boldsymbol{s}^4 \cdot \boldsymbol{\nabla})\boldsymbol{u} \end{pmatrix}. \tag{25}$$

In equations (25) p is the pressure, ρ is the density, and $\boldsymbol{s}^k$ is a unit vector orthogonal to $\boldsymbol{m}^k$, $\boldsymbol{m}^2$ is aligned with the velocity vector, $\boldsymbol{m}^3 = \boldsymbol{m}^2$ and $\boldsymbol{m}^4 = -\boldsymbol{m}^2$. By using natural coordinates (ξ, η) as the independent variables, equations (23) become:

$$\frac{\partial U}{\partial t} = -R\left(A_V \frac{\partial V}{\partial \xi} + B_V \frac{\partial V}{\partial \eta}\right). \tag{26}$$

Equations (26) will be used to derive a general formulation of linear matrix FS schemes.

On the other hand, for supersonic flow, it is convenient to recast equations (26) into a symmetric form. One such form can be easily obtained by performing the change of variables:

$$\partial Q = \begin{pmatrix} \partial p/\rho c \\ \partial u_\xi \\ \partial u_\eta \\ \partial p - c^2 \partial \rho \end{pmatrix} = L_Q \partial V, \tag{27}$$

to give:

$$\frac{\partial Q}{\partial t} = -\left(A_Q \frac{\partial Q}{\partial \xi} + B_Q \frac{\partial Q}{\partial \eta}\right). \tag{28}$$

In equations (27) ∂u_ξ and ∂u_η are the differentials of the velocity components in the (ξ, η) coordinate system, and the symmetric Jacobians A_Q and B_Q read:

$$A_Q = \begin{bmatrix} u_\xi & c & 0 & 0 \\ c & u_\xi & 0 & 0 \\ 0 & 0 & u_\xi & 0 \\ 0 & 0 & 0 & u_\xi \end{bmatrix}, \qquad B_Q = \begin{bmatrix} 0 & 0 & c & 0 \\ 0 & 0 & 0 & 0 \\ c & 0 & 0 & 0 \\ 0 & 0 & 0 & 0 \end{bmatrix}. \tag{29}$$

Furthermore, the matrix P introduced by van Leer *et al.* (van Leer 91) is considered:

$$P = \frac{1}{u_\xi}\begin{bmatrix} \frac{\chi}{\beta^2}M^2 & -\frac{\chi}{\beta^2}M & 0 & 0 \\ -\frac{\chi}{\beta^2}M & \frac{\chi}{\beta^2}+1 & 0 & 0 \\ 0 & 0 & \chi & 0 \\ 0 & 0 & 0 & 1 \end{bmatrix}, \tag{30}$$

where:

$$\beta = \sqrt{|\, M^2 - 1\,|}, \quad \chi = \beta \max(M, 1), \tag{31}$$

M being the Mach number. Premultiplying equations (28) by the matrix P and introducing the following set of variables:

$$\partial W = \begin{pmatrix} \beta \partial p/\rho c + M \partial u_\eta \\ \beta \partial p/\rho c - M \partial u_\eta \\ \partial p/\rho c + M \partial u_\xi \\ \partial p - c^2 \partial \rho \end{pmatrix} = L_W \partial Q, \tag{32}$$

the governing equations can be finally written as:

$$\frac{\partial U}{\partial t} = -R\left[R_P\left(A_W\frac{\partial W}{\partial \xi} + B_W\frac{\partial W}{\partial \eta}\right)\right]. \tag{33}$$

In equations (33), $R_P = L_Q^{-1}P^{-1}L_W^{-1}$ and the Jacobian matrices A_W and B_W read:

$$A_W = L_W P A_Q L_W^{-1} = \begin{bmatrix} \chi\nu^+ & \chi\nu^- & 0 & 0 \\ \chi\nu^- & \chi\nu^+ & 0 & 0 \\ 0 & 0 & 1 & 0 \\ 0 & 0 & 0 & 1 \end{bmatrix}, \tag{34}$$

$$B_W = L_W P B_Q L_W^{-1} = \begin{bmatrix} \frac{\chi}{\beta} & 0 & 0 & 0 \\ 0 & -\frac{\chi}{\beta} & 0 & 0 \\ 0 & 0 & 0 & 0 \\ 0 & 0 & 0 & 0 \end{bmatrix}, \tag{35}$$

where:

$$\nu^+ = \frac{M^2 - 1 + \beta^2}{2\beta^2}, \; \nu^- = \frac{M^2 - 1 - \beta^2}{2\beta^2}. \tag{36}$$

It is noteworthy that the two Jacobian matrices A_W and B_W are still symmetric and, more importantly, they diagonalize for supersonic flow ($\nu^- = 0$). Therefore, for such a case, the Euler system is equivalent to a set of four preconditioned scalar advection equations, according to its space hyperbolic nature. For subsonic flow, the first two equations form an elliptic sub-system, whereas both the third and fourth equations remain scalar advection equations (hyperbolic-elliptic splitting) (Mesaros 95, Paillère 95).

3.2 Multidimensional upwind methods

In recent years different approaches have been used to extend the FS schemes developed for scalar equations to systems. In particular, for the Euler equations, such an extension is straightforward in the case of one space variable, insofar as the system is equivalent to a set of advection equations written in terms of characteristic variables. In the case of two and three space variables, the decoupling of the Euler equations is not generally possible, since the Jacobians do not commute. Moreover, the basic design criteria of multidimensional upwind schemes for systems are different from those for scalar equations: the solutions of the scalar equations are TVD so that a scheme satisfying such a condition is monotonicity preserving; the only functional known to be bounded for systems is the energy of the solution. Friedrichs (Friedrichs 54) has shown that for a symmetric hyperbolic linear system,

$$\frac{\partial U}{\partial t} + \sum_{s=1}^{d} A_s \frac{\partial U}{\partial x_s} = 0, \tag{37}$$

the L^2 norm of the solution has bounded growth. Furthermore, the discrete solution itself,

$$U_j^{n+1} = \sum_k C_k U_{j+k}^n, \tag{38}$$

satisfies the same condition if the matrices C_k satisfy the following properties:

1. C_k are symmetric nonnegative;
2. $\sum_k C_k = I$;
3. $C_k = 0$, except for a finite set of k;
4. C_k depend Lipschitz continuously on x_s.

The schemes currently employed to solve the Euler equations have not been designed so as to fulfill such conditions. Most of them are based on some simplifications which allow to exploit the techniques developed for the scalar case. The Flux Difference Splitting (FDS) method (Roe 86a), for example, is based on the solution of one Riemann problem in each grid direction, thus reducing the multidimensional problem to 2 or 3 one-dimensional ones. Another approach consists in solving each characteristic equation separately, with the difficulty of handling the coupling terms. Both approaches have some disadvantages: the former incorrectly models wave propagation phenomena which are oblique to the grid; the latter generally faces convergence problems, due to an unphysical treatment of the coupling terms.

Liu and Lax (Liu 96) proposed a novel scheme for solving the Euler equations which is energy bounded. The scheme is first developed for the one-dimensional case and then extended "dimension by dimension" to the multidimensional one, thus resulting very similar to classical non compact methods based on the TVD property (Yee 87).

One of the first attempts to design a genuinely multidimensional compact scheme for the solution of the Euler system is due to Hughes and Mallet (Hughes 86b) who proposed a generalization of the scalar SUPG scheme for smooth flows (Hughes 86a). The scheme was designed, without considering the energy boundness, so as to: i) reduce correctly to the one-dimensional system case; ii) be equivalent to the scalar scheme for a multidimensional advection equation; iii) reduce to the scalar scheme for each uncoupled component of multidimensional diagonalizable systems. The authors also provide a rigorous mathematical convergence study for such a scheme, which is very satisfactory for computing smooth flows. In order to better control the behaviour of the solution in high-gradient regions, the same authors propose the addition of a discontinuity capturing term (Hughes 86c) obtained by an heuristic generalization of the one previously used in the scalar SUPG scheme (Hughes 86a).

More recently, Sidilkover (94) extended the positive linearity preserving scheme for scalar equations discussed before to the multidimensional Euler system. As in the scalar case, the scheme produces higher-order-accurate steady-state solutions, and combines very effectively with a Gauss–Seidel relaxation procedure embedded in a multigrid strategy.

Finally, a new class of matrix schemes was introduced generalizing the FS scalar schemes to the case of systems (van der Weide 96, Catalano 97a). Such schemes will be described in more details in the following, with reference to equations (26).

3.3 Linear matrix FS schemes

The conservative flux balance over each triangle T is defined as the fluctuation, $\Phi_{U,T}$,

$$\Phi_{U,T} = -\int_T \left(A\frac{\partial U}{\partial x} + B\frac{\partial U}{\partial y} \right) \, \mathrm{d}S, \tag{39}$$

so that equations (26) can be integrated over a triangle to give:

$$\int_T \frac{\partial U}{\partial t} \mathrm{d}S = \Phi_{U,T} = -\int_T \left[R \left(A_V \frac{\partial V}{\partial \xi} + B_V \frac{\partial V}{\partial \eta} \right) \right] \mathrm{d}S. \tag{40}$$

In the numerical method, a discrete counterpart of equations (40) is needed, which requires a conservative linearization of the governing equations in order to compute shocks correctly. By generalizing the linearization introduced by Roe (Roe 81) for the one-dimensional Euler equations, Deconinck *et al.* (Deconinck 93a) assumed the parameter vector $Z = \sqrt{\rho}(1, u, v, H)^{\tau}$ to vary bilinearly over each triangle. In such a way, the use of cell-averaged values for all variables, evaluated analytically using the average value of Z, guarantees conservation. The discrete fluctuation is then expressed as:

$$\Phi_{U,T} = -\bar{R} \left(\bar{A}_V \overline{\frac{\partial V}{\partial \xi}} + \bar{B}_V \overline{\frac{\partial V}{\partial \eta}} \right) S_T = \bar{R} \Phi_T. \tag{41}$$

The fluctuation Φ_T is then rewritten in terms of appropriate fluxes through the edges of each triangle (van der Weide 96, Catalano 97a) as:

$$\Phi_T = -\sum_{j=1}^{3} \frac{l_j}{2} \bar{\boldsymbol{A}}_V \cdot \boldsymbol{n}_j V_j = -\sum_{j=1}^{3} K_j V_j, \tag{42}$$

where

$$K_j = \frac{1}{2} l_j \left(\bar{A}_V n_{\xi,j} + \bar{B}_V n_{\eta,j} \right). \tag{43}$$

Due to the hyperbolic nature of the system, K_j can be written as

$$K_j = (\bar{R}_K \bar{\Lambda}_K \bar{L}_K)_j = (\bar{R}_K \bar{\Lambda}_K^+ \bar{L}_K)_j + (\bar{R}_K \bar{\Lambda}_K^- \bar{L}_K)_j = K_j^+ + K_j^-. \tag{44}$$

In equations (44), $\bar{R}_{K,j}$ and $\bar{L}_{K,j}$ are the right and left eigenvector matrices of K_j, whereas $\bar{\Lambda}_{K,j}^+$ and $\bar{\Lambda}_{K,j}^-$ are the corresponding positive and negative eigenvalue matrices. Introducing the following vectors:

$$\begin{aligned} V_{\text{in}} &= \left(\textstyle\sum_{j=1}^3 K_j^- \right)^{-1} \left(\textstyle\sum_{j=1}^3 K_j^- V_j \right), \\ V_{\text{out}} &= \left(\textstyle\sum_{j=1}^3 K_j^+ \right)^{-1} \left(\textstyle\sum_{j=1}^3 K_j^+ V_j \right), \end{aligned} \tag{45}$$

linear matrix schemes can be obtained by extending the definitions of the scalar fluctuation splitting schemes described in the previous section. The matrix N scheme is obtained as:

$$\Phi_j^N = -K_j^+ \left[V_j - V_{\text{in}}\right], \tag{46}$$

whereas the matrix LDA scheme reads:

$$\Phi_j^{LDA} = -K_j^+ \left[V_{\text{out}} - V_{\text{in}}\right]. \tag{47}$$

Employing an Euler explicit time integration scheme for the space discretized equations (40), one has:

$$U_i^{n+1} - U_i^n = \frac{\Delta t}{S_i} \sum_T \bar{R}\Phi_{T,i} = -\frac{\Delta t}{S_i} \sum_T \bar{R}\mathcal{C}_T \mathcal{V}_T, \tag{48}$$

where the sum is extended to all triangles sharing node i. In equations (48) $\mathcal{V}_T = [V_1, V_2, V_3]^T$ is the vector containing the unknowns for each triangle and $\mathcal{C}_T$ is the convection matrix. Computing the element energy matrix, Barth *et al.* (Barth 96) have shown that the N scheme is energy bounded. Moreover, like their scalar counterparts, the N scheme is only first-order-accurate, whereas the LDA scheme is LP and thus is suitable for computing smooth flows.

3.4 A nonlinear FS scheme

The optimal nonlinear FS scheme should be linearity preserving, capable of providing monotone solutions to the Euler equations in the case of discontinuous flow, and robust in the range of Mach number $0 \leq M \leq 2$. Previously proposed schemes, based on a scalar limiting of the residual (Catalano 97c) have not been able to provide a satisfactory convergence to steady-state, especially in the case of transonic flow. On the other hand, such a difficulty does not seem to be due to the employed limiter function, which guarantees full convergence for the case of scalar advection equations with or without discontinuities (Struijs 91). Therefore, a different method was proposed (Catalano 97d): namely, to extend the level line approach to the Euler system. Without changing the steady-state residual of the Euler equations, a nonlinear Jacobian is introduced into equations (33), namely,

$$\bar{\boldsymbol{A}}^{nl} = \bar{\boldsymbol{A}}_W + \Omega \bar{\boldsymbol{A}}_\perp, \qquad \bar{\boldsymbol{A}}_\perp \cdot \overline{\boldsymbol{\nabla} W} = 0. \tag{49}$$

A general approach would require constructing a matrix $\bar{\boldsymbol{A}}_\perp$, as proposed in (Hughes 86c) in the context of Petrov–Galerkin finite elements:

$$\bar{\boldsymbol{A}}_\perp = \bar{\boldsymbol{A}}_W - \bar{\boldsymbol{A}}_\parallel, \tag{50}$$

where

$$\bar{\boldsymbol{A}}_{\parallel}^{\tau} = \bar{\boldsymbol{A}}_{W}^{\tau} \frac{\boldsymbol{\nabla} W \, (\boldsymbol{\nabla} W)^{\tau}}{(\boldsymbol{\nabla} W)^{\tau} \, \boldsymbol{\nabla} W}, \tag{51}$$

and then finding the matrix Ω capable of making the scheme based on the nonlinear Jacobian $\bar{\boldsymbol{A}}^{nl}$ energy stable and linearity preserving. Unfortunately, an extremely difficult analytical problem arises, for which no solution in closed form could be found. Therefore, a simpler route was undertaken, namely, the following matrices $\bar{\boldsymbol{A}}_{\perp}$ and Ω were designed:

$$\bar{\boldsymbol{A}}_{\perp}^{\tau} = \left[\begin{bmatrix} -W_{\eta}^1 & 0 & 0 & 0 \\ 0 & -W_{\eta}^2 & 0 & 0 \\ 0 & 0 & -W_{\eta}^3 & 0 \\ 0 & 0 & 0 & -W_{\eta}^4 \end{bmatrix}, \begin{bmatrix} W_{\xi}^1 & 0 & 0 & 0 \\ 0 & W_{\xi}^2 & 0 & 0 \\ 0 & 0 & W_{\xi}^3 & 0 \\ 0 & 0 & 0 & W_{\xi}^4 \end{bmatrix} \right], \tag{52}$$

$$\Omega = \begin{bmatrix} \omega^1 & & & & & & & \\ & \omega^2 & & & & & & \\ & & \omega^3 & & & 0 & & \\ & & & \omega^4 & & & & \\ & & & & \omega^1 & & & \\ & & 0 & & & \omega^2 & & \\ & & & & & & \omega^3 & \\ & & & & & & & \omega^4 \end{bmatrix}. \tag{53}$$

Notice that $\bar{\boldsymbol{A}}_{\perp}$ is the tensor of the normals to each component of the gradient of W and Ω is a diagonal matrix, so that for supersonic flow the proposed nonlinear matrix scheme identically recovers the corresponding scalar PSI scheme; this is a very desirable feature, since the Jacobians in equations (33) diagonalize in the case of supersonic flow. In the case of subsonic flow, when such a diagonalization is not possible, the coefficients ω^i are still computed using the scalar procedure. In conclusion, the nonlinear matrix scheme proposed in (Catalano 97d) reads:

$$\Phi_{U,j} = \overline{R}\,\overline{R}_P \left(\nu^- K_j^{nl+} \left[W_{\text{out}}^{nl} - W_{\text{in}}^{nl} \right] - \nu^+ K_j^{nl+} \left[W_j - W_{\text{in}}^{nl} \right] \right), \tag{54}$$

where K_j^{nl}, W_{in}^{nl}, W_{out}^{nl} are computed using the Jacobian $\bar{\boldsymbol{A}}^{nl}$. In equations (54), the first term ensures that the scheme is linearity preserving when the Jacobians are not diagonal ($\nu^- = -1$, $\nu^+ = 0$), whereas the second term allows to recover the scalar PSI scheme of (Struijs 91) when the Jacobians commute ($\nu^- = 0$, $\nu^+ = 1$). Such a scheme is very accurate at all flow regimes, but does not converge at stagnation points, where the inverse of the preconditioning matrix used to obtain the hyperbolic-elliptic splitting becomes ill-conditioned (Mesaros 95). In spite of the considerable effort by several groups, who have proposed different ways to fix such a problem (see, e.g., (Lee 97)), none of them is completely satisfactory.

3.5 Hybrid approach

In order to exploit the positive features of the FS schemes discussed before, while side-stepping their drawbacks, the authors proposed a hybrid method, which combines the linear matrix LDA scheme for subsonic flows and the nonlinear scheme previously described for supersonic flows (De Palma 98b). In the former, case the characteristic variables V are employed in order to avoid ill-conditioning at low Mach number regions, so that the fluctuation to node j can be written using equations (47). In the latter case, using the set of characteristic variables W allows to recast the Euler system into an equivalent set of four scalar advection equations, which are discretized using equations (54) with $\nu^+ = 1$ and $\nu^- = 0$. The hybrid FS scheme is obtained by using either equations (47) or equations (54), depending on the value of the local (cell averaged) Mach number, M, namely:

$$\Phi_{U,j} = \overline{R}\left(\nu^- K_j^+ \left[V_{\text{out}} - V_{\text{in}}\right] - \overline{R}_P \nu^+ K_j^{nl+} \left[W_j - W_{\text{in}}^{nl}\right]\right). \qquad (55)$$

The scheme given in equations (55) has been shown to be accurate and robust for transonic flows with weak shocks (De Palma 98a). However, it experienced some conservation difficulty in the presence of strong shocks, probably due to the use of two different sets of equations to evaluate the residual contributions to nodes located inside a captured discontinuity. Moreover, one may reasonably argue that it is inappropriate to use the LDA scheme for evaluating the residual at those cells having $M < 1$ and containing a shock, where a monotone scheme should be used instead. An improved version of the hybrid approach is then obtained by splitting the fluctuation in such cells by means of the matrix N scheme using the characteristic variables W:

$$\Phi_{U,j} = -\overline{R}\,\overline{R}_P K_j^{l+}\left(W_j - W_{\text{in}}^l\right), \qquad (56)$$

where K_j^l and W_{in}^l are computed using the Jacobian $\bar{\boldsymbol{A}}_W$. In order to pick out these special cells, disregarding those where the transition from supersonic flow conditions to subsonic ones is smooth enough to be properly handled by the LDA scheme, it is necessary to characterize them uniquely. By a careful analysis of local flow properties across a normal shock, it is concluded that "shock cells" must contain: i) at least one supersonic node; ii) at least one subsonic node with its local Mach number lower than 0.9. At every step of the computational process, all cells with $M < 1$ ($\nu^+ = 0$, $\nu^- = -1$) are scanned and, for all "shock cells", the scheme switches from equations (55) to equations (56) to distribute the cell residual.

The residual of the Euler equations at vertex j is then evaluated by collecting all contributions coming from the surrounding triangles, as:

$$\left(\frac{\partial U}{\partial t}\right)_j = \frac{1}{S_j}\sum_T \Phi_{U,j}^T. \qquad (57)$$

Equations (57) are integrated in time by means of an explicit three stage Runge–Kutta scheme with coefficients $c_1 = .132$, $c_2 = .377$, combined with a standard FAS multigrid V-cycle (Brandt 82), using full-weighting collection and bilinear prolongation.

Standard characteristic boundary conditions are imposed at inflow and outflow points, whereas isentropic simple radial equilibrium is imposed at solid walls (Catalano 95).

3.6 Results for inviscid flows

Three well documented transonic flows have been employed to assess the accuracy and robustness of the proposed methodology.

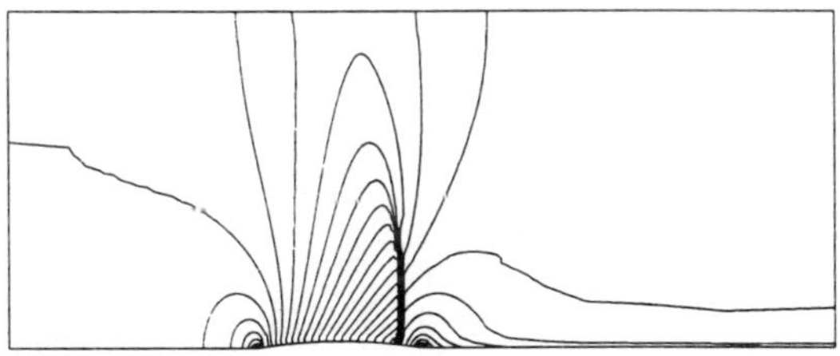

Fig. 5. Mach number contours ($\Delta M = 0.03$).

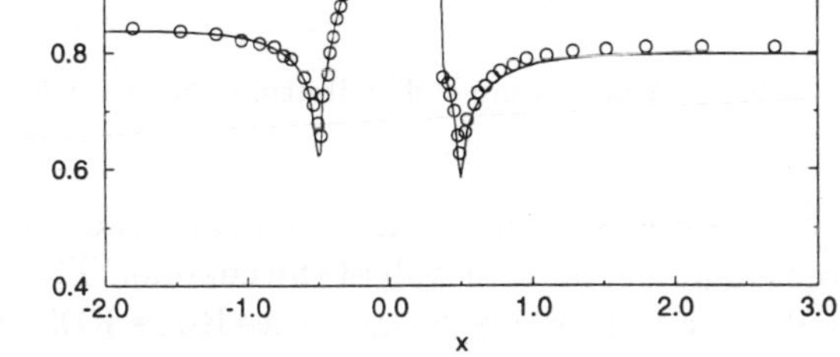

Fig. 6. Mach number distributions at the wall.

Flow inside a channel with a 4.2% thickness circular bump and isentropic outlet Mach number equal to 0.85 has been computed at first, using a stretched grid with 80×32 quadrilateral cells divided into triangles by means of equally oriented diagonals, and a CFL number equal to 0.2. Figure 5 shows

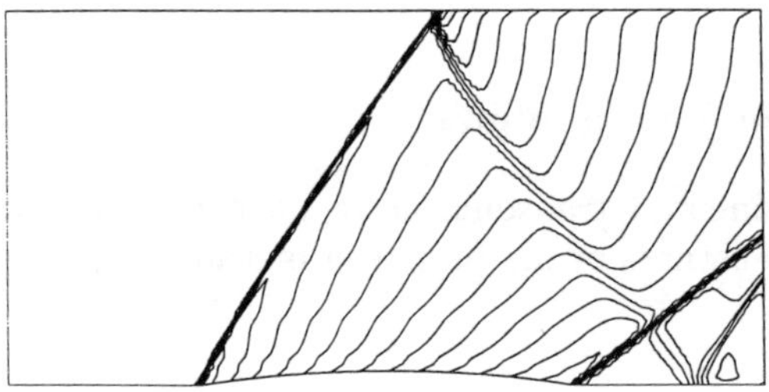

Fig. 7. Mach number contours ($\Delta M = 0.05$).

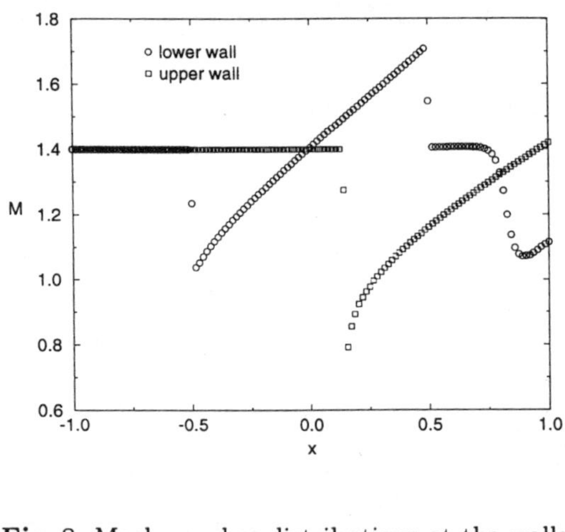

Fig. 8. Mach number distributions at the walls.

the Mach number contours, whereas figure 6 shows the corresponding Mach number distribution at the bottom wall of the channel. The solution obtained on the same grid using a second-order-accurate Roe's FDS scheme (Roe 86a) is also shown, for comparison. The present solution is monotone and computes the maximum Mach number, and thus the strength of the normal shock at the wall, more accurately than Roe's scheme.

In order to better assess the shock capturing capability of the scheme, the flow inside a channel with a 4% thickness circular bump and inlet Mach number $M_i = 1.4$ has been computed, the CFL number being equal to 0.3. Figure 7 shows the Mach number contours obtained on a 128×48 grid with equally oriented diagonals. The Mach number distributions at the lower and upper walls are also given in figure 8. The shocks are captured monotonically and the overall solution is typical of a second-order-accurate method with limiter. Moreover, a subsonic region is present behind the λ-shock at the upper wall, where the scheme switches among the three residual-distribution schemes, without any difficulty.

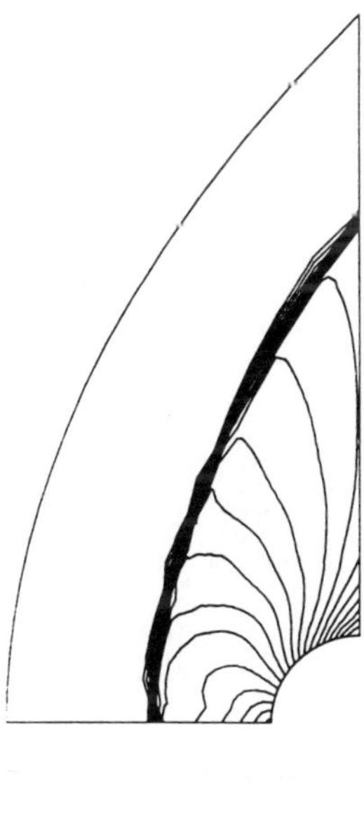

Fig. 9. Mach number contours for $M_\infty = 2$ ($\Delta M = 0.1$).

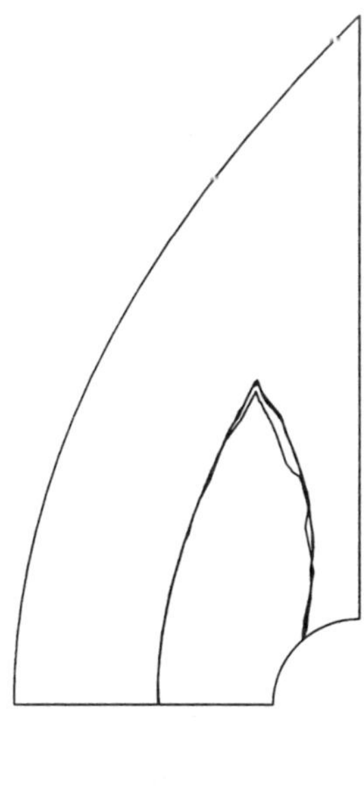

Fig. 10. Sonic lines for $M_\infty = 2$ using three different grids.

Finally, to provide a more thorough validation of the scheme with respect to its capabilities of capturing shocks monotonically and to cross sonic lines smoothly, supersonic flow past a circular cylinder has been considered as an ideal test case, both a shock of variable intensity and a sonic line being present in the flow field. Satisfactory results have been obtained using a CFL number equal to 0.3 for two values of the free stream Mach number, M_∞. A value $M_\infty = 2$ has been considered at first. Figure 9 shows the Mach number contours obtained on a 32×16 grid with equally oriented diagonals, whereas figure 10 shows the sonic line computed on three different grids containing 32×16, 64×32 and 128×64 cells, respectively. The solution is smooth

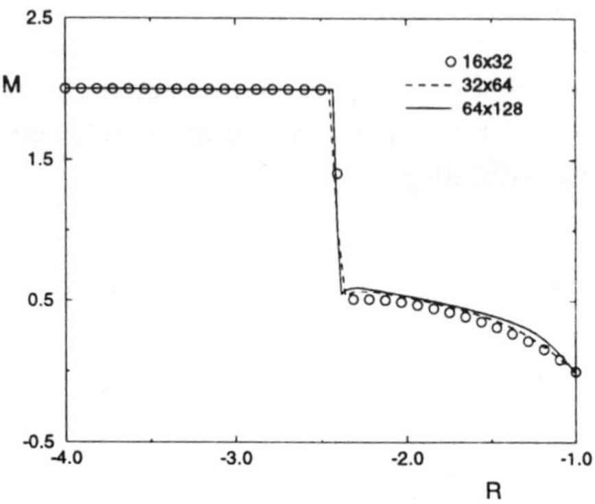

Fig. 11. Mach number distributions along the symmetry line.

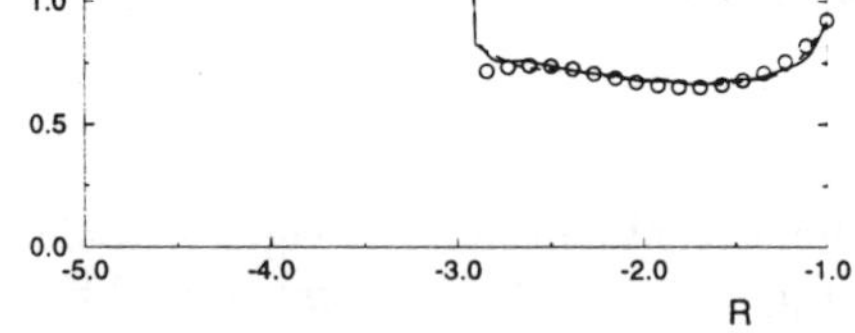

Fig. 12. Mach number distributions along the radial line at 45°.

in the region after the shock and, in particular, across the sonic line, where the method switches between two different sets of equations and residual-distribution schemes. The Mach number distributions along the horizontal (symmetry) line and the radial line inclined at 45 degrees are also given in figure 11 and 12, respectively. Three sets of results are plotted, which refer to the three grids, and demonstrate grid convergence. In all cases, the shock is captured without spurious oscillations, indicating that the switching strategy between the two sets of equations and the three residual-distribution schemes is quite effective.

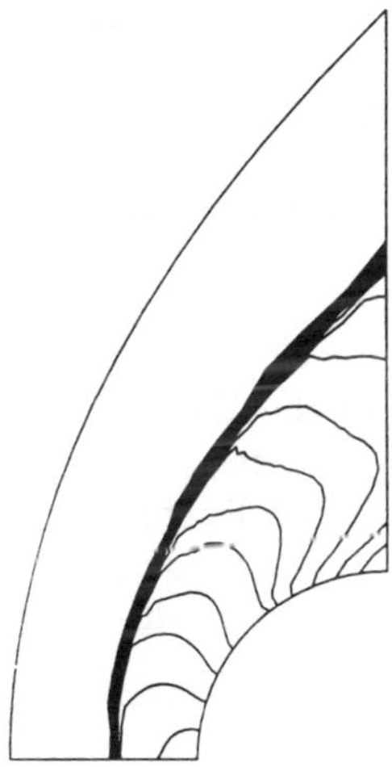

Fig. 13. Mach number contours for $M_\infty = 6$ ($\Delta M = 0.2$).

The very severe $M_\infty = 6$ case (see the carbuncle phenomenon (Liou 95)) has been then considered. Figures 13 and 14 show the Mach number contours and the Mach number distribution along the symmetry line, obtained on a uniform 32×16 grid with equally oriented diagonals. The present results agree well with those of (Liou 95) and demonstrate that the scheme maintains its qualities also for such an extreme flow case. It is noteworthy that the total number of *special cells*, where the lower-order N scheme is employed, is very small, so that the global accuracy of the solution can be considered second-order. Moreover, a residual drop of about six orders of magnitude is obtained starting from rest, for $M_\infty = 2$, and from an initial solution with linearly varying M, for $M_\infty = 6$, demonstrating the robustness of the present approach. Finally, the present results provide some minor improvement with respect to those of (De Palma 98b), where a problem dependent sensor was used to switch from the LDA scheme to the N one.

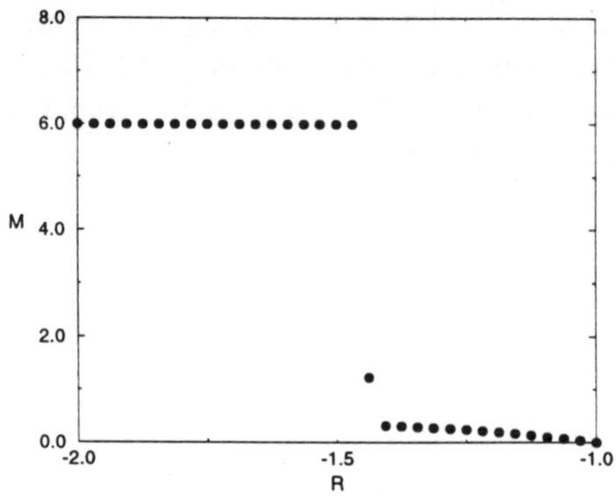

Fig. 14. Mach number distribution along the symmetry line.

In conclusion, for the case of the Euler equations, the proposed scheme excels previous state of the art FS schemes with respect to accuracy and, more importantly, robustness in the full range of Mach numbers ($0 \leq M \leq 6$). The proposed method can thus be extended with confidence to the Navier–Stokes equations, no difficulty caused by the presence of viscous layers being anticipated.

4 Navier–Stokes equations

4.1 Numerical method

The compressible Navier–Stokes equations are written in vector form as

$$\frac{\partial U}{\partial t} = -\boldsymbol{\nabla} \cdot (\mathcal{F}^i - \mathcal{F}^v) = \mathrm{Res}^i(U) + \mathrm{Res}^v(U), \tag{58}$$

$$\mathcal{F}^i = (F^i, G^i), \qquad \mathcal{F}^v = (F^v, G^v), \tag{59}$$

where U is the vector of the conservative variables and the flux vectors in the x and y directions are given as:

$$\begin{aligned} F^i &= (\rho u,\ p + \rho u^2,\ \rho uv,\ \rho uH)^T, \\ F^v &= (0,\ \tau_{xx},\ \tau_{xy},\ b_x)^T, \\ G^i &= (\rho v,\ \rho uv,\ p + \rho v^2,\ \rho vH)^T, \\ G^v &= (0,\ \tau_{yx},\ \tau_{yy},\ b_y)^T. \end{aligned} \tag{60}$$

In equation (60), the components of the viscous fluxes, F^v, G^v, are given as:

$$
\begin{aligned}
\tau_{xx} &= \mu\left[2\partial u/\partial x - (2/3)(\partial u/\partial x + \partial v/\partial y)\right],\\
\tau_{xy} &= \tau_{yx} = \mu\left[\partial u/\partial y + \partial v/\partial x\right],\\
\tau_{yy} &= \mu\left[2\partial v/\partial y - (2/3)(\partial u/\partial x + \partial v/\partial y)\right],\\
b_x &= u\tau_{xx} + v\tau_{xy} + \mathrm{k}(\partial T/\partial x),\\
b_y &= u\tau_{yx} + v\tau_{yy} + \mathrm{k}(\partial T/\partial y),
\end{aligned} \tag{61}
$$

where μ and k are the (laminar+eddy) dynamic viscosity and thermal conductivity, respectively. The system of governing equations is finally closed by assuming perfect gas with constant specific heats and laminar viscosity varying according to Sutherland's law.

The viscous terms are discretized using a standard Galerkin finite element scheme. Consider the generic node i of the computational domain and the area S_i composed by all triangles sharing node i. The standard procedure used in finite element analysis is applied to the divergence of the viscous fluxes, to provide:

$$
\int_{S_i} \omega_i \nabla \cdot \boldsymbol{\mathcal{F}}^v \mathrm{d}S = -\int_{S_i} \nabla\omega_i \cdot \boldsymbol{\mathcal{F}}^v \mathrm{d}S - \oint_{\partial S_i} \omega_i \boldsymbol{\mathcal{F}}^v \cdot \boldsymbol{n}\, \mathrm{d}s, \tag{62}
$$

where ω_i is the linear tent function and $\boldsymbol{n}$ is the inward unit vector normal to the boundary of S_i, ∂S_i. The second term in the right-hand-side of equation (62) is zero since $\omega_i = 0$ on ∂S_i, so that:

$$
\int_{S_i} \omega_i \nabla \cdot \boldsymbol{\mathcal{F}}^v \mathrm{d}S = -\sum_T \left[\frac{l_{iT}\boldsymbol{n}_{iT}}{2S_T} \cdot \int_{S_T} \boldsymbol{\mathcal{F}}^v \mathrm{d}S\right]. \tag{63}
$$

In equation (63), the sum is extended over all triangles meeting in i, and the subscript iT refers to the edge of the triangle T opposite to the vertex i.

For the present viscous flow computations, the following solid wall boundary conditions are employed. The velocity is prescribed, whereas the pressure and the temperature are computed by linear extrapolation from the first two interior grid-points (the grid being always orthogonal at the wall).

4.2 Results for viscous flows

Laminar flow past a NACA 0012 airfoil with $M_\infty = 2$, $10°$ angle of attack and Reynolds number, based on the chord of the airfoil and freestream flow conditions, equal to 1000, has been considered at first to test the methodology independently of turbulence modeling effects. The wall temperature has been assumed to be equal to the freestream total temperature. Two C-meshes have been employed, with 128×32 and 256×64 quadrilateral cells subdivided using equally oriented diagonals, and the far field boundary located at 12 chords from the airfoil. The pressure and skin friction coefficient (C_p, C_f) distributions along the profile are shown in figures 15 and 16, respectively. The corresponding values of the lift and drag coefficients are $C_L = 0.315$, $C_D = 0.252$, for the coarser mesh, and $C_L = 0.315$, $C_D = 0.259$, for the

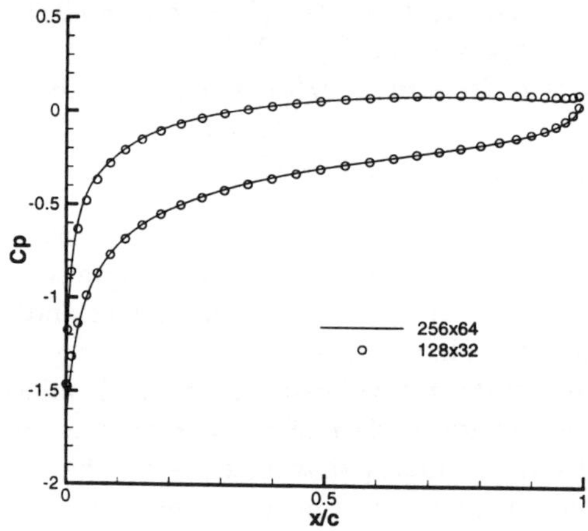

Fig. 15. Pressure coefficient distributions at the wall.

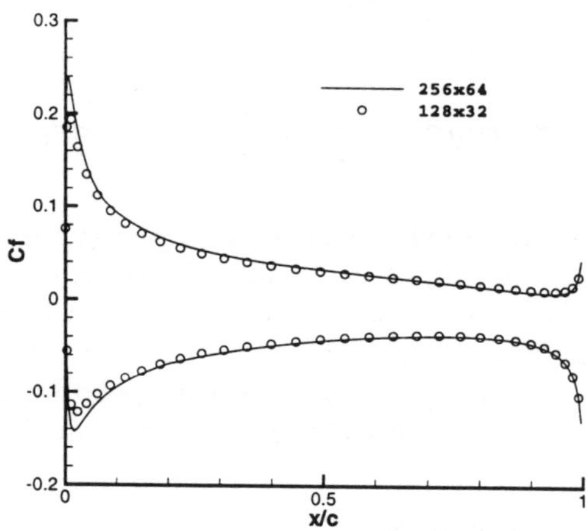

Fig. 16. Skin friction coefficient distributions at the wall.

finer one. Such values agree with the results given in (Bristeau 87). An overall description of the flowfield is also provided in figure 17, which shows the Mach number contours obtained on the finer grid. From these results it appears that the scheme is very accurate and the 256 $\times$ 64 grid solution can be considered grid converged.

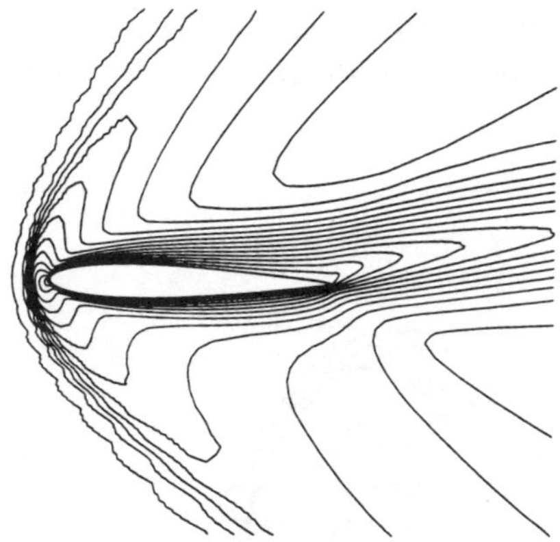

Fig. 17. Mach numebr contours ($\Delta M = 0.1$).

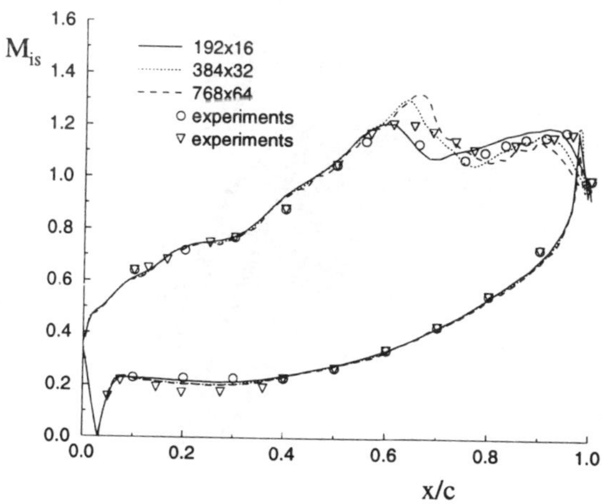

Fig. 18. Isentropic Mach number distributions at the wall.

Transonic flow through the VKI-LS59 cascade has been finally considered as a severe turbulent test case. The isentropic exit Mach number and the Reynolds number based on the blade chord and exit conditions are equal to 1 and 7.44×10^5, respectively (Kiock 86). The simple algebraic Baldwin-Lomax (Baldwin 78) turbulence model has been used, which has already proven satisfactory for such a flow (Arnone 92). Solutions have been computed on three non-periodic C-grids having 192×16, 384×32, and 768×64

cells, respectively. The average y^+ at the first grid-points away from the blade is equal to 1 for the finest grid. Figure 18 shows the resulting isentropic Mach

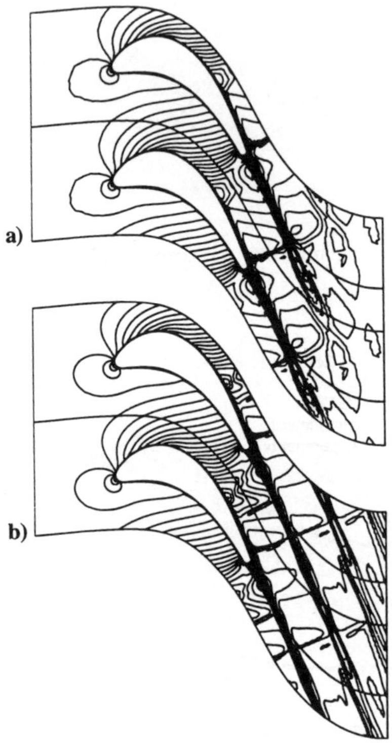

Fig. 19. Mach number contours ($\Delta M = 0.05$): a) coarsest grid; b) finest grid.

number distributions along the blade, together with two sets of experimental data (Kiock 86). These have been obtained in two different wind tunnels and mainly differ in the values of the maximum isentropic Mach number. The numerical solutions agree very well with the experimental results and show a trend towards grid convergence. The Mach number contours obtained on the coarsest and finest grids are also given in figure 19, for completeness. The complex structure of the flow is very well captured by the present method even on the coarsest mesh. The loss coefficient of the cascade (Kiock 86) has been finally computed, the resulting values (0.0737, 0.0568, and 0.0512) agreeing well with the experimental range (0.05-0.057).

5 Conclusions and future work

A genuinely multidimensional upwind hybrid fluctuation splitting scheme has been developed for the 2D steady-state compressible Euler and Navier–Stokes equations. The proposed approach employs the nonlinear scalar PSI scheme at supersonic cells, and either the monotone matrix N scheme or the linearity preserving matrix LDA one at subsonic cells, according to the local nature (discontinuous or smooth) of the flow. A clever switching strategy allows to choose the optimal scheme, locally, so that, for the first time, a fluctuation splitting method is obtained which is quasi second-order-accurate, captures shocks monotonically, and converges without any difficulty at subsonic, supersonic, and transonic flow regimes. In the future, the proposed approach will be combined with two-equation turbulence models and later extended to three space dimensions.

6 Acknowledgements

The authors are grateful to professor Roe who paved the way for a challenging and rewarding research.

The present research has been supported by M.U.R.S.T.

References

ARNONE A., LIOU M.-S., POVINELLI L. A., Navier-Stokes solution of transonic cascade flows using nonperiodic C-type grids. *Journal of Propulsion and Power,* **8**, *No. 2, pp. 410-417,* 1992.

BALDWIN B. S., LOMAX H., Thin layer approximation and algebraic model for separated turbulent flows. *AIAA Paper 78-0257,* 1978.

BARTH T.J., VAN DER WEIDE E., DECONINCK H., ABGRALL R., Matrix distribution schemes and energy stability for hyperbolic systems with application to compressible flows. *6th International Conference on Hyperbolic Problems, Hong Kong,* 1996.

BRANDT A., Guide to multigrid development. *Lecture Notes in Mathematics,* **960**, *Springer Verlag, pp. 220-312,* 1982.

BRISTEAU M. O., GLOWINSKI R., PERIAUX J., VIVIAND A. (EDS.), Numerical simulation of compressible Navier-Stokes flows. *Notes on Numerical Fluid Mechanics,* **18**, *Vieweg, Braunschweig,* 1987.

CATALANO L. A., DE PALMA P., NAPOLITANO M., PASCAZIO G., Cell-vertex adaptive Euler method for cascade flows. *AIAA Journal,* **33**, *No. 12, pp. 2299-2304,* 1995.

CATALANO L. A., DE PALMA P., NAPOLITANO M., PASCAZIO G., A critical analysis of multi-dimensional upwinding for the Euler equations. *Computers and Fluids,* **25**, *No. 1, pp. 29-38,* 1996.

CATALANO L. A., DE PALMA P., PASCAZIO G., NAPOLITANO M., Matrix fluctuation splitting schemes for accurate solutions to transonic flows. *Lecture Notes in Physics,* **490**, *Springer Verlag, pp. 328-333,* 1997a.

CATALANO L. A., DE PALMA P., NAPOLITANO M., PASCAZIO G., A very efficient local-adaptive multigrid method based on a simple-wave decomposition of the Euler equations. *Notes on Numerical Fluid Mechanics,* **57**, *Vieweg, Braunschweig, pp. 187-219,* 1997b.

CATALANO L. A., DE PALMA P., NAPOLITANO M., PASCAZIO G., Genuinely multi-dimensional upwind methods for accurate and efficient solutions of compressible flows. *Notes on Numerical Fluid Mechanics,* **57**, *Vieweg, Braunschweig, pp. 221-250,* 1997c.

CATALANO L. A., DE PALMA P., PASCAZIO G., NAPOLITANO M., A contribution to multidimensional upwinding and fluctuation splitting: nonlinear matrix schemes. *AIAA Paper 97-2031,* 1997d.

DECONINCK H., ROE P. L., STRUIJS R., A multidimensional generalization of Roe's flux difference splitter for the Euler equations. *Computers and Fluids,* **22**, *No. 2/3, pp. 215-222,* 1993a.

DECONINCK H., STRUIJS R., PAILLÈRE H., CATALANO L. A., DE PALMA P., NAPOLITANO M., PASCAZIO G., Development of cell-vertex multidimensional upwind solvers for the compressible flow equations. *CWI-Quarterly,* **6**, *No. 1, pp. 1-28,* 1993b.

DE PALMA P., PASCAZIO G., NAPOLITANO M., A hybrid fluctuation splitting scheme for transonic inviscid flows. *4th ECCOMAS CFD Conference, Athens,* 1998a.

DE PALMA P., PASCAZIO G., NAPOLITANO M., A validation study of a hybrid fluctuation splitting scheme for transonic inviscid flows. *Lecture Notes in Physics,* **490**, *Springer Verlag, pp. 373-378,* 1998b.

FRIEDRICHS K. O., Symmetric hyperbolic linear differential equations. *Commun. Pure Appl. Math.* **7**, *pp. 345-392,* 1954.

HUGHES T. J. R., MALLET M., MIZUKAMI A., A new finite element formulation for computational fluid dynamics: II. Beyond SUPG. *Comp. Meth. Appl. Mech. Engrg.,* **54**, *pp. 341-355,* 1986a.

HUGHES T. J. R. AND MALLET M., A new finite element formulation for computational fluid dynamics: III. The generalized streamline operator for multi-dimensional advective-diffusive systems. *Comp. Meth. Appl. Mech. Engrg.,* **58**, *pp. 305-328,* 1986b.

HUGHES T. J. R. AND MALLET M., A new finite element formulation for computational fluid dynamics: IV. A discontinuity-capturing operator for multidimensional advective-diffusive systems *Comp. Meth. Appl. Mech. Engrg.,* **58**, *pp. 329-336,* 1986c.

KIOCK R., LETHAUS F., BAINES N. C., SIEVERDING C. H., The transonic flow through a plane turbine cascade as measured in four European wind tunnels. *Journal of Engineering for gas turbines and power,* **108**, *No. 2, pp. 277-285,* 1986.

Liu X. and Lax P. D., Positive schemes for solving multi-dimensional hyperbolic systems of conservation laws. *CFD Journal, **5**, No. 2, pp. 133-156*, 1996.

Lee D., van Leer B., Lynn J. F., A local Navier-Stokes preconditioner for all Mach and Reynolds numbers. *AIAA Paper 97-2024*, 1997.

Liou M., Wada Y., A quest towards ultimate numerical flux schemes. *CFD Review, M. Hafez and K. Oshima editors, Wiley, pp. 251-278*, 1995.

Mesaros L. M., Roe P. L., Multidimensional fluctuation splitting schemes based on decomposition methods. *AIAA Paper 95-1699*, 1995.

Mizukami A., Hughes T. J. R., A Petrov–Galerkin finite element method for convection-dominated flows: an accurate upwinding technique for satisfying the maximum principle. *Comp. Meth. Appl. Mech. Engrg., **50**, pp. 181-193*, 1985.

Paillère H., Carette J-C., Deconinck H., Multidimensional upwind and SUPG methods for the solution of the compressible flow equations on unstructured grids. *VKI LS 1994-05, Computational Fluid Dynamics, von Karman Institute, Belgium*, 1994.

Paillère H., Multidimensional upwind residual distribution schemes for the Euler and Navier-Stokes equations on unstructured grids. *Ph.D Thesis, Université Libre de Bruxelles, Belgium*, 1995.

Roe P. L., Approximate Riemann solvers, parameter vectors and difference schemes. *J. Comput. Phys., **43**, No. 2, pp. 357-372*, 1981.

Roe P. L., Characteristic-based schemes for the Euler equations. *Ann. Rev. Fluid Mech., **18**, pp. 337-365*, 1986a.

Roe P. L., Discrete models for the numerical analysis of time-dependent multidimensional gas dynamics. *J. Comput. Phys., **63**, No. 2, pp. 458-476*, 1986b.

Roe P. L., Linear advection schemes on triangular meshes. *CoA Report No. 8720, Cranfield Institute of Technology*, 1987.

Roe P. L., "Optimum" upwind advection on a triangular mesh. *ICASE Report No. 90-75*, 1990.

Sidilkover D., A genuinely multidimensional upwind scheme and efficient multigrid solver for the compressible Euler equations. *ICASE Report No. 94-84*, 1994.

Sidilkover D., Roe P. L., Unification of some advection schemes in two dimensions. *ICASE Report No. 95-10*, 1995.

Struijs R., Deconinck H., Roe P. L., Fluctuation splitting schemes for the 2D Euler equations. *VKI LS 1991-01, Computational Fluid Dynamics, von Karman Institute, Belgium*, 1991.

van der Weide E., Deconinck H., Positive matrix distribution schemes for hyperbolic systems, with applications to the Euler equations. *3rd ECCOMAS CFD Conference, Paris*, 1996.

van Leer B., Lee W. T., Roe P. L., Characteristic time-stepping or local preconditioning of the Euler equations. *AIAA Paper 91-1552*, 1991.

Yee H. C., Construction of explicit and implicit symmetric TVD schemes and their applications. *J. Comput. Phys., **68**, pp. 151-179*, 1987.

Some Recent Developments in Kinetic Schemes based on Least Squares and Entropy Variables[1]

Deshpande S. M.

AR & DB Centre for Excellence in Aerospace CFD
Department of Aerospace Engineering
Indian Institute of Science, Bangalore, INDIA
Email: suresh@aero.iisc.ernet.in

Keywords: Boltzmann equation, KFVS, Least Squares Kinetic Upwind Method, Entropy variables, Euler Equations, Moment method strategy

Abstract: Least squares kinetic upwind method (LSKUM) uses least squares formulae to approximate spatial derivatives in the Boltzmann equation and then uses moment method strategy to get a numerical method for the Euler equations of gas dynamics. The second order LSKUM is obtained through defect correction which involves use of first order formula with a modified difference. Further simplification can be achieved by use of entropy variables. The modified difference becomes a difference between two modified Maxwellians when entropy variables are used thus leading to q-LSKUM. The state update for q-LSKUM is just like the state update for first order LSKUM because of the difference between modified Maxwellians. The LSKUM based entropy variables has been tested against standard 2-D examples indicating superior performance of q-LSKUM compared to usual LSKUM.

1 Introduction

Kinetic numerical schemes are being continuously developed in the CFD Laboratory of Indian Instiute of Science, Bangalore for the past 15 years by various investigators. There are also other research groups elsewhere in the world who are contributing a great deal to the development of kinetic schemes. Deshpande [1,2] has reviewed many of these developments. The present article mainly focuses on the recent developments in kinetic schemes pursued in the CFD Laboratory. Deshpande[3] has put forward the idea of 'moment method strategy' using which a numerical scheme for the solution of Euler equations of inviscid gas dynamics or Navier-Stokes (NS) equations can be obtained by taking suitable moments of a numerical scheme for the Boltzmann equation of kinetic theory of gases. We start with the Boltzmann equation.

$$\frac{\partial f}{\partial t} + v_1 \frac{\partial f}{\partial x_1} + v_2 \frac{\partial f}{\partial x_2} + v_3 \frac{\partial f}{\partial x_3} = 0 \tag{1}$$

without collision term. Here $f = f(t, x_1, x_2, x_3, v_1, v_2, v_3) = f(t, \vec{x}, \vec{v})$ is a velocity distribution function, v_1, v_2, v_3 are molecular or particle velocites along

[1]This paper was presented in "Solutions of PDE" Conference in honour of Prof. Roe on the occassion of his 60th birthday, July 1998, Arachaon, France

co-ordinates directions x_1, x_2, x_3. This is a linear hyperbolic equation for a scalar f and a lot of theoretical work has been done for such equations. Introduce the moment function vector Ψ given by

$$\Psi = \begin{bmatrix} I \\ v_1 \\ v_2 \\ v_3 \\ I + \frac{v^2}{2} \end{bmatrix}, \qquad v^2 = v_1^2 + v_2^2 + v_3^2 \tag{2}$$

where we have introduced the internal energy variable I for having a specified ratio γ of specific heats. The distribution function f then obviously becomes a function of I also, that is, $f = f(t, \vec{x}, \vec{v}, I)$. The fluid dynamical variables $\rho =$ mass density, $\vec{u} =$ fluid velocity and $e =$ total energy/volume $= \frac{p}{\gamma-1} + \frac{1}{2}\rho\mathbf{u}^2$, and $p =$ pressure are related to f as given below

$$\rho = \int_{-\infty}^{+\infty} \int \int \int_0^{\infty} f dv_1 dv_2 dv_3 dI = \int_{R^3} \int_{R^+} f d^3v dI$$

$$\rho\vec{u} = \int_{R^3} \int_{R^+} \vec{v} f d^3v dI, \qquad \rho e = \int_{R^3} \int_{R^+} \left(I + \frac{v^2}{2} \right) f d^3v dI \tag{3}$$

The conserved vector U is given by

$$U = \begin{bmatrix} \rho \\ \rho u_1 \\ \rho u_2 \\ \rho u_3 \\ \rho e \end{bmatrix} \tag{4}$$

The vector U is realted to the Maxwellian F by the equation

$$U = \int_{R^3} \int_{R^+} \psi F d^3v dI = (\psi, F) \tag{5}$$

The Maxwellian distribution F is defined by

$$F = \frac{\rho}{I_0} \left(\frac{\beta}{\pi} \right)^{\frac{3}{2}} exp \left[-\beta(\vec{v} - \vec{u})^2 - \frac{I}{I_0} \right] \tag{6}$$

where β=1/2(RT), R =Gas constant, T=temperature and I_0= internal energy= $\frac{5-3\gamma}{2(\gamma-1)3}$. Note that I_0=0 for γ=5/3. The Euler equations in the conservation form

$$\frac{\partial U}{\partial t} + \frac{\partial}{\partial x_1}(GX) + \frac{\partial}{\partial x_2}(GY) + \frac{\partial}{\partial x_3}(GZ) = 0 \tag{7}$$

are moments of (1),that is

$$\int_{R^3} \int_{R^+} \psi \left(\frac{\partial F}{\partial t} + \vec{v}.\frac{\partial F}{\partial x} \right) d^3v dI = 0 \tag{8}$$

The fluxes are given by

$$
\begin{aligned}
GX &= \int_{R^3}\int_{R^+} v_1 \psi F d^3v dI \\
GY &= \int_{R^3}\int_{R^+} v_2 \psi F d^3v dI \\
GZ &= \int_{R^3}\int_{R^+} v_3 \psi F d^3v dI
\end{aligned}
\tag{9}
$$

The moment method strategy precisely expliots the eq.(8),that is, start with a numerical scheme for

$$
\frac{\partial F}{\partial t} + \vec{v}.\frac{\partial F}{\partial x} = 0 \tag{10}
$$

and take ψ-moments to obtain a mapped numerical scheme for (7).The flux vector splitting [3] based on $v_1 > 0$ or $v_1 < 0$ or $v_2 > 0$ or $v_2 < 0$ and $v_3 > 0$ or $v_3 < 0$ gives us

$$
\begin{aligned}
GX^{\pm} &= \int_{R^3}\int_{R^+} \frac{v_1 \pm |v_1|}{2} \psi F d^3v dI \\
GY^{\pm} &= \int_{R^3}\int_{R^+} \frac{v_2 \pm |v_2|}{2} \psi F d^3v dI \\
GZ^{\pm} &= \int_{R^3}\int_{R^+} \frac{v_3 \pm |v_3|}{2} \psi F d^3v dI
\end{aligned}
\tag{11}
$$

and corresponding kinetic flux vector split (KFVS) Euler equations are

$$
\begin{aligned}
\frac{\partial}{\partial t}(U) &+ \frac{\partial}{\partial x_1}(GX^+) + \frac{\partial}{\partial x_1}(GX^-) + \frac{\partial}{\partial x_2}(GY^+) + \frac{\partial}{\partial x_2}(GY^-) \\
&+ \frac{\partial}{\partial x_3}(GZ^+) + \frac{\partial}{\partial x_3}(GZ^-) = 0
\end{aligned}
\tag{12}
$$

Two dimensional cell-centered as well as cell vertex based finite volume codes based on KFVS with reconstruction for higher order accuracy have been developed by Mathur and Deshpande [4] and thoroughly tested for very low subsonic (William's airfoil), subsonic, transonic and supersonic flows around airfoils. Deshpande et al[5] have developed 3D cell-centered finite volume code called BHEEMA based on KFVS and this code is being routinely used in DRDL, Hyderabad for computing transonic, superonic flows around several flight vehicles.

2 Least Squares Kinetic Upwind Method (LSKUM)

In many practical configurations, the mesh plays a very important role in determining the solution accuracy. Flow solvers are known to lose their accuracy if the mesh is too much distorted or of poor quality. It would be very nice to have a flow solver which is capable of working on any type of mesh-structured, unstructured, chimera, dragon, stacked and even one of poor quality. A grid fault tolerant scheme will be very valuable. The design environment requires capability of calculating flow quickly with engineering accuracy. In such situations, the grid is generated quickly and it may not be of good quality, that is, it may involve distorted hexahedra or tetrahedra, cells or finite volumes could be highly skewed having very acute or obtuse angles in some regions, also the cell volumes may vary rapidly in some pockets. This generally happens when grid stacking is employed for generating quasi-3D meshes. There is always a certain degree of trade-off between the quality of the grid and the flow-solver employed. If the flow-solver is not robust enough, then very good quality grid is essential. The other possibility is to have robust flow-solver capable of working on bad grids. We started work in this direction to find out whether kinetic schemes could be developed further to satisfy this need. The Least squares Kinetic Upwind Method (LSKUM) developed by Ghosh, Mandal and Deshpande [6,7,8,9] is a step in this direction. To illustrate the basic idea behind the method, consider a 2D linear advection equation

$$\frac{\partial f}{\partial t} + v_1 \frac{\partial f}{\partial x} + v_2 \frac{\partial f}{\partial y} = 0 \tag{13}$$

We keep in mind that when $f = F =$ Maxwellian, the equation (13) gives the 2-D Euler equations if suitable moments are taken. The central problem then is to determine discrete approximations to derivatives $\frac{\partial f}{\partial x}, \frac{\partial f}{\partial y}$ at point or a node P_o using data at the neighbouring points as shown in fig. 1. The distribution of points around P_o is completely arbitrary. We can define the set of neighbours (also called stencil of P_o) of P_o by

$$N(P_o) = \{ \text{ all } P_i \text{ such that } d(P_o, P_i) < h\} \tag{14}$$

where $d(P_o, P_i) =$ Euclidean distance between P_o and P_i and h=characteristic linear dimension of the neighbourhood $N(P_o)$ to be specified by the user. Exact definition of the neighbourhood is not important to LSKUM as long a P_i are local to P_o. The discrete approximation to the derivatives $\left(\frac{\partial f}{\partial x}\right)_o, \left(\frac{\partial f}{\partial y}\right)_o$ are determined by using least squares. For this purpose we introduce the notations

$$\Delta f_i = f_i - f_0, \quad \Delta x_i = x_i - x_0, \quad \Delta y_i = y_i - y_0$$

using Taylor series, we obtain

$$\Delta f_i = \left(\frac{\partial f}{\partial x}\right)_0 \Delta x_i + \left(\frac{\partial f}{\partial y}\right)_0 \Delta y_i + \ldots\ldots, \quad \text{for i} = 1, 2, 3 \ldots\ldots n \tag{15}$$

where n=number of nodes in $N(P_o)$

For n $\geq$ 3 the equation (15) leads to classical over determined problem for the derivatives $\left(\frac{\partial f}{\partial x}\right)_o, \left(\frac{\partial f}{\partial y}\right)_o$ we can write (15) in the matrix form

$$A_n(Df) = \Delta f \tag{16}$$

where

$$\begin{aligned} Df &= \begin{bmatrix} f_{xo} \\ f_{yo} \end{bmatrix}, \qquad \Delta f = [\Delta f_1, \Delta f_2 ....\Delta f_n]^T \\ A_n &= \begin{bmatrix} \Delta x_1 & \Delta y_1 \\ \Delta x_2 & \Delta y_2 \\ \dots & \dots \\ \dots & \dots \\ \Delta x_n & \Delta y_n \end{bmatrix} \end{aligned} \tag{17}$$

One of the well known methods to solve (16) is to use least squares principle, that is minimise the sum of squares norm

$$\|e\|^2 = \|A_n Df - \Delta f\|^2 \tag{18}$$

we then get

$$Df = (A_n^T A_n)^{-1}(A_n^T \Delta f) \tag{19}$$

Here$A_n^T A_n$ is a 2x2 square matrix and for two dimenstions its inversion is easy. We can explicitly write the solution as

$$\begin{aligned} f_{xo} = \left(\frac{\partial f}{\partial x}\right)_o &= \frac{\|\Delta y\|^2(\Delta f, \Delta x) - (\Delta x, \Delta y)(\Delta f, \Delta y)}{\|\Delta x\|^2\|\Delta y\|^2 - (\Delta x, \Delta y)^2} \\ f_{yo} = \left(\frac{\partial f}{\partial y}\right)_o &= \frac{\|\Delta x\|^2(\Delta f, \Delta y) - (\Delta x, \Delta y)(\Delta f, \Delta x)}{\|\Delta x\|^2\|\Delta y\|^2 - (\Delta x, \Delta y)^2} \\ \text{where } \|\Delta x\|^2 &= \sum_{i=1}^{n} \Delta x_i^2, \quad \|\Delta y\|^2 = \sum_{i=1}^{n} \Delta y_i^2 \end{aligned} \tag{20}$$

and the inner product is defined by

$$(\Delta a, \Delta b) = \sum_{i=1}^{n} \Delta a_i \Delta b_i \tag{21}$$

upwinding can be enforced in the above frame work by recognizing that the exact solution of (13) is

$$f(t + \Delta t, x_o, y_o, v_1, v_2) = f(t, x_o - v_1 \Delta t, y_o - v_2, \Delta t, v_1, v_2) \tag{22}$$

This exact solution indicates that the propagation of information from any node P_i (contributing node) to the receiving node (P_o) depends upon the location of P_i relative to P_o and signs of v_1, v_2. If $v_1 > 0$ then the node P_i will influence the solution at P_o provided P_i lies to the left of $P_o(i.e.\Delta x_1 < 0)$ and by similar arguments it follows that P_i will influence the solution at P_o for $v_1 < O$,if $\Delta x_i > 0$. This property called signal propagation property needs to be taken into account while developing an upwind scheme. We accomplish upwinding by dividing stencil of points or equivalently by introducing weights. We first note that the norm defined by (18) can be modified as

$$\|e\|^2 = \sum_{i=1}^{n} w_i\{\Delta f_i - \Delta x_i f_{xo} - \Delta y_i f_{yo}\}^2$$

where w_i are positive weights. The formulae (20) then become

$$f_{xo} = \frac{(\sum w_i \Delta y_i^2)(\sum w_i \Delta f_i \Delta x_i) - (\sum w_i \Delta x_i \Delta y_i)(\sum w_i \Delta f_i \Delta y_i)}{\sum w_i \Delta x_i^2)(\sum w_i \Delta y_i^2) - (\sum w_i \Delta x_i \Delta y_i)^2}$$

$$f_{yo} = \frac{(\sum w_i \Delta x_i^2)(\sum w_i \Delta f_i \Delta y_i) - (\sum w_i \Delta x_i \Delta y_i)(\sum w_i \Delta f_i \Delta x_i)}{(\sum w_i \Delta x_i^2)(\sum w_i \Delta y_i^2) - (\sum w_i \Delta x_i \Delta y_i)^2} \tag{23}$$

These are the weighted least squares approximations to the spatial derivatives of f at the node P_o in terms of the data at the neighbouring nodes P_i . The weights required for upwinding approximations are then given by

$$w_i = \begin{cases} 1, for \Delta x_i > 0, v_1 < 0 \\ 1, for \Delta x_i < 0, v_1 > 0 \\ 1, for \Delta y_i > 0, v_2 < 0 \\ 1, for \Delta y_i < 0, v_2 > 0 \\ 0, \quad otherwise \end{cases} \tag{24}$$

In other words we reject the nodes from the set $N(P_o)$ which are not upwind of P_o. We can also interpret this choice of weights with reference to Courant-Issacson-Rees upwind scheme which is a forerunner of many modern upwind schemes. First we cast (13) in the form

$$\frac{\partial f}{\partial t} + \frac{v_1 + |v_1|}{2}\frac{\partial f}{\partial x} + \frac{v_1 - |v_1|}{2}\frac{\partial f}{\partial x} + \frac{v_2 + |v_2|}{2}\frac{\partial f}{\partial y} + \frac{v_2 - |v_2|}{2}\frac{\partial f}{\partial y} = 0 \tag{25}$$

The second and the fourth terms in (25) are present respectively for $v_1 > 0$ and $v_2 > 0$, they are zero for $v_1 < 0$ and $v_2 < 0$. For the later possibility, the third and fifth terms in (25) are present. Thus, while obtaining a discrete approximation for the derivative in the second term we use weight $w_i = 1$ for nodes with $\Delta x_i < 0$ and $v_1 > 0$.

The LSKUM for the 2D Euler equations can now be obtained by taking ψ moments of (25) with $f = F$. We then have

$$\begin{aligned} \int_{R^+} dI \int_R \int_R dv_1 dv_2 \psi \frac{\partial F}{\partial t} &+ \int_{R^+} dI \int_R \int_R dv_1 dv_2 \left\{\frac{v_1 + |v_1|}{2}\frac{\partial F}{\partial x}\psi\right\}_{N_1(P_o)} \\ &+ \int_{R^+} dI \int_R \int_R dv_1 dv_2 \left\{\frac{v_1 - |v_1|}{2}\frac{\partial F}{\partial x}\psi\right\}_{N_2(P_o)} \\ &+ \int_{R^+} dI \int_R \int_R dv_1 dv_2 \left\{\frac{v_2 + |v_2|}{2}\frac{\partial F}{\partial x}\psi\right\}_{N_3(P_o)} \\ &+ \int_{R^+} dI \int_R \int_R dv_1 dv_2 \left\{\frac{v_1 - |v_1|}{2}\frac{\partial F}{\partial x}\psi\right\}_{N_4(P_o)} = 0 \end{aligned} \tag{26}$$

Here the substencils $N_1(P_o), N_2(P_o), N_3(P_o), N_4(P_o)$ are subsets of $N(P_o)$ and are defined by

$$
\begin{aligned}
N_1(P_o) &= \{\text{all nodes to the left of y-axis}\} \\
&= \{P_4, P_5, P_6, P_7\} \\
N_2(P_o) &= \{\text{all nodes to the right of y-axis}\} \\
&= \{P_8, P_1, P_2, P_3\} \\
N_3(P_o) &= \{\text{all nodes below x-axis}\} \\
&= \{P_5, P_6, P_7, P_8\} \\
N_4(P_o) &= \{\text{all nodes above x-axis}\} \\
&= \{P_1, P_2, P_3, P_4\}
\end{aligned}
$$

Fig.2 shows these various substencils. The two dimensional Maxwellian F is given by

$$
F = \frac{\rho\beta}{\pi I_0}\exp\left[-\beta\left(v_1-u_1\right)^2-\beta\left(v_2-u_2\right)^2-\frac{I}{I_0}\right]
$$

$$
I_0 = \frac{2-\gamma}{(\gamma-1)\beta}, \quad \beta = \frac{1}{2RT}, \quad \psi = \left[1, v_1, v_2, I + \frac{v_1^2+v_2^2}{2}\right]^T \tag{27}
$$

After performing the integrations in (26), we get

$$
\begin{aligned}
\frac{\partial U}{\partial t} &+ \left\{\frac{\|\Delta y\|^2\left(\Delta GX^+, \Delta x\right) - (\Delta x, \Delta y)\left(\Delta GX^+, \Delta y\right)}{\|\Delta x\|^2\|\Delta y\|^2 - (\Delta x, \Delta y)^2}\right\}_{N_1(P_o)} \\
&+ \left\{\frac{\|\Delta y\|^2\left(\Delta GX^-, \Delta x\right) - (\Delta x, \Delta y)\left(\Delta GX^-, \Delta y\right)}{\|\Delta x\|^2\|\Delta y\|^2 - (\Delta x, \Delta y)^2}\right\}_{N_2(P_o)} \\
&+ \left\{\frac{\|\Delta x\|^2\left(\Delta GY^+, \Delta y\right) - (\Delta x, \Delta y)\left(\Delta GY^+, \Delta x\right)}{\|\Delta x\|^2\|\Delta y\|^2 - (\Delta x, \Delta y)^2}\right\}_{N_3(P_o)} \\
&+ \left\{\frac{\|\Delta x\|^2\left(\Delta GY^-, \Delta y\right) - (\Delta x, \Delta y)\left(\Delta GY^-, \Delta x\right)}{\|\Delta x\|^2\|\Delta y\|^2 - (\Delta x, \Delta y)^2}\right\}_{N_4(P_o)}
\end{aligned} \tag{28}
$$

The inner products in each of the above curly brackets in (28) are to be taken over the nodes belonging to the respective substencils shown at the bottom of the closing curly brackets. The split fluxes $GX^{\pm}, GY^{\pm}$ are the usual KFVS formulae and these are given by

$$
\begin{bmatrix} GX^{\pm}(1) \\ GX^{\pm}(2) \\ GX^{\pm}(3) \\ GX^{\pm}(4) \end{bmatrix} = \begin{bmatrix} \rho\left\{u_1\frac{A^{\pm}(S_1)}{2} \pm \frac{B(S_1)}{2\sqrt{\pi\beta}}\right\} \\ \rho\left\{\left(u_1^2+\frac{p}{\rho}\right)\frac{A^{\pm}(S_1)}{2} \pm u_1\frac{B(S_1)}{2\sqrt{\pi\beta}}\right\} \\ \rho\left\{u_1u_2\frac{A^{\pm}(S_1)}{2} \pm \frac{B(S_1)}{2\sqrt{\pi\beta}}\right\} \\ \rho\left\{u_1\left(\frac{\gamma p}{\rho(\gamma-1)}+\frac{u_1^2+u_2^2}{2}\right)\frac{A^{\pm}(S_1)}{2} \pm \left(\frac{(\gamma+1)p}{\rho(\gamma-1)}+\frac{u_1^2+u_2^2}{2}\right)\frac{B(S_1)}{2\sqrt{\pi\beta}}\right\} \end{bmatrix} \tag{29}
$$

$$\begin{bmatrix} GY^{\pm}(1) \\ GY^{\pm}(2) \\ GY^{\pm}(3) \\ GY^{\pm}(4) \end{bmatrix} = \begin{bmatrix} \rho\left\{u_2 \frac{A^{\pm}(S_2)}{2} \pm \frac{B(S_2)}{2\sqrt{\pi\beta}}\right\} \\ \rho\left\{u_1 u_2 \frac{A^{\pm}(S_2)}{2} \pm \frac{B(S_2)}{2\sqrt{\pi\beta}}\right\} \\ \rho\left\{\left(u_2^2 + \frac{p}{\rho}\right) \frac{A^{\pm}(S_2)}{2} \pm u_2 \frac{B(S_2)}{2\sqrt{\pi\beta}}\right\} \\ \rho\left\{u_2\left(\frac{\gamma p}{\rho(\gamma-1)} + \frac{u_1^2+u_2^2}{2}\right) \frac{A^{\pm}(S_2)}{2} \pm \left(\frac{(\gamma+1)p}{\rho(\gamma-1)} + \frac{u_1^2+u_2^2}{2}\right) \frac{B(S_2)}{2\sqrt{\pi\beta}}\right\} \end{bmatrix} \tag{30}$$

In the above formulae, speed ratios S_1, S_2 are given by
$S_1 = u_1\sqrt{\beta}, S_2 = u_2\sqrt{\beta}, A^{\pm}(S) = 1 \pm erf(S), B(S) = exp(-S^2)$
we note that $\Delta GX^{\pm}, GY^{\pm}$ in (28) have the meaning $\Delta GX^{\pm} = GX_i^{\pm} - GX_0^{\pm}$ and $GX^{\pm}, GY^{\pm}$, respectively stand for the column vectors defined by (29) and (30) at the node i. The state update formula based on LSKUM is finally obtained by using any time marching scheme for (28). Ghosh[10] has also developed yet another version of LSKUM using quadrantwise splitting, that is, the integrals with respect to v_1, v_2 are split over four quadrants $(v_1 > 0, v_2 > 0), (v_1 > 0, v_2 < 0), (v_1 < 0, v_2 > 0)$ and $(v_1 < 0, v_2 < 0)$. The quadrantwise splitting LSKUM is found to contain more numerical diffusion than x-y splitting described above and is sometimes useful because of this fact. Deshpande, Ghosh and Mandal [8,10] have also developed LSKUM with rotation for reducing the inherent numerical diffusion. In the least squares formulation, employing x-y splitting, the rotation of coordinate directions (from the point of view of minimal numerical diffusion) is possible. Even though this question has not been answered completely satisfactorily as yet, it has been found that streamline upwinding considerably reduces diffusion. We will come back to the question later in this paper.

The first order LSKUM contains unacceptably large numerical diffusion - a characteristic property of many other first-order upwind numerical methods. It is therefore essential to reduce the diffusion by developing higher order LSKUM which is done by two steps scheme described in the next section.

3 Second Order LSKUM

The basic idea behind second-order LSKUM can be illustrated with reference to 1-D problem. The least squares approximations to the derivate F_{xo} at the node o is given by

$$F_{x0}^{(1)} = \frac{\sum \Delta x_i \Delta F_i}{\sum \Delta x_i^2} \tag{31}$$

where $\Delta x_i = x_i - x_o, \Delta F_i = F_i - F_o$ as before. The superscript(1) on F_{xo} indicates that the formula (31) is first-order accurate in x. Consider Taylor expansion

$$\Delta F_i = \Delta x_i \left(\frac{\partial F}{\partial x}\right)_o + \frac{\Delta x_i^2}{2}\left(\frac{\partial^2 F}{\partial x^2}\right)_0 + H.O.T \tag{32}$$

where H.O.T = higher -order-terms. We note that

$$\Delta F_{xi} = \Delta x_i \left(\frac{\partial^2 F}{\partial x^2} \right)_0 + H.O.T \tag{33}$$

This suggest that ΔF_{xi} can be used to cancel the second derivate term in (32) that is define the modified

$$\Delta \tilde{F}_i = \Delta F_i - \frac{\Delta x_i}{2} \Delta F_{xi}^{(1)} \tag{34}$$

where

$$\Delta F_{xi}^{(1)} = F_{xi}^{(1)} - F_{xo}^{(1)} \tag{35}$$

which is the difference between two first order accurate derivatives of F at two neighbouring node i and o. The eqn. (34) gives the defect correction step. The difference $\Delta F_{xi}^{(1)}$ can be obtained by using the first-order formula (31), and it can then be used to determine the modified difference $\Delta \tilde{F}_i$ defined by (34). Because of the construction of $\Delta \tilde{F}_i$, we have

$$\Delta \tilde{F}_i = \Delta x_i \left(\frac{\partial F}{\partial x} \right)_o + O\left(\Delta x_i^3\right)$$

Hence the second- order accurate formulae for F_{xo} is given by

$$F_{xo}^{(2)} = \frac{\sum \Delta x_i \Delta \tilde{F}_i}{\sum \Delta x_i^2} \tag{36}$$

we therefore get a 2-step second-order accurate formula for F_{xo} and sometimes this is called defect correction procedure. In the first step determine $F_{xo}^{(1)}$ by using (31), the modified difference $\Delta \tilde{F}_i$ is then determined by (34) which requires $F_{xo}^{(1)}, F_{xi}^{(1)}$. The second step is given by (36) which makes use of the modified differences. The extension of the above 2-step formula to two-dimensional problem is straightforward. We first determine first-order accurate derivatives $F_{xo}^{(1)}, F_{yo}^{(1)}$ by the least squares expression

$$\begin{aligned} F_{xo}^{(1)} &= \frac{\sum \Delta y_i^2 \sum \Delta F_i \Delta x_i - \sum \Delta x_i \Delta y_i \sum \Delta F_i \Delta y_i}{DET} \\ F_{yo}^{(1)} &= \frac{\sum \Delta x_i^2 \sum \Delta F_i \Delta y_i - \sum \Delta x_i \Delta y_i \sum \Delta F_i \Delta y_i}{DET} \\ DET &= (\sum \Delta x_i^2)(\sum \Delta y_i^2) - (\sum \Delta x_i y_i)^2 \end{aligned} \tag{37}$$

Determine the modified differences

$$\Delta \tilde{F}_i = \Delta F_i - \frac{\Delta x_i}{2} \Delta F_{xi}^{(1)} - \frac{\Delta y_i}{2} \Delta F_{yi}^{(1)} \tag{38}$$

$$\Delta F_{xi}^{(1)} = F_{xi}^{(1)} - F_{xo}^{(1)}, \qquad \Delta F_{yi}^{(1)} = F_{yi}^{(1)} - F_{yo}^{(1)} \tag{39}$$

The second step consists in determining the second-order accurate derivatives $F_{xo}^{(2)}, F_{yo}^{(2)}$ by

$$F_{xo}^{(2)} = \frac{\sum \Delta y_i^2 \sum \Delta \tilde{F}_i \Delta x_i - \sum \Delta x_i \Delta y_i \sum \Delta \tilde{F}_i \Delta y_i}{DET}$$

$$F_{xo}^{(2)} = \frac{\sum \Delta x_i^2 \sum \Delta \tilde{F}_i \Delta y_i - \sum \Delta x_i \Delta y_i \sum \Delta F_i \Delta y_i}{DET} \tag{40}$$

An interesting property of the above 2-step formulae is the repeated use of the first-order formula which has been found to be very robust. It is worth noting that because of Cauchy-Schwartz inequality

$$\left(\sum \Delta x_i^2\right)\left(\sum \Delta y_i^2\right) \geq \left(\sum \Delta x_i y_i\right)^2$$

the determinant $DET \geq 0$. It is zero only if all the points (x_i, y_i) in the stencil fall on a straight line which is impossible for any 2-D grid. However it sometimes does happen that the nodes fall approximately on a straight line, they are in a sense contained in a long thin pencil and then the determinant DET can become quite small leading to local blowing up of residue or physical variable becoming negative. The remedy then consists in locally modifying the stencil. Every second-order scheme develops wiggles (spurious oscillations in flow variables near shocks) unless some limiter is employed. In case of LSKUM it is easier to use slope limiter. Consider the node o and let i be any node in the stencil of nodes used in obtaining $F_{xo}^{(2)}, F_{yo}^{(2)}$. Let

$$F_{min} = min\{F_i, i \in N(P_0)\}, \qquad F_{max} = max\{F_i, i \in N(P_o)\} \tag{41}$$

Spurious oscillations in solutions appear if the criterion

$$F_{min} \leq F_o + \Delta x_i \left(\frac{\partial F^{(2)}}{\partial x}\right)_o + \Delta y_i \left(\frac{\partial F^{(2)}}{\partial y}\right)_o \leq F_{max} \tag{42}$$

is violated by any node $i \in N(P_o)$. The solution to the problem consists in using a slope limiter Θ such that

$$F_{min} \leq F_o + \Theta\left\{\Delta x_i \left(\frac{\partial F^{(2)}}{\partial x}\right)_o + \Delta y_i \left(\frac{\partial F^{(2)}}{\partial y}\right)_o\right\} \leq F_{max} \tag{43}$$

We can always choose a suitable value of Θ so that (41) holds good for all the nodes in the stencil. In the present formulation slope limiters are applied to ψ -moments of (43) and not to F.

The LSKUM described above for 2D flows can be easily extended to 3D flows. We first determine first order derivatives by

$$\begin{bmatrix} f_{xo}^{(1)} \\ f_{yo}^{(1)} \\ f_{zo}^{(1)} \end{bmatrix} = (A_n^T A_n)^{-1} A_n^T \begin{bmatrix} \Delta f_1 \\ \Delta f_2 \\ \vdots \\ \Delta f_n \end{bmatrix} \tag{44}$$

where

$$A_n = \begin{bmatrix} \Delta x_1 & \Delta y_1 & \Delta z_1 \\ \Delta x_2 & \Delta y_2 & \Delta z_2 \\ \vdots & \vdots & \vdots \\ \vdots & \vdots & \vdots \\ \Delta x_n & \Delta y_n & \Delta z_n \end{bmatrix}$$

and n = number of points in $N(P_o)$. Obviously $A_n^T A_n$ is a 3x3 matrix The second step of the two-step defect correction method is to define modified differences by

$$\Delta \tilde{f}_i = \Delta f_i - \left(\frac{\Delta x_i}{2} \Delta f_{xi}^{(1)} + \frac{\Delta y_i}{2} \Delta f_{yi}^{(1)} + \frac{\Delta z_i}{2} \Delta f_{zi}^{(1)} \right) \tag{45}$$

where as before we have

$$\Delta f_{xi}^{(1)} = f_{xi}^{(1)} - f_{xo}^{(1)}, \Delta f_{yi}^{(1)} = f_{yi}^{(1)} - f_{yo}^{(1)}, \Delta f_{zi}^{(1)} = f_{zi}^{(1)} - f_{zo}^{(1)} \tag{46}$$

The second order accurate derivatives $f_{xo}^{(2)}, f_{yo}^{(2)}, f_{zo}^{(2)}$ are then given by

$$\begin{bmatrix} f_{xo}^{(2)} \\ f_{yo}^{(2)} \\ f_{zo}^{(2)} \end{bmatrix} = (A_n^T A_n)^{-1} A_n^T \begin{bmatrix} \Delta \tilde{f}_1 \\ \Delta \tilde{f}_2 \\ \vdots \\ \Delta \tilde{f}_n \end{bmatrix} \tag{47}$$

Ramesh et. al. and Ramesh[11, 12] have developed 3D LSKUM for computing 3D flows such as blast wave propagation in air, supersonic flowaround a hemisphere and transonic flow around ONERA M6 wing. Again a few minor problems can arise while computing 3D flows. The grid chosen can be very bad in some pockets, that is, all the nodes in a neighbourhood of some point P_o may fall in a long and thin pencil or on a disc. In such cases $A_n^T A_n$ is close to a singular matrix leading to residue divergence or physical variables (ρ, p or T) becoming negative. The remedy then consists in locally modifying the stencil of such troublesome nodes.

4 LSKUM based on Entropy Variables: q-LSKUM

Recently Ghosh, Mathur and Deshpande [13] have used entropy variables in the framework of cell centered finite volume formulation based on KFVS. The results obtained for 2D flows studied are very encouraging, shocks are sharper and q-KFVS code takes much less CPU time. It is highly worthwhile extending these ideas to LSKUM. Some comments on LSKUM considered so far are required to understand why q-LSKUM needs to be developed. The first order LSKUM uses ΔF_i which is a difference between two Maxwellians. The first order LSKUM has been thoroughly studied by Ghosh, Ramesh and Deshpande [7,10,11] on a variety of problems and it is found to possess some advantages and some disadvantages. It always gives smooth contours has very good convergence property (residue falls through several decades) and under a suitable Courant condition it is found to be positivity preserving, that is, the physical variables density, pressure and temperature never become negative. Unfortunately, it is highly dissipative and does not give results with acceptable accuracy unless very very fine grid is chosen. Second order LSKUM described before certainly yields accurate results but requires limiters to avoid pre and post shock spurious wiggles. The second order LSKUM like many other second order numerical schemes sometimes gives local stagnation pressure p_o more than the free stream stagnation pressure $p_{o\infty}$ even for cases where such a rise in p_o is theoretically not possible. Evidently, there are some local pockets in flow region where entropy decrease takes place thus leading to local violation of entropy condition. Another feature of second order LSKUM is that $\Delta \tilde{F}_i$ used in the second step is not a difference between two Maxwellians. We note that for 2D case (eqns. (38), (39))

$$\begin{aligned}
\Delta \tilde{F}_i &= \Delta F - \left(\frac{\Delta x_i}{2} \Delta F_{xi}^{(1)} + \frac{\Delta y_i}{2} \Delta F_{yi}^{(1)} \right) \\
&= (F_i - F_o) - \left\{ \frac{\Delta x_i}{2} \left(F_{xi}^{(1)} - F_{xo}^{(1)} \right) \frac{\Delta y_i}{2} \left(F_{yi}^{(1)} - F_{yo}^{(1)} \right) \right\} \\
&= \left(F_i - \frac{\Delta x_i}{2} F_{xi}^{(1)} - \frac{\Delta y_i}{2} F_{yi}^{(1)} \right) - \left(F_o - \frac{\Delta x_o}{2} F_{xo}^{(1)} - \frac{\Delta y_o}{2} F_{yo}^{(1)} \right) \quad (48)
\end{aligned}$$

Thus $\Delta \tilde{F}_i$ is in general not even a difference between two non-negative velocity distributions. This is a significant difference between the structure of first order LSKUM and second order LSKUM. An interesting question can be asked now: Is it possible to develop a modified second order LSKUM which has the same structure as the first order LSKUM and therefore expecting that it has all the nice characteristics of the first order LSKUM. One possible way to achieve this objective at least partially is to use entropy variables called q-variables introduced by Deshpande [14]. Let us briefly study the q-variables. Deshpande showed that 1-D Euler equations

$$\frac{\partial U_i}{\partial t} + \frac{\partial G_i}{\partial x} = 0, \quad i = 1, 2, 3 \tag{49}$$

can be cast in the symmetric hyperbolic form

$$P \frac{\partial q_i}{\partial t} + B \frac{\partial q_i}{\partial x} = 0 \tag{50}$$

if we can find two scalars M_o, M_1 such that

$$U_i = \frac{\partial M_O}{\partial q_i}, \qquad G_i = \frac{\partial M_1}{\partial q_i} \tag{51}$$

The matrices P and B are given by

$$P = \left[\frac{\partial^2 M_0}{\partial q_i \partial q_j}\right] \qquad B = \left[\frac{\partial^2 M_1}{\partial q_i \partial q_j}\right] \tag{52}$$

which are obviously Hessians of scalars M_0 and M_1. Deshpande [14] further showed that the desired functions accomplishing symmetrisation are

$$M_0 = \rho, \qquad M_1 = \rho u \tag{53}$$

and the q-variables are defined by

$$q_1 = ln\rho + \frac{ln\beta}{\gamma - 1} - \beta u^2, \quad q_2 = 2u\beta, \quad q_3 = -2\beta \tag{54}$$

The appearence of the q-variables in the above formulation becomes obvious when we recognize that (eq 31 of Deshpande [14])

$$d(lnF) = \frac{dF}{F} = dq_1 + v dq_2 + \left(I + \frac{v^2}{2}\right) dq_3 + \frac{5 - 3\gamma}{3 - \gamma}(I_0 - I) dq_3 \tag{55}$$

we further note that

$$\rho = \int_R \int_{R^+} F dv dI \qquad \text{and hence}$$

$$\begin{aligned} dM_0 = d\rho &= \int_R \int_{R^+} dF dv dI \\ &= \int_R \int_{R^+} F \left[dq_1 + v dq_2 + \left(I + \frac{v^2}{2}\right) dq_3\right] dv dI \end{aligned} \tag{56}$$

This relation follows from (55) because

$$\int_R \int_{R^+} (I - I_o) F dv dI = 0 \tag{57}$$

The relation (51) between U and q_i then follows from (56). Proceeding similarly, we get

$$\begin{aligned} dM_1 &= d(\rho u) = \int_R \int_{R^+} v dF dv dI \\ &= \int_R \int_{R^+} v F \left[dq_1 + v dq_2 + \left(I + \frac{v^2}{2}\right) dq_3\right] dv dI \end{aligned} \tag{58}$$

and hence follows the relation

$$G_i = \frac{\partial M_1}{\partial q_i} \tag{59}$$

Similar analysis can be carried out for 2D Euler equations

$$\frac{\partial U}{\partial t} + \frac{\partial}{\partial x}(GX) + \frac{\partial}{\partial y}(GY) = 0 \tag{60}$$

for which we need q_1, q_2, q_3, q_4 and functions M_0, M_1, M_2 to carry out symmetrisation. These are given by

$$U_i = \frac{\partial M_0}{\partial q_i}, \quad GX_i = \frac{\partial M_1}{\partial q_i} \quad GY_i = \frac{\partial M_2}{\partial q_i} \quad M_0 = \rho, \quad M_1 = \rho u_1, \quad M_2 = \rho u_2 \tag{61}$$

$$q_1 = ln\rho + \frac{ln\beta}{\gamma - 1} - \beta\left(u_1^2 + u_2^2\right), \quad q_2 = 2u_1\beta \quad q_3 = -2u_2\beta \quad q_4 = -2\beta \tag{62}$$

The Matrices P, B_1 and B_2 are given by

$$P = \left[\frac{\partial^2 M_0}{\partial q_i \partial q_j}\right] \quad B_1 = \left[\frac{\partial^2 M_1}{\partial q_i \partial q_j}\right] \quad B_2 = \left[\frac{\partial^2 M_2}{\partial q_i \partial q_j}\right] \tag{63}$$

and the corresponding symmetric hyperbolic form is

$$P\frac{\partial q}{\partial t} + B_1\frac{\partial q}{\partial x} + B_2\frac{\partial q}{\partial y} = 0 \tag{64}$$

where the column vector q is given by

$$q = \begin{bmatrix} ln\rho + \frac{ln\beta}{\gamma-1} - \beta\left(u_1^2 + u_2^2\right) \\ 2\beta u_1 \\ 2\beta u_2 \\ -2\beta \end{bmatrix} \tag{65}$$

Again the q-variables appear in the above formulation because of the relation

$$d(lnF) = \frac{dF}{F} = dq_1 + v_1 dq_2 + + v_2 dq_3 + \left(I + \frac{v^2}{2}\right) dq_4 + \frac{3-2\gamma}{2-\gamma}(I_0 - I) dq_4 \tag{66}$$

The importance of q-variables to us is primarily due to two reasons. First Deshpande [14] has shown that the Boltzmann H-function (which appears in the famous H-theorm) is related to q-variables by

$$H(F) = \int_R \int_{R^+} \left(\sum q_i \frac{\partial F}{\partial q_i} - F\right) dv dI \tag{67}$$

and hence q are called entropy variables. Secondly, if we take two sets of q-variables denoted by q' and q'' corresponding to two Maxwellians F' and F'' then the linearly interpolated q-variables

$$\tilde{q} = (1 - \Theta)\, q' + \Theta q'', \quad 0 \leq \Theta \leq 1 \tag{68}$$

give us another Maxwellian $\tilde{F}$ which is a geometric mean of the Maxwellians F' and F''. The formulae (65) give q_1, q_2, q_3 and q_4 in terms of the fluid variables

p, u_1, u_2 and β. Similarly, the latter variables can be obtained in terms of the former through the relations

$$\beta = -\frac{q_4}{2} \qquad u_1 = -\frac{q_2}{q_4} \qquad u_2 = -\frac{q_1}{q_4} \qquad p = \exp\left(\frac{\left[q_1 - \frac{q_2^2+q_3^2}{2q_4}\right]}{-\left(\frac{q_4}{2}\right)^{\gamma-1}}\right) \tag{69}$$

Hence any maxwellian F can be determined uniquely in terms of either the usual variables p, u_1, u_2, β or in terms of the entropy variables, and we denote the Maxwellian so determined by $F(q)$ to emphasize that it is obtained from q.

With this background we come back to (48) which suggests quite strongly q-LSKUM. We note that first order LSKUM uses

$$\Delta F_i = F(q_i) - F(q_o) \tag{70}$$

The idea now is to obtain

$$\Delta \tilde{F}_i = F(\tilde{q}_i) - F(\tilde{q}_o) \tag{71}$$

with modified entropy variables $\tilde{q}_i$ and $\tilde{q}_o$, so that the use of $\Delta\tilde{F}_i$ instead of ΔF_i in the first order LSKUM formula yields the second order q-LSKUM. Following Dauhoo, Ghosh. Ramesh and Deshpande [15], we define modified entropy variables

$$\begin{aligned} \tilde{q}_i &= q_i - \frac{\Delta x_i}{2} q_{xi}^{(1)} - \frac{\Delta y_i}{2} q_{yi}^{(1)} \\ \tilde{q}_o &= q_o - \frac{\Delta x_i}{2} q_{xo}^{(1)} - \frac{\Delta y_{(i)}}{2} q_{yo}^{(1)} \end{aligned} \tag{72}$$

For these tilde variables we can now obtain

$$\Delta \tilde{F}_i = F(\tilde{q_i}) - F(\tilde{q_o}) \tag{73}$$

and then get the second order formulae. The definitions of the tilde variables are in fact suggested by (48).The second order formulae for space derivatives are then given by

$$F_{xo}^{(2)} = \frac{\sum \Delta y_i^2 \sum \Delta\tilde{F}_i \Delta x_i - \sum \Delta x_i \Delta y_i \sum \Delta\tilde{F}_i \Delta y_i}{DET} \tag{74}$$

$$F_{yo}^{(2)} = \frac{\sum \Delta x_i^2 \sum \Delta\tilde{F}_i \Delta y_i - \sum \Delta x_i \Delta y_i \sum \Delta\tilde{F}_i \Delta y_i}{DET}$$

$$DET = (\sum \Delta x_i^2)(\sum \Delta y_i^2) - (\sum \Delta x_i \Delta y_i)^2 \tag{75}$$

The second order q-LSKUM is obtained by taking ψ moments of

$$\begin{aligned} &\tfrac{\partial F}{\partial t} + \tfrac{v_1+|v_1|}{2}\{F_{xo}^{(2)}\}_{N_1(P_o)} + \tfrac{v_1-|v_1|}{2}\{F_{xo}^{(2)}\}_{N_2(P_o)} + \tfrac{v_2+|v_2|}{2}\{F_{yo}^{(2)}\}_{N_1(P_O)} \\ &\qquad + \tfrac{v_2-|v_2|}{2}\{F_{yo}^{(2)}\}_{N_4(P_o)} = 0 \end{aligned} \tag{76}$$

Obviously we can use a suitable time marching scheme to obtain desired time accuracy. Thus q-LSKUM uses update formula very similar in structure to the first order LSKUM update formula and hence as mentioned before is expected to have many nice properties like upwinding, validity of entropy condition and positivity preservation. These expectations are indeed borne out by the computations of Dauhoo et. al. [15]. Dauhoo and Deshpande [17] are working on yet another interesting idea of rotated q-LSKUM. The LSKUM update formula (76) is valid in any frame x,y and we can ask at this stage an interesting question: which is the optimal frame from the point of view having minimum numerical dispersion or minimum numerical disspation? One possibility is to locally rotate the frame to natural coordinate system s,n with s along the local streamline and n perpendicular to s-direction. For enforcing streamline upwinding dividing the stencil is required.

The stencil for node P_o is split into $N^+(P_o), N^-(P_o)$ defined by

$$\begin{aligned} N^+(P_o) &= \{ \text{ For all } P_i \text{ such that} \Delta : s_i = s_i - s_o < 0 \} \\ N^-(P_o) &= \{ \text{ For all } P_i \text{ such that} \Delta : s_i = s_i - s_o \geq 0 \} \end{aligned} \tag{77}$$

Let v_s, v_n be the particle velocities along s,n and let us u_s, u_n be corresponding fluid velocities. The rotated second order q-LSKUM will obviously require the entropy variables in rotated frame, that is,

$$\bar{q} = \begin{bmatrix} ln\rho + \frac{ln\beta}{\gamma-1} - \beta u^2 \\ 2\beta u_s \\ 2\beta u_n \\ -2\beta \end{bmatrix} \qquad u^2 = u_1^2 + u_2^2 = u_n^2 + u_s^2 \tag{78}$$

and this can be easily obtained from q given by (65) in the global Cartesian frame. The corresponding tilde variables can now be defined by

$$\begin{aligned} \tilde{\bar{q}}_i &= \bar{q}_i - \frac{\Delta s_i}{2} \bar{q}_{si}^{(1)} - \frac{\Delta n_i}{2} \bar{q}_{ni}^{(1)} \\ \tilde{\bar{q}}_o &= \bar{q}_o - \frac{\Delta s_i}{2} \bar{q}_{so}^{(1)} - \frac{\Delta n_i}{2} \bar{q}_{no}^{(1)} \end{aligned} \tag{79}$$

For these tilde variables we require the first order derivatives $\bar{q}_{si}^{(1)}, \bar{q}_{si}^{(1)}, \bar{q}_{si}^{(1)}\ \bar{q}_{si}^{(1)}$ and these can be determined from the first order least squares formulae. In respect of choice of stencils for obtaining these derivatives several possibilities arise. For example, we can choose the full stencil $N(P_o)$, that is,

$$\begin{aligned} \bar{q}_{so}^{(1)} &= \left\{ \frac{\|\Delta n\|^2 (\Delta\bar{q}, \Delta s) - (\Delta s, \Delta n)(\Delta\bar{q}, \Delta n)}{DET} \} \right\}_{N(P_o)} \\ \bar{q}_{no}^{(1)} &= \left\{ \frac{\|\Delta s\|^2 (\Delta\bar{q}, \Delta n) - (\Delta s, \Delta n)(\Delta\bar{q}, \Delta s)}{DET} \} \right\}_{N(P_o)} \end{aligned} \tag{80}$$

$$DET = \left\{ \sum \Delta s_i^2 \sum \Delta n_i^2 - \left(\sum \Delta s_i \Delta n_i \right)^2 \right\}_{N(P_o)} \tag{81}$$

Yet another possibility is to use $N^{\pm}(P_o)$ for $\bar{q}_{so}$ and we use full stencil for determining $\bar{q}_{so}$. The streamline upwind q-LSKUM requires Maxwellians $F(\bar{\bar{q}}_i)$ and$F(\bar{\bar{q}}_o)$which can be easily obtained from $\bar{\bar{q}}_i, \bar{\bar{q}}_o$ and these in turn are given by (79). Once these Maxwellians are obtained we proceed to get the modified difference

$$\Delta \tilde{\tilde{F}}_i = F(\bar{\bar{q}}_i) - F(\bar{\bar{q}}_o) \tag{82}$$

The state update formula for second order rotated q-LSKUM is obtained by taking ψ moments of

$$\frac{\partial F}{\partial t} + \frac{v_s + |v_s|}{2}\left(F_{so}^{(2)}\right)_{N^-(P_o)} + \frac{v_s - |v_s|}{2}\left(F_{so}^{(2)}\right)_{N^+(P_o)} + v_n\left(F_{no}^{(2)}\right)_{N^+(P_o)} = 0 \tag{83}$$

It is worth noting that full stencil is used for the derivative $F_{no}^{(2)}$ while streamline upwinding is employed for the derivative $F_{so}^{(2)}$. A few technical points need to be mentioned. The tilde entropy variables defined by (79) or by (72) are linear combinations of entropy variables at neighbouring nodes. It is quite possible therefore to encounter a situation in which the tilde variables can go beyond the minimum or the maximum. More importantly the $\tilde{\beta}$ corresponding to $\tilde{q}_4$ or $\tilde{\tilde{q}}_4$ can become negative. It is therefore necessary to introduce slope limiter to preserve positivity of $\tilde{\beta}$. The relations (69) show that ρ is always positive if q_4 is negative irrespective of signs of q_1, q_2 and q_3. No separate slope limiter is required to preserve positivity of density.

5 Results and Discussions

In this section we present some typical results for some latest application of LSKUM and q-LSKUM. Recently Ramesh, Anandhnarayan and Deshpande [12, 16] have made an attempt to exploit the power of LSKUM for arbitrary grids.

Continuing along the same lines Ramesh and Deshpande [12] have considered the case in which a body fitted mesh (structured grid) around airfoil overlap with a background Cartesian mesh. In the overlap region we consider all the points from both node P_o is the set of neighbouring points and LSKUM uses this information to obtain discrete approximation of spatial derivatives. The connectivity $N(P_o)$ is required for all P_o in the computational domain. For purely structured, Cartesian or unstructured grids, connectivity generation is relatively easy. In case of such grids use is made of the implicit data structure associated with those grids. However for the Chimera mesh, it is not so easy because the points from different types of grids are combined together. Ramesh [12] has developed a quad tree code which acts as a preprocessor on the distribution of the points and its output is the connectivity $N(P_o)$ for each Po . Generally $N(P_o)$ contains the addresses of nodes in the nieghbourhood of P_o . The flowsolver then operates on the set of points with the associated connectivity data for each point.

Table 1: Comparison of q-LSKUM & LSKUM

Method	C_1	C_d	P_{ocomp} / P_{otheo}
q-LSKUM	0.315	-0.0027	0.9
LSKUM	0.3817	0.006	1.024

Fig 3 shows the structured grid overlapping with Cartesian mesh for NACA-0012 airfoil. Fig 4 shows the computed pressure contours for flow past NACA-0012 Airfoil for M = 0.85 and $\alpha = 1.0$ for the above case. In fact it can be seen that in the overlap region also the contour's are continuous and smooth. The smoothness of contours is most probably due to relatively higher density of grid points in the overlap region.

Another important point about the present approach (LSKUM on overlap grid) is that there are no boundaries such as inner boundary of overlap domain or outer bounday of overlap domain, the flowsolver is same throughout the entire computational domain. Consequently, there is no data transfer across boundaries and no need to interpolate. All that we require are a set of nodes or points and their connectivity sets which in the present approach are generated by quad tree preprocessor. This is one of the most attractive features of LSKUM.

Fig 5 shows an another application of chimera mesh by Dhokrikar et al [16] for the Biplane configuration. In this case a structured grid is generated over the upper airfoil and another structured grid over the lower airfoil which lies within the grid of the upper airfoil. The quadtree pre-processor is then run on the set of nodes obtained by combining these two grids(with nodes which lie inside the bodies being deleted) to obtain the connectivity data as output. The flow solver then operates on this set of nodes with the connectivity data. Fig 6 shows the pressure contours for transonic flow past this configuration for M = 0.85 and $\alpha = 1.0$ All the essential features are satisfactorily computed.

The q-LSKUM has been applied [17] to subsonic lifting case of NACA 0012 airfoil with $M_\infty = 0.63$ and $\alpha = 2.0^o$ An unstructured grid of 4074 nodes has been employed. The C_p- distribution for this test case is shown in Fig 7. Table (1) shows the superior performance of q-LSKUM. Here p_{ocomp} = computed total pressure at the leading edge stagnation point and p_{otheo} = $p_{o\infty}$ = stagnation pressure in free stream. Referring to the comment made before it is worth noting that p_{ocomp} for q-LSKUM is lower than p_{otheo} unlike for LSKUM which obviously locally violates entropy condition. Also, for this subsonic lifting case C_d is exactly zero and C_d predicted by q-LSKUM is closer to the exact value. Evidently, more computations and testing is required for confirming the above claim of superior performance.

6 Acknowledgement

I am thankful to Prof. M. Hafez for giving me an opportunity to present a paper in this symposium held to honor Prof. P.L.Roe on his 60th birthday. Prof. Roe visited NAL, Bangalore in 1988 to participate in CAARC conference on CFD and CFD community in Bangalore was tremendously impressed with his insight in various aspects of CFD. I am thankful to R. Harish, R.Tyagaraju, and Radha of CFD Centre for giving all the help in preperation of this paper. A part of the work reported here was supported by Indo-French Centre, Delhi sponsored project IFCPAR/CEFIPRA 1601-1 and I am deeply indebted to IFCPAR/CEFIPRA for their continued support.

References

[1] SM Deshpande *Applications of kinetic schemes to all types of meshes* , Invited paper in 7th International Conference on Hyperbolic problems, Theory, Numerics, Applications, ETH Zurich, February 9-13, 1998.

[2] SM Deshpande *Kintetic Flux Splitting Schemes*, CFD Review 1995 eds. M.Hafez, K.Oshima, John Wiley & Sons, pp 161-181, Sept. 1995.

[3] SM Deshpande *Kinetic theory based new upwind methods for Inviscid compressible flows* AIAA-86-0275.

[4] JS Mathur, SM Deshpande, June 1996 *Reconstruction on unstructured grids using an upwind kinetic method* 15th International Conference in Numerical Methods in Fluid Dynamics (ICNMFD), Monterey, California, USA.

[5] SM Deshpande, S. Sekar, M.Nagarathinam, R.Krishnamurthy, PK Sinha and PS Kulkarni *A 3-dimensional upwind Euler Solver using kinetic flux vector splitting method* proceedings of the 13th International Conference on Numerical Methods in Fluid Dynamics, Rome, 6-10 July 1992. Lecture notes in Physics (Berlin:Springer-Verlag)414

[6] Deshpande SM, presented a paper on *Kinetic Upwind Method for Inviscid Gas Dynamics in specialists* CFD meeting organised by commonwealth Advisory Aeronautical Research Council (CAARC) at NAL, Bangalore, Dec 1988.

[7] AK Ghosh and SM Deshpande *A robust least squares kinetic upwind scheme for Euler equations on arbitrary meshes* Proceedings of 14th ICNMFD, Bangalore, July 11-15, 1994.

[8] SM Deshpande, AK Ghosh & JC Mandal *Least squares weak upwind method for Euler equations* Fluid mech. Rep.89 FM 4, Dept. of Aerospace Engg. Indian Institute of Science, Bangalore.

[9] AK Ghosh and SM Deshpande *Least Squares Kinetic Upwind Method for Inviscid Compressible Flows* 12th AIAA CFD Conference, San Diego, AIAA 95-1735, June 1995.

[10] AK Ghosh, Ph.D thesis *Robust least squares kinetic upwind method for inviscid compressible flows* Department of Aerospace Engineering, Indian Institute of Science, Bangalore 1996.

[11] V.Ramesh, AK Ghosh & SM Deshpande, Sept 1997 *Computational of three dimensional Inviscid compressible flows using least squares kinetic upwind method (LSKUM)* Fluid Mechanics Report 97 FM 8, Dept. of Aerospace Engineering, Indian Institute of Scicnce, Bangalore.

[12] V.Ramesh, Ph.D thesis, Department of Aerospace Engineering, IISc, Bangalore (Under preparation)

[13] AK Ghosh, JS Mathur and SM Deshpande *q-KFVS scheme - A new higher order kinetic method for Euler equations* 16th International conference in Numerical Methods in Fluid Dynamics (ICNMFD) held at Arcachon, France.

[14] SM Deshpande*On the Maxwellian Distribution, Symmetric Form and the Entropy conservation for the Euler Equations* NASA Technical paper 2583, August 1986.

[15] MZ Dauhoo, AK Ghosh, V.Ramesh and SM Deshpande *q-LSKUM A new higher order kinetic method for Euler equations using entropy variables* ISCFD 99, September 1999, Bremen, Germany

[16] D.Dhokrikar, K. Anandhanarayanan, V. Ramesh and SM Deshpande *A grid free approach for 2-D Euler computations using Least Squares Kinetic Upwind Method* (ACFD3) Dec 7-11, 1998, Bangalore.

[17] MZ Dauhoo, Ph.D thesis, University of Mauritius (Under preparation)

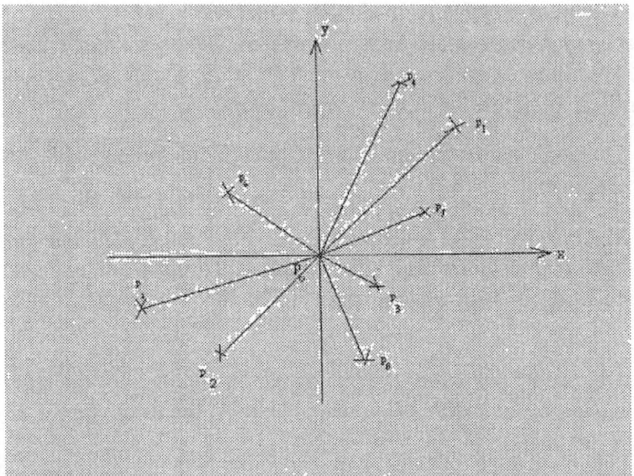

Figure 1.

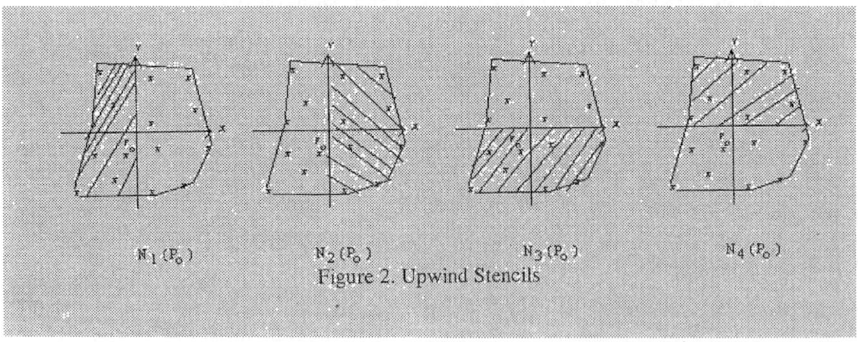

Figure 2. Upwind Stencils

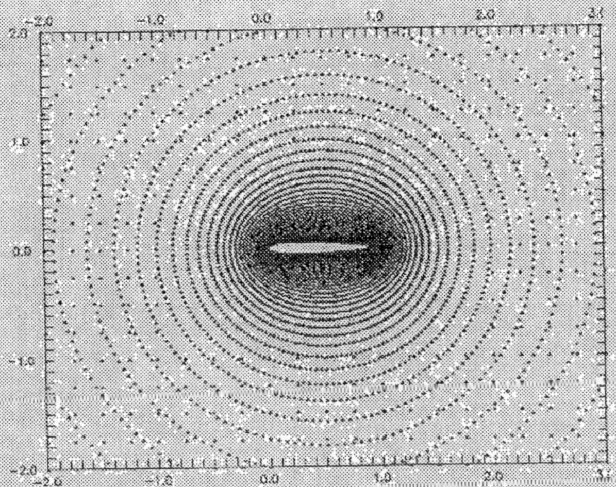

Figure. 3 Overlapped grid

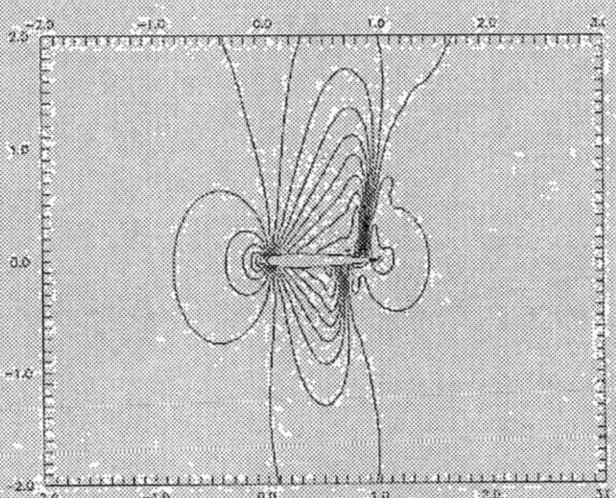

Figure 4. Pressure contours
M=0.85, α=1.0

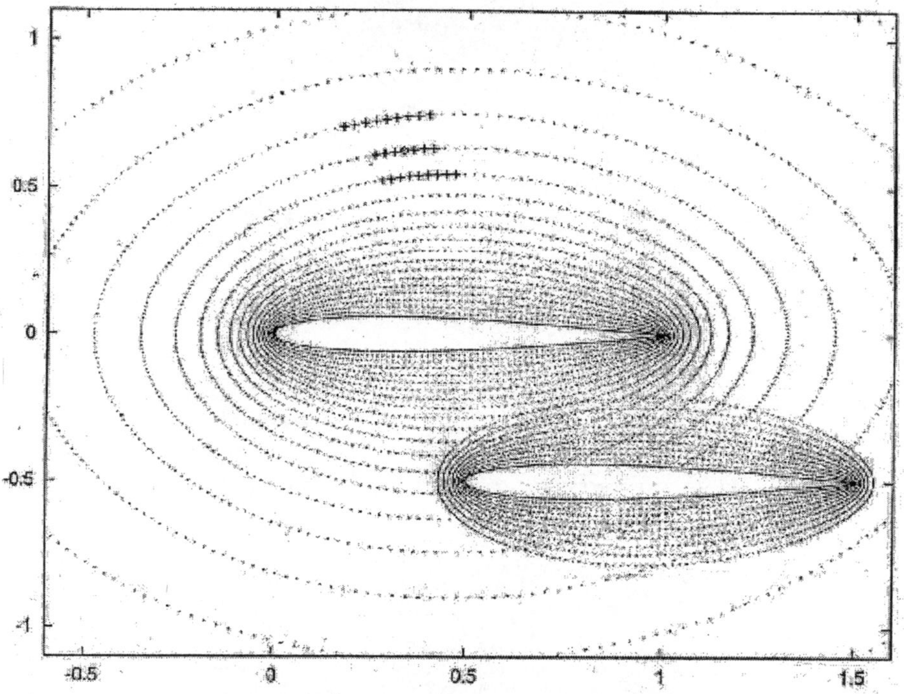

Figure 5: Chimera mesh for Biplane configuration

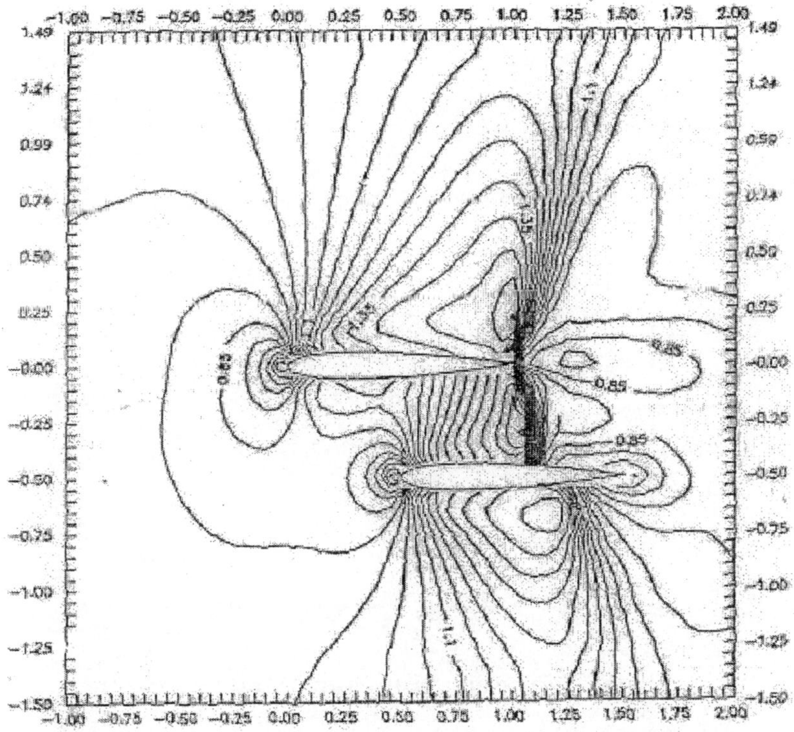

Figure 6 : Pressure contours : Biplane configuration
$M_\infty = 0.85, \alpha = 0°$

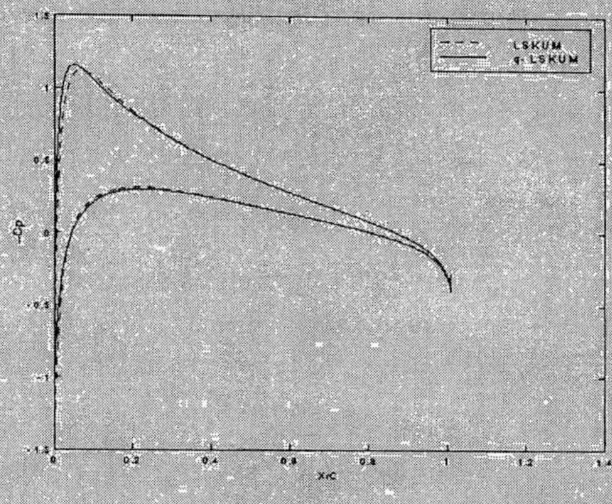

Figure 7. Pressure Distribution M=0.63, α=2.0

Difference Approximation for Scalar Conservation Law.
– Consistency with Entropy Condition from the viewpoint of Oleĭnik's E-condition –

AISO, Hideaki

Computational Sciences Division, National Aerospace Laboratory,
Jindaiji-Higashi-machi 7-44-1 Chofu TOKYO 182 8522 JAPAN,
E-mail: aiso@nal.go.jp

Key Words: Conservation Law, Entropy Condition, Entropy Inequiality, Oleĭnik's E-Condition, Consistency of Difference Approximation

Dedication. Among the contribution by Professor P.L. Roe, the extension of upwind difference concept by the averaging matrix [11] is an essential idea for all the CFD researchers and users. The idea is a classic but not yet old. Combined with the entropy fix technique of adding small numerical viscosity, it is still a base on which we consider or develop difference schemes for various problems. We revisit and analyze the numerics of difference schemes including this basic difference scheme in the scalar case. The author dedicates the paper to the 60 year-birthday of Professor P.L. Roe.

Abstract In usual a method to discretize partial differential equation is required to have some consistency with the solution to original differential equation. When we solve hyperbolic conservation laws numerically by difference approximation, it is very important to know how consistent with the entropy solution the difference approximation or numerical result is. In the case of scalar conservation law, we know that a difference approximation converges to the entropy solution if it satisfies some condition which is described in terms of numerical viscosity coefficients. (See [1, 2, 10, 13].) In these works, the consistency with entropy condition is an important subject and is discussed through the numerical entropy inequality or cell entropy inequality, which is a discrete analogue of the entropy inequality that is employed in the definition of entropy condition by Lax. (See [6, 12].) On the other hand, the results on convergence may lack practical information on the numerical behavior of difference approximation, where the numerical behavior means the behavior of numerical result which is calculated at some finite values of difference increments (Δx, Δt etc.). But information on numerical behavior is important when we are interested in the quality of numerical computation.

We here discuss the consistency with entropy condition from the viewpoint of Oleĭnik's E-condition. Although the discussion is not applied so widely as that based on the numerical entropy inequality, the result gives more information on the numerical behavior of difference approximations. It helps us discuss the quality of numerical computation.

1 Introduction

We are concerned with difference approximation for the initial value problem of scalar conservation law

$$\begin{cases} u_t + f(u)_x = 0, \ -\infty < x < \infty, 0 < t < \infty \\ u(x,0) = u_0(x), \ -\infty < x < \infty, \end{cases} \tag{1}$$

where the flux function f is strictly convex, *i.e.*

$$f'' > c \tag{2}$$

for some fixed positive number c, and the initial function u_0 is of bounded variation. Difference approximation is considered in the following viscous form

$$u_i^{n+1} = u_i^n - \frac{\lambda}{2}\{f(u_{i+1}^n) - f(u_{i-1}^n)\} + \frac{\lambda}{2}\left\{a_{i+\frac{1}{2}}^n(u_{i+1}^n - u_i^n) - a_{i-\frac{1}{2}}^n(u_i^n - u_{i-1}^n)\right\}, \tag{3}$$

where each u_i^n is supposed as an approximate value for u at the node $(i\Delta x, n\Delta t)$ and the ratio $\lambda = \dfrac{\Delta t}{\Delta x}$ of the temporal difference increment Δt to the spatial one Δx is fixed so that it should satisfy the general CFL condition

$$\lambda \sup_{m \leq s \leq M} |f'(s)| \leq 1, \tag{4}$$

where m and M are lower and upper bounds, respectively, for $\{u_i^n\}_{i,n\in\mathbf{Z},n\geq 0}$, [1] *i.e.*

$$m < u_i^n < M, \ \forall i, n. \tag{5}$$

To analyze the difference approximation's consistency with entropy solution, we usually discuss two kinds of consistency, the consistency with weak solution and the consistency with entropy condition. This comes from the essential property of solutions to scalar conservation laws. (See [8, 12] etc.)

It is well known that the global existence of smooth solution to a scalar conservation law may be violated even with smooth initial data because of the nonlinearity of flux function f. Then we employ the concept of weak solution to avoid the inconvenience and obtain the existence of solution, but on the other hand we lose the uniqueness of solution because the equation may admit infinitely many different weak solutions. To obtain a "unique" solution, which should be physically relevant as well, we have to employ an additional

[1] In this article we are interested only in difference approximations of the form (3) satisfying the so called TVD condition

$$\left|\frac{f(u_{i+1}^n) - f(u_i^n)}{u_{i+1}^n - u_i^n}\right| \leq a_{i+\frac{1}{2}}^n \leq \frac{1}{\lambda}.$$

In the case the difference approximations have the following property as well.

$$\inf_{i\in\mathbf{Z}} u_i^n \leq \inf_{i\in\mathbf{Z}} u_i^{n+1} \leq \sup_{i\in\mathbf{Z}} u_i^{n+1} \leq \sup_{i\in\mathbf{Z}} u_i^n, \ n = 0, 1, 2, 3, \cdots$$

Therefore, it suffices to suppose the condition that only the initial data $\{u_i^0\}_{i\in Z}$ satisfy

$$m < u_i^0 < M, \ \forall i,$$

instead of the condition (5).

condition called the entropy condition. Then a unique solution, called the entropy solution, is determined as a weak solution that satisfies the entropy condition.

In view of the situation above, the convergence of difference approximation to entropy solution is usually guaranteed through the following two steps of discussion. The first step is to show the consistency with weak solution, which is discussed through some kind of compactness or stability. The compactness from TV(Total Variation)-Stability is usually used. (See [7, 8]) The first step of analysis can only guarantee the convergence (strictly saying, accumulation) to some weak solution and still admits the possibility of convergence to a weak solution violating the entropy condition. Then we need the second step of analysis, where we discuss the consistency with entropy condition to guarantee that the weak solution of difference approximation's convergence limit is really the entropy solution. The works [1, 3, 10] give typical examples of this kind of discussion. In these works the consistency with entropy condition is discussed through the numerical entropy inequality

$$\begin{aligned} U(u_i^{n+1}) - U(u_i^n) + & \frac{\lambda}{2}\{F(u_{i+1}^n) - F(u_{i-1}^n)\} \\ & - \frac{\lambda}{2}\left\{A_{i+\frac{1}{2}}^n(u_{i+1}^n - u_i^n) - A_{i-\frac{1}{2}}^n(u_i^n - u_{i-1}^n)\right\} \leq 0 \end{aligned} \tag{6}$$

or some other equivalent derived from the entropy inequality

$$U(u)_t + F(u)_x \leq 0 \tag{7}$$

that a weak solution is required to satisfy in the definition of entropy condition by Lax [6], where (U, F) is an arbitrary pair of a convex function U and another function F (not necessarily convex) satisfying $F' = U'f'$.

With the discussion above, we have already determined rather wide classes of difference approximation which converges to the entropy solution. For example, the following result is obtained in [1]. This is one of the best possible results of this kind.

Theorem 0 *The difference approximation (3) converges to the entropy solution if each numerical viscosity coefficient $a_{i+\frac{1}{2}}^n$ satisfies the inequality*

$$\max\left\{a^{MR}(u_i^n, u_{i+1}^n), \frac{\epsilon}{\lambda} sgn(u_{i+1}^n - u_i^n)\right\} \leq a_{i+\frac{1}{2}}^n \leq a^{LF}, \tag{8}$$

where

$$a^{MR}(u_i^n, u_{i+1}^n) = \left|\int_0^1 f'(u_i^n + (u_{i+1}^n - u_i^n)\theta)d\theta\right| = \left|\frac{f(u_{i+1}^n) - f(u_i^n)}{u_{i+1}^n - u_i^n}\right| \tag{9}$$

and

$$a^{LF} = \frac{1}{\lambda} \tag{10}$$

are the numerical viscosity coefficients of Murmann-Roe scheme and Lax-Friedrichs scheme, respectively.

We note that the convergence of difference approximation is considered in the following manner. For each $\Delta = (\Delta x, \Delta t)$ satisfying $\dfrac{\Delta t}{\Delta x} = \lambda$, a family

$\{u_i^n\}_{i,n\in\mathbf{Z},n\geq 0}$ of approximate values are obtained from the numerical initial data $\{u_i^0\}$ and the difference approximation. Let the approximate solution $u_\Delta(x,t)$ be a function defined by

$$u_\Delta(x,t) = u_i^n \text{ if } (i-\tfrac{1}{2})\Delta x < x < (i+\tfrac{1}{2})\Delta x \text{ and } n\Delta t \leq t < (n+1)\Delta t, \quad (11)$$

for each Δ and the corresponding family $\{u_i^n\}_{i,n\in\mathbf{Z},n\geq 0}$. Then we discuss the convergence of $u_\Delta(x,t)$ as $\Delta \longrightarrow (0,0)$ with the following assumption of consistency with the initial condition;

$$u_\Delta(\cdot,0) \longrightarrow u_0. \quad (12)$$

The convergence is discussed in the topology of L^1_{loc}.

We remember that difference approximations for scalar conservation law are usually understood in the concept of finite volume method. We understand that each interval $((i-\frac{1}{2})\Delta x, (i+\frac{1}{2})\Delta x)$ is a finite volume and that the value of u over the finite volume at the time $t = n\Delta t$ is represented by u_i^n. It comes from the essential concept of conservation law. In fact the equation (1) means the conservation of u within any finite volume including the import and export of u through the boundary of volume by the flux $f(u)$. The entropy inequality (7) governs the violation of conservation of entropy U at any finite volume if it happens. Then it is natural to discuss the consistency with entropy condition through the numerical entropy inequality which also implies the finite volume concept. In view of the observation above and the successful results [1, 10], we may say that the numerical entropy inequality is a substantial tool to analyze the consistency with entropy condition and to obtain a theoretical proof.

But, when we are interested in the quality of numerical results calculated by difference approximation, discussion on the convergence to "right" solution is not enough. Numerical calculation is conducted with some finite values of difference increments ($\Delta x, \Delta t$ etc.). It means that we need to know the behavior of numerical result that is calculated as a solution to the equation of difference approximation with some finite difference increments. Let us call it the numerical behavior of difference approximation.

Now we remember another way to define the entropy condition. The entropy condition is defined through Oleĭnik's E-condition as well, without using the concept of entropy inequality (7) or entropy U. Oleĭnik's E-condition requires that some positive constant E should exist so that a weak solution $u = u(x,t)$ to the problem (1) should satisfy the inequality

$$\frac{u(x+h,t) - u(x,t)}{h} \leq \frac{E}{t} \quad (13)$$

for any $x \in \mathbf{R}, h > 0, t > 0$. In [8], which is the first work to prove the existence and uniqueness of solution to scalar conservation laws, Oleĭnik discussed the entropy condition of this form to show the uniqueness. We also note that the optimal value of constant E for the inequality (13) is the supremum of $\dfrac{1}{|f''|}$.

The E-condition is regarded to govern the decay of an inverse (physically irrelevant) shock into rarefaction or the temporal change of profile of rarefaction wave. We here remember that it is still difficult to obtain a good profile of rarefaction wave in numerical computation and to know the exact position of an edge (beginning point or ending point) of rarefaction, even though various techniques of capturing shocks have been developed after the introduction of TVD (Total Variation Diminishing) concept. (See [4, 5, 9] etc.) Therefore we expect that the analysis based on the E-condition gives more precise information on the numerical behavior of difference approximation.

The analysis is not applied to so wide a class of difference approximation, compared with that based on the numerical entropy inequality. But it gives more information on the numerical behavior of difference approximation of some class as well as gives an alternative proof of the convergence to entropy solution. The discussion includes Murmann-Roe scheme with Harten's entropy fix, which is still regarded as a base of scheme development.

2 Results

The statement of basic result is the following.

THEOREM 1 *Let ϵ be an arbitrary fixed positive number less than $\frac{1}{2}$ and $Q_\epsilon(s)$, $-\dfrac{1}{\lambda} \leq s \leq \dfrac{1}{\lambda}$ be a function defined by*

$$Q_\epsilon(s) = \begin{cases} \dfrac{\epsilon}{\lambda} \ \textit{if } |s| \leq \dfrac{\epsilon}{\lambda}, \\ s \ \textit{ otherwise.} \end{cases} \tag{14}$$

Each $q^n_{i+\frac{1}{2}}$ is given by

$$q^n_{i+\frac{1}{2}} = \int_0^1 f'(u^n_i + (u^n_{i+1} - u^n_i)\theta)d\theta = \begin{cases} \dfrac{f(u^n_{i+1}) - f(u^n_i)}{u^n_{i+1} - u^n_i}, \ u^n_i \neq u^n_{i+1} \\ |f'(u^n_i)| = |f'(u^n_{i+1})|, \ u^n_i = u^n_{i+1} \end{cases} \tag{15}$$

Suppose that each numerical viscosity coefficient $a^n_{i+\frac{1}{2}}$ in (3) is given by

$$a^n_{i+\frac{1}{2}} = Q(q^n_{i+\frac{1}{2}}) \tag{16}$$

or

$$a^n_{i+\frac{1}{2}} = \begin{cases} Q(q^n_{i+\frac{1}{2}}), \ u^n_{i+1} > u^n_i \\ \left|q^n_{i+\frac{1}{2}}\right|, \ \textit{otherwise.} \end{cases} \tag{17}$$

Then the following estimates are obtained.

(E1) Some positive constants E' and N exist so that

$$\frac{u_{i+1}^n - u_i^n}{\Delta x} < \frac{E'}{n\Delta t} \tag{18}$$

holds for $n > N$. Furthermore E' can be as small as $\frac{4}{c}$

(E2) Some positive constants $\bar{E}$ exists so that

$$\frac{u_{i+1}^n - u_i^n}{\Delta x} < \frac{\bar{E}}{n\Delta t} \tag{19}$$

holds.

We easily remember that the function Q_ϵ is introduced in the idea of entropy fix [4, 5]. ϵ means the so called additional viscosity coefficient for entropy fix. We note that the optimal value of E' in the estimate (E1) is not as small as the optimal of E in the Oleĭnik's E-condition (13).

The estimate (E2) immediately gives the following corollary.

COROLLARY 2 *If the difference approximation (3) satisfies the assumption in the theorem above, it converges to the entropy solution.*

The fact is already proved because the case is included in Theorem 0, the main theorem of [1]. But we here obtain an alternative proof without the use of entropy inequality, numerical entropy inequality or cell entropy inequality.

We give a rough sketch of proof of the theorem. The estimate (E1) is rather basic and the other (E2) is naturally derived from (E1). The following lemma plays an important role in the proof of theorem.

LEMMA 3 *We determine $z^n, n \in \boldsymbol{Z}, n \geq 0$ by*

$$z^n = \max\{\max_{i\in\boldsymbol{Z}}\{u_{i+1}^n - u_i^n\}, 0\}. \tag{20}$$

Then we obtain the following estimate.

$$z^{n+1} - z^n \leq \max\left\{-\epsilon z^n, -\frac{\lambda}{4}(z^n)^2\right\}. \tag{21}$$

The proof of estimate (21) is too complicated to be shown in a limited space. But the idea is almost straightforward. What is complicated is that the discussion can not be unified but that many different cases should be discussed separately. First we write $u_{i+1}^{n+1} - u_i^{n+1}$ in the following form.

$$\begin{aligned} u_{i+1}^{n+1} - u_i^{n+1} &= \frac{\lambda}{2} a_{i-\frac{1}{2}}^n (u_i^n - u_{i-1}^n) + \frac{\lambda}{2}\{f(u_i^n) - f(u_{i-1}^n)\} + (1 - a_{i+\frac{1}{2}}^n)(u_{i+1}^n - u_i^n) \\ &+ \frac{\lambda}{2} a_{i+\frac{3}{2}}^n (u_{i+2}^n - u_{i+1}^n) - \frac{\lambda}{2}\{f(u_{i+2}^n) - f(u_{i+1}^n)\}. \\ &= \frac{\lambda}{2}(a_{i-\frac{1}{2}}^n + q_{i-\frac{1}{2}}^n)(u_i^n - u_{i-1}^n) + (1 - a_{i+\frac{1}{2}}^n)(u_{i+1}^n - u_i^n) \\ &+ \frac{\lambda}{2}(a_{i+\frac{3}{2}}^n - q_{i+\frac{3}{2}}^n)(u_{i+2}^n - u_{i+1}^n) \end{aligned} \tag{22}$$

Using the relation, we estimate $u_{i+1}^{n+1} - u_i^{n+1}$ by $u_i^n - u_{i-1}^n$, $u_{i+1}^n - u_i^n$ and $u_{i+2}^n - u_{i+1}^n$. More precisely speaking, our interest is on the relation of $(u_{i+1}^{n+1} - u_i^{n+1})^+$ to $(u_i^n - u_{i-1}^n)^+$, $(u_{i+1}^n - u_i^n)^+$ and $(u_{i+2}^n - u_{i+1}^n)^+$, where we mean $\max\{s, 0\}$ by s^+. We progress the estimate by discussing each of the following cases.

Case 1. $u_i^n - u_{i-1}^n, u_{i+1}^n - u_i^n, u_{i+2}^n - u_{i+1}^n > 0$
Case 2a. $u_i^n - u_{i-1}^n, u_{i+1}^n - u_i^n > 0, u_{i+2}^n - u_{i+1}^n \leq 0$
Case 2b. $u_i^n - u_{i-1}^n \leq 0, u_{i+1}^n - u_i^n, u_{i+2}^n - u_{i+1}^n > 0$
Case 3. $u_i^n - u_{i-1}^n > 0, u_{i+1}^n - u_i^n \leq 0, u_{i+2}^n - u_{i+1}^n > 0$
Case 4a. $u_i^n - u_{i-1}^n > 0, u_{i+1}^n - u_i^n, u_{i+2}^n - u_{i+1}^n \leq 0$
Case 4b. $u_i^n - u_{i-1}^n, u_{i+1}^n - u_i^n \leq 0, u_{i+2}^n - u_{i+1}^n > 0$
Case 5. $u_i^n - u_{i-1}^n \leq 0, u_{i+1}^n - u_i^n > 0, u_{i+2}^n - u_{i+1}^n \leq 0$
Case 6. $u_i^n - u_{i-1}^n, u_{i+1}^n - u_i^n, u_{i+2}^n - u_{i+1}^n \leq 0$

Each case is again divided into several subcases. For example, Case 1 is divided into several subcases according to the relation among $q_{i-\frac{1}{2}}^n, q_{i+\frac{1}{2}}^n, q_{i+\frac{3}{2}}^n$ and $\pm\epsilon$.

Main machinery used in the estimate consists of two kinds. One is the strict convexity of flux function f. The other is what comes from the small additional viscosity for entropy fix, which we see as $\frac{\epsilon}{\lambda}$ in the definition (14) of function Q_ϵ.

First we show an example in which the strict convexity of f is applied to the estimate. Let us consider the subcase of $\epsilon < q_{i-\frac{1}{2}}^n < q_{i+\frac{1}{2}}^n < q_{i+\frac{3}{2}}^n$ belonging to Case 1. In this subcase we easily observe

$$u_{i+1}^{n+1} - u_i^{n+1} = \lambda q_{i-\frac{1}{2}}^n (u_i^n - u_{i-1}^n) + (1 - \lambda q_{i+\frac{1}{2}}^n)(u_{i+1}^n - u_i^n). \tag{23}$$

$q_{i-\frac{1}{2}}^n (u_i^n - u_{i+1}^n)$ is estimated as follows

$$\begin{aligned} q_{i-\frac{1}{2}}^n (u_i^n - u_{i-1}^n) &= f'(u_i^n)(u_i^n - u_{i-1}^n) - \tfrac{1}{2} f''(\theta)(u_i^n - u_{i-1}^n)^2 \\ &\leq f'(u_i^n)(u_i^n - u_{i-1}^n) - \tfrac{1}{2} c (u_i^n - u_{i-1}^n)^2 \end{aligned} \tag{24}$$

with some θ between u_{i-1}^n and u_i^n, We observe

$$q_{i+\frac{1}{2}}^n (u_{i+1}^n - u_i^n) \geq f'(u_i^n)(u_{i+1}^n - u_i^n) + \tfrac{1}{2} c (u_{i+1}^n - u_i^n)^2 \tag{25}$$

in a similar manner as well. Then we obtain the following estimate.[2]

$$\begin{aligned} u_{i+1}^{n+1} - u_i^{n+1} &\leq \lambda f'(u_i^n)(u_i^n - u_{i-1}^n) + \{1 - \lambda f'(u_i^n)\}(u_{i+1}^n - u_i^n) \\ &\quad - \frac{c\lambda}{2}\{(u_i^n - u_{i-1}^n)^2 + (u_{i+1}^n - u_i^n)^2\} \\ &\leq \max\{(u_i^n - u_{i-1}^n)^+, (u_{i+1}^n - u_i^n)^+\} \\ &\quad - \frac{c\lambda}{4}[\max\{(u_i^n - u_{i-1}^n)^+, (u_{i+1}^n - u_i^n)^+\}]^2. \end{aligned} \tag{26}$$

From the general CFL condition (4), we obtain

$$\left|f'(u_{i+1}^n) - f'(u_i^n)\right| \leq \frac{2}{\lambda} \tag{27}$$

for every i, n, and

$$|f'(u_{i+1}^n) - f'(u_i^n)| = |f''(\theta)(u_{i+1}^n - u_i^n)| > c|u_{i+1}^n - u_i^n| \tag{28}$$

is a direct conclusion of the mean value theorem, where θ is some value between u_i^n and u_{i+1}^n. Then we obtain

$$|u_{i+1}^n - u_i^n| < \frac{2}{c\lambda} \tag{29}$$

for every i, n. Observing

$$s - \frac{c\lambda}{4}s^2 = -\frac{c\lambda}{4}\left(s - \frac{2}{c\lambda}\right)^2 + \frac{1}{c\lambda}, \tag{30}$$

we conclude

$$u_{i+1}^{n+1} - u_i^{n+1} \leq z^n - \frac{c\lambda}{4}(z^n)^2. \tag{31}$$

Second we show an example of estimate which is derived from the machinery of entropy fix. The example is the subcase of $u_i^n - u_{i-1}^n, u_{i+2}^n - u_{i+1}^n \leq 0, u_{i+1}^n - u_i^n > 0, -\epsilon < q_{i+\frac{1}{2}}^n < \epsilon$ belonging to Case 5. In this subcase we easily obtain

$$u_{i+1}^{n+1} - u_i^{n+1} \leq (1 - \epsilon)(u_{i+1}^n - u_i^n) \leq (1 - \epsilon)z^n. \tag{32}$$

In all the other cases or subcases the estimate is obtained through either of the two kinds of machinery shown above. Some cases or subcases are tricky, and the most tricky is Case 3. But careful division into subcases and straightforward estimate in each subcases lead us to the estimate that we prove.

Once we prove the estimate (21), it is not so difficult to observe that

$$z^n < w(n) \tag{33}$$

for every n, where $w = w(s)$ is a function determined by

$$\begin{cases} \dfrac{dw}{ds} = \max\left\{-\epsilon w, -\dfrac{c\lambda}{4}w^2\right\} \\ w(0) = \dfrac{2}{c\lambda}. \end{cases} \tag{34}$$

---

[2] It is possible to obtain the estimate

$$u_{i+1}^{n+1} - u_i^{n+1} \leq \max\{(u_i^n - u_{i-1}^n)^+, (u_{i+1}^n - u_i^n)^+\} - \frac{c\lambda}{2}[\max\{(u_i^n - u_{i-1}^n)^+, (u_{i+1}^n - u_i^n)^+\}]^2.$$

in this subcase, but in some other subcases we have to replace the coefficient $\frac{c\lambda}{2}$ by $\frac{c\lambda}{4}$ because of some technical reason. Therefore we here employ the estimate (25).

We also observe

$$w(s) \le \max\left\{ \frac{2}{c\lambda} e^{-\epsilon s}, \frac{4}{c\lambda}\frac{1}{s} \right\}. \tag{35}$$

Then we obtain

$$u_{i+1}^n - u_i^n \le z^n < \max\left\{ \frac{2}{c\lambda} e^{-\epsilon n}, \frac{4}{c\lambda}\frac{1}{n} \right\}. \tag{36}$$

If n is large enough, (36) yields

$$u_{i+1}^n - u_i^n < \frac{4}{c\lambda}\frac{1}{n} = \frac{4\Delta x}{cn\Delta t}$$

and then

$$\frac{u_{i+1}^n - u_i^n}{\Delta x} < \frac{\frac{4}{c}}{n\Delta t}, \tag{37}$$

which is the estimate (E1). We easily find some constant $\bar{E}$ satisfying

$$e^{\epsilon s} < \frac{2\bar{E}}{s} \tag{38}$$

for $s > 0$. This completes the estimate (E2).

We here note that similar discussion employing the strict convexity of flux function is given in [8] to prove a similar inequality

$$\frac{u_{i+1}^n - u_{i-1}^n}{2\Delta x} < \frac{\frac{2}{c}}{n\Delta t} \tag{39}$$

for Lax-Friedrichs scheme of difference approximation. The manner in which we apply the strict convexity to obtain the inequalities (24) and (25) is almost the same as that in [8]. It seems natural to use the strict convexity of flux function for such estimates because the evolution of each nonlinear wave like rarefaction or shock is essentially governed by the strict convexity of flux function. But we need the other kind of machinery derived from the constant ϵ for entropy fix. We may understand that the latter machinery is more "numerical" than the former.

The corollary is proved as follows. The difference approximation satisfies the so called TVD condition

$$a^{MR}(u_i^n, u_{i+1}^n) \le a_{i+\frac{1}{2}}^n \le a^{LF}, \tag{40}$$

which means that the convergence limit of any convergent subfamily of approximate solutions u_Δ is a weak solution to the initial value problem (1). (See [7], [12] or [1].) But the approximate solutions satisfy (E2), which is exactly Oleĭnik's E-condition. Then the convergence limit is unique and it is the entropy solution to (1). (See [8].)

We also note that similar discussion is possible even if we replace the function $Q_\epsilon(s)$ in Theorem 1 by another function $Q(s)$, $-\frac{1}{\lambda} \le s \le \frac{1}{\lambda}$ satisfying the following (C1)-(C4), where η is an arbitrary fixed positive number less than $\frac{1}{2}$.

(C1) Q is an even function, *i.e.* $Q(-s) = Q(s)$.
(C2) $Q(s) = |s|$ if $|s| \geq \dfrac{1}{2\lambda}$.
(C3) Q is a convex function.
(C4) $Q(0) = \dfrac{\eta}{\lambda}$.

The proof is obtained from the fact that $Q(s)$ is an convex combination of the family $\{Q_\epsilon\}_{0 \leq \epsilon < \frac{\lambda}{2}}$ of functions Q_ϵ determined by (14)

For example, a function

$$Q(s) = \begin{cases} \dfrac{1}{2}\left(\dfrac{\alpha}{2} + \dfrac{s^2}{2\alpha}\right) & \text{if } |s| \leq \alpha, \\ |s| & \text{otherwise} \end{cases} \tag{41}$$

(α is a positive constant smaller than $\frac{\lambda}{2}$) satisfies (C1)-(C4) and is often used for practical computation. (See [4, 5].)

3 Discussion

We remember that the difference approximation in Theorem 1 or that with Q_ϵ replaced by Q satisfying (C1)-(C4) satisfies the assumption of Theorem 0 as well. It means that the convergence of such difference approximations to the entropy solution is already proved in [1]. But, on the contrary, it is well known that the difference approximation gives some strange numerical behavior like "bridge"[3] especially when ϵ is very small. This can be regarded as a kind of impropertv in numerical interpretation of rarefaction. Of course Theorem 0 means that we may avoid or neglect such deterioration when difference increments become small enough. But if we are interested in the quality of numerical computation, we can not neglect such numerical behavior of difference approximation and it is required to analyze it in some way. Our discussion in this article is an answer to this problem.

[3]Even if we use difference approximation converging to the entropy solution, we may have strange numerical phenomena which are not seemed to be physically relevant especially where the rarefaction happens around a sonic point (a point where $f' = 0$). This is called "bridge". Here is an example. (The line shows the exact solution and the set of dots • shows the numerical result. The "bridge" is circled.)

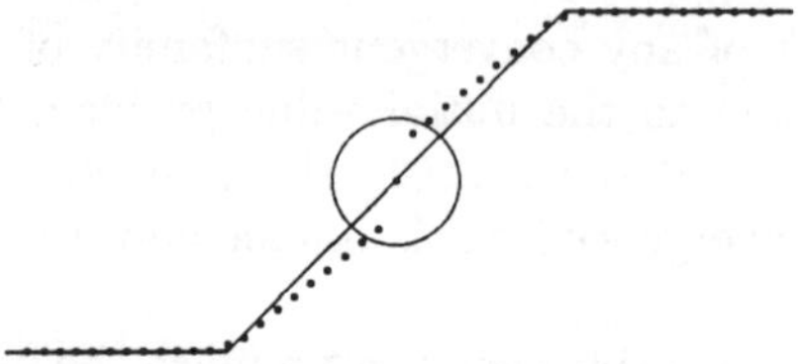

It is well known that this phenomenon is noticeable when we employ Harten's entropy-fix technique with the additional viscosity very small.

Theorem 1 implies us that improper numerical phenomena like "bridge" might be noticeable if $\frac{u_{i+1}^n - u_i^n}{\Delta x}$ is not governed by $\frac{4}{cn\Delta t}$ but by the exponential order of n. And we easily observe that $\frac{u_{i+1}^n - u_i^n}{\Delta x}$ is governed by $\frac{4}{cn\Delta t}$ if n is large enough and that it is easy to determine a lower bound N for such n. We note that the validity of the estimate

$$\frac{u_{i+1}^n - u_i^n}{\Delta x} < \frac{4}{cn\Delta t}$$

depends only on n but not on $n\Delta t$. It implies that the "bridge"-like numerical improperty may happen only when n is small and that it can be avoided any fixed time $T = n\Delta t > 0$ by taking difference increments that are small enough so that $n = T/\Delta t$ should be larger than N. In other words, it means that the kind of numerical impoperty is noticeable only within a so called initial layer which is a region where t is near 0 in the half plane $\{(x,t)|x, t \in \boldsymbol{R}, t \geq 0\}$. Therefore, if we do the computation to know the behavior of solution as $t \longrightarrow \infty$, the choice of value of ϵ (or η) is not so important.

The difference between the optimal values of constants E in Oleĭnik's E-condition (13) and E' in the estimate (E1) might have some effect to the formation of "bridge"-like numerical improperty. We can not exclude it, but the effect is not so strong.

Extension of this discussion to the systems' case is not straightforward because we do not have any analogue of Oleĭnik's E-condition in the case of systems of conservation law. But we might expect that "bridge"-like numerical improperty would occur soon after a spatial discontinuity breaks down to form a rarefaction wave.

Our discussion is applied rather to rarefatcion waves than to shock waves. The quality of numerical result around rarefation waves has not been discussed so much nor so carefully, while that around shock waves has been discussed in many works and various methods to capture shocks have been developed. Of course we easily imagine why the situation is as above. For example, the failure in computing rarefaction waves seems to bring less damage to the computation while the failure in computing shock waves easily makes much noticeable inconvenience like oscilation. Or, if too large numerical viscosity brings the smearing effect, the deterioration of numerical results are much more easily recognized around shock waves than around rarefaction waves. But the importance to obtain a good numerical interpertation of rarefaction waves should be respected as well. For example, it is very important to know where the rarefaction begins or ends in some problems of high-speed aerodynamics.

Anyway we still need to analyze numerical behavior of difference approximation for conservation laws to improve the quality of numerical computation.

References

[1] H. Aiso. Admissibility of difference approximation for scalar conservation laws. *Hiroshima Math. J.*, 23(1):15–61, 1993.

[2] H. Aiso. A General Class of Higher Order-Accurate Difference Approximations for Scalar Conservation Laws Converging to the Entropy Solution. *Computational Fluid Dynamics '96 (Proceedings of Third ECCOMAS Computational Fluid Dynamics Conference.)*, 937–943 , 1996.

[3] B. Engquist and S. Osher. Stable and entropy satisfying approximations for transonic flow calculations. *Math. Comp.*, 34:45–75, 1980.

[4] A. Harten. High resolution schemes for hyperbolic conservation laws. *J. Comput. Phys.*, 49:357–393, 1983.

[5] A. Harten. On a class of high resolution total-variation-stable finite-difference schemes. *SIAM J. Numer. Anal.*, 21(1):1–23, 1984.

[6] P. D. Lax. Shock waves and entropy. In *Contributions to nonlinear functional analysis*, pages 603–634. Academic Press, NewYork, 1971.

[7] A. Y. LeRoux. A numerical conception of entropy for quasi-linear equations. *Math. Comp.*, 31:848–872, 1977.

[8] O. Oleĭnik. Discontinuous solutions of nonlinear differential equations. *Uspekhi Mat. Nauk.(N.S.)*, 12:3–73, 1957. English transl. in *Amer. Math. Soc. Transl., Ser. 2, vol.* 26, 95-172.

[9] S. Osher and S. Chakravarthy. High resolution schemes and the entropy condition. *SIAM J. Numer. Anal.*, 21(5):955–984, October 1984.

[10] S. Osher and E. Tadmor. On the convergence of difference approximations to scalar conservation laws. *Math. Comp.*, 50:19–51, 1988.

[11] P. L. Roe. Approximate Riemann solvers, parameter vectors and difference schemes. *J. Comput. Phys.*, 43:357–372, 1981.

[12] J. Smoller. *Shock waves and reaction-diffusion equations.* Springer-Verlag, NewYork, 1982.

[13] E. Tadmor. Numerical viscosity and the entropy condition for conservative difference schemes. *Math. Comp.*, 43:369–382, 1984.

Lessons Learned from the Blast Wave Computation Using Overset Moving Grids - Grid Motion Improves the Resolution -

Kozo Fujii

The Institute of Space and Astronautical Science
3-1-1, Yoshinodai, Sagamihara, Kanagawa, 229-8510, JAPAN

Abstract: This paper describes some of the issues that are obtained from the computational efforts to develop an accurate computational tool for the purpose of establishing a better estimation method of the safety distance for the blast wave. Two approaches are considered since this type of simulations require high level of grid resolution. One is an overset zonal method and the other is an unstructured solution adaptive grid method. The former approach is taken and three-dimensional simulations are carried out with sufficient accuracy. During the process of the simulation, it is found that the grid motion improves the accuracy of the solution. The feature of the moving grids is confirmed by another applications such as a shock tube problem and a vortex transport problem.

1 Introduction

Investigation of blast wave propagation is one of the important research topics of unsteady shock wave behavior [1,2]. How the blast wave behaves when propagating on the ground surface is of considerable importance for many practical applications such as an accidental explosion of chemical or nuclear plants or rocket propellants. Volcanic eruptions may be another application. In these problems, accurate estimations of the blast wave strength are required because it is the key factor of the damage.

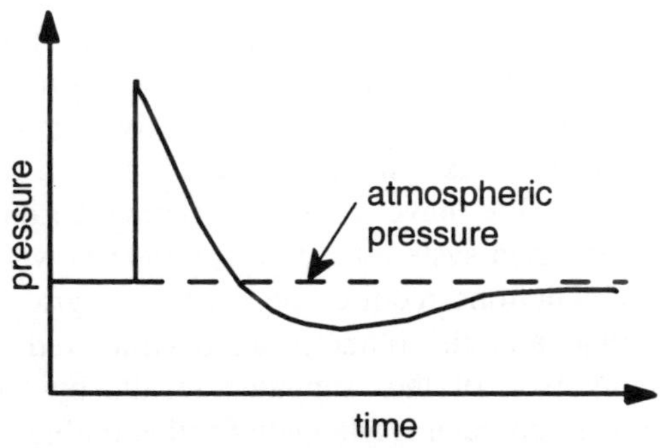

Fig. 1 Pressure history over the ground surface due to the blast wave

Figure 1 shows the schematic picture of the blast wave propagation. This figure shows the time history of the pressure observed at a certain distance from the point of explosion. The pressure stays at the atmospheric value until the frontal shock wave reaches the point of measurement. After an abrupt increase of the pressure due to the shock wave, the pressure decreases gradually and becomes lower than the atmospheric value due to the following expansion and eventually comes back to the atmospheric value. The rapid pressure rise is called "overpressure" and it defines the strength of the blast wave. Figure 2 shows the history of the instantaneous pressure profiles in space. As the blast wave propagates, the overpressure decays under a power law in general. One of the criteria for the estimation of the "safety distance" is the strength of the overpressure. Conventionally, the empirical method based on the one-dimensional theory and the experimental data by Kingery and Pannill [3] is widely used for the evaluation of the "safety distance". In this theory, the overpressure simply depends on the amount of released energy and the distance, and other factors such as the ground surface geometry and/or atmospheric conditions are only empirically considered. When considering the diffraction, reflection and other interactions that may occur on the shock wave passing over some geometries, it is imagined that ground surface geometry may change the strength of the blast wave. Discretization methods easily handle such effects and therefore developing such methods may improve the accuracy of the estimation of "safety distance". Computatiönal efforts using the Lagrangean approach exist but they are restricted to the primitive simulations under the assumption of point symmetry or axisymmetry [4,5].

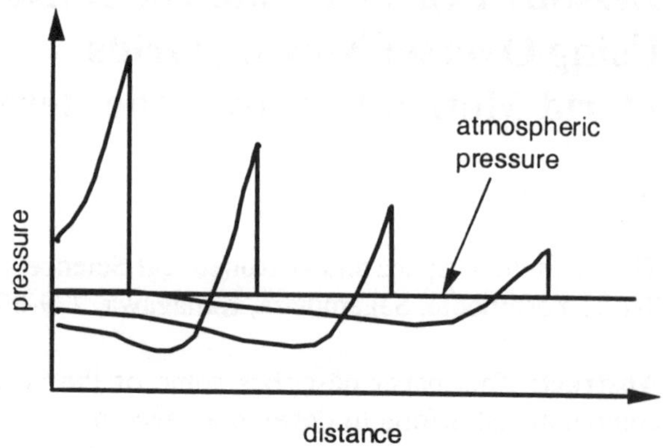

Fig. 2 Instantaneous pressure profiles of the blast wave

We have adopted a computational method using an overset zonal moving grid system but two approaches were initially considered since this type of simulations requires high level of grid resolution. One is an overset zonal method and the other is an unstructured solution-adaptive grid method. The comparison of the computed results by these two methods, indicate that the former approach is adequate for this problem. Main reason was that we found the grid motion improves the accuracy of the solution during the course of the simulation. The main purpose of the present paper is to see the effect of moving grids and discuss the feature based on the several computational examples. The results indicate that the simple grid motion may improve some of the unsteady flow simulations with a small modification of the computer program. In the next section, the simulation results using an overset moving gird system is shown and the reason for the choice will be discussed in the following sections.

2 Blast wave simulation using an overset moving grid system

Euler equations are considered here because the viscous effect on the ground surface would change the flow field left over by the frontal shock wave but would not change the strength of the frontal shock wave. The equations are written in the generalized coordinate system where ground surface is one of the boundaries. The FSA (Fortified Solution Algorithm) interface method [6] is used to transfer the information between each zone. The details of the formulation as well as the merits of the FSA zonal method compared to the conventional CHIMERA approach is discussed in Ref. 6.

The convective terms are discretized using the flux difference splitting scheme with Roe's approximate Riemann Solver [7]. Third-order upwind biased differencing is achieved using the MUSCL interpolation with the van Albada's flux limiter[8] although ENO-type of scheme might work better. Since the computation requires time accuracy, two-step Runge-Kutta time integration scheme is adopted.

The grid system is composed of three zonal grid components. One is the base grid which covers the whole region. Second is the centered grid which is prepared to avoid the topological singularity in the center of the explosion. Third one is the locally-defined moving grid covering the blast wave region. This grid moves at the speed of sound. The grid could move with the exact speed of the frontal shock wave by monitoring the speed of the shock wave, but the speed is fixed to the speed of sound as the solution can be improved by the grid speed not necessarily the same as the shock wave, as indicated in the research to be shown in the latter part of the present paper. Note that the speed of the shock wave is close to the speed of sound except very near the point of explosion where the shock wave is very strong. The base grid is fixed and local moving grids are defined for all the directions to see the effect of the ground surface geometry. To achieve this within the restriction of the memory size of 256 MB, the computational region is divided into 8 parts and eight computations are carried out with the local moving fine-grid region located only in one part. Figure 3 shows one instantaneous shot of the base, centered, and all the eight moving grids together. Each computation uses roughly 0.5 million grid points in total. The computations were carried out on Fujitsu VPP500/1PE and

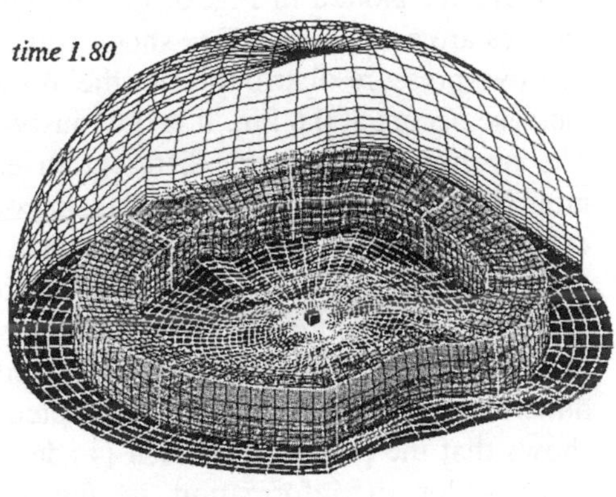

Fig. 3 Computational Grid at t=1.80

the required computer time was roughly 20 hours.

One thing to note here is the consideration for the deformation of the computational grids. Geometric Conservation Law (GCL) should be satisfied when computational grids are deforming. In the present paper, all the metrics and Jacobians are computed considering the geometrical conservation[9] both in time and space to satisfy the GCL and no special treatment was necessary.

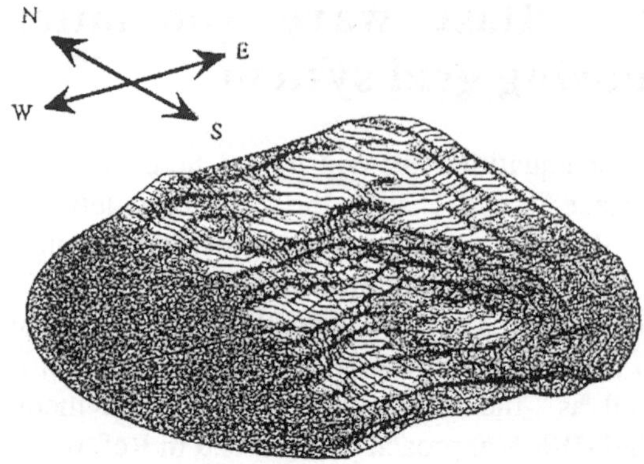

Fig. 4 Ground surface configuration

The ground surface geometry is taken from a real physical model which is shown in Fig. 4. The west area (left side) is the sea. The point of explosion is the center and is located about 200 m above the sea level. There is a small hill in the east direction (right) from the point of explosion along the line from north to south. The blast wave propagating to the east direction goes over this hill of 300 m high. The energy core of 100 atm pressure and 2000K temperature with the core radius of 17 m is assumed. This corresponds to the realistic case with the released energy of 55 tons TNT equivalence. Outside the core, atmospheric conditions are given. The pressure is 1 atm and the temperature is 300K there. Density gradient in the horizontal direction is not considered. The region to be simulated is about 2.5 km from the center of explosion because the strength of the blast wave is estimated to become less than 0.2 psi there. Note that the criteria of the safety distance is 0.2 psi.

The computed overpressure distributions in the WEST, EAST, NORTH, SOUTH are plotted in Fig. 5. In this figure, some of the instantaneous pressure profiles are also plotted. The shock wave is crisply captured in all the directions. The overpressure distribution in the WEST direction shows that strong pressure increase occurs at 0.6 km. It is obviously due to the diffraction of the shock wave. The overpressure is very different in each direction and it indicates that the ground surface geometry effect is important and should be considered for an accurate estimation of the safety distance of the blast wave.

The regions in which the overpressure become larger than 2-10% of the atmospheric pressure are plotted in Fig. 6. The region strongly depends on the ground surface geometry as expected from Fig. 5. Although the three dimensional simulations were conducted only for a specific example, the result shows that the present approach is adequate for the blast wave simulations and gives a lot of information to the estimation of "safety distance". More discussions on the simulated results including two-dimensional simulations can be found in Ref. 10.

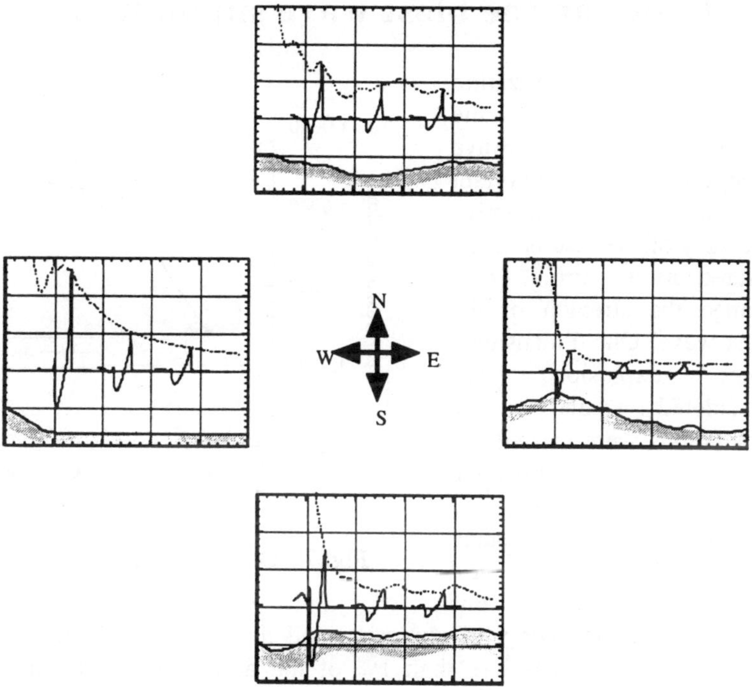

Fig. 5 Computed overpressure distributions over the ground surface

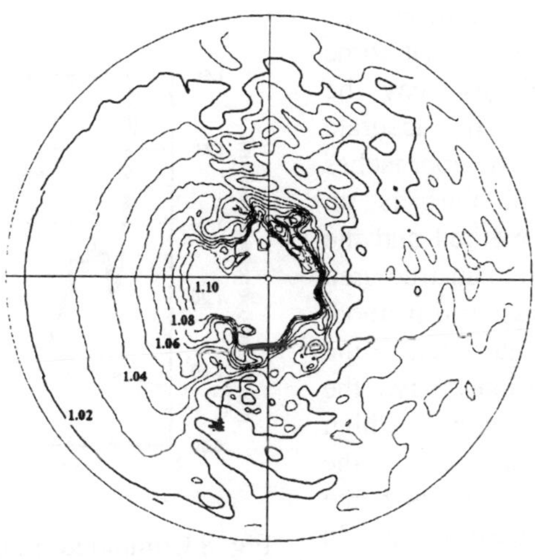

Fig. 6 Computed overpressure distributions

3 Issues on the blast wave simulations

Although discretization methods are adequate for the simulation of blast wave propagation, it is important to check the quantitative accuracy of the results as the discretization errors may change the strength of the blast wave. One-dimensional point explosion was considered and the simulations were carried out prior to the practical computations. Figure 7 shows the computed overpressure versus grid resolution. In this computation, the energy of explosion and the distance are defined where the overpressure becomes less than 1% of the atmospheric pressure because the overpressure for the "safety distance" is conventionally considered to be 0.2 psi which is roughly 1.4 % of the atmospheric pressure. The horizontal axis is the number of grid points in the direction of the blast wave propagation. Even with 5000 grid points, the asymptotic value is not yet obtained and the result indicates that more than 6000 grid points are required.

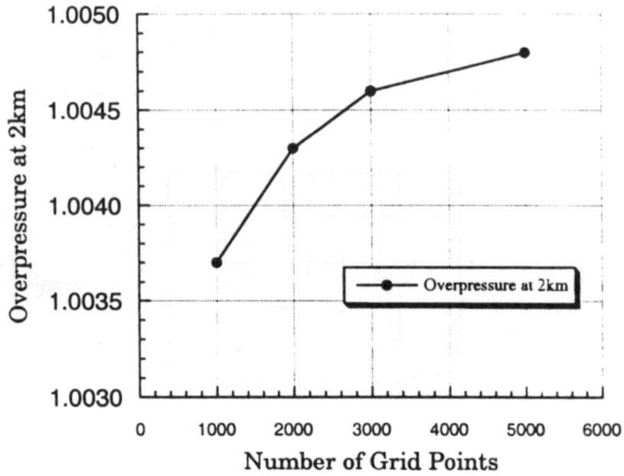

Fig. 7 Overpressure .vs. grid resolution

The one-dimensional simulations showed that some special techniques may be required for an accurate simulation. Two approaches are considered. One is an overset zonal method and the other is an unstructured solution-adaptive grid method. Since we are interested in the strength of the blast wave, the region that requires high accuracy is restricted in the certain area that moves at the speed of the blast wave. In the overset zonal method, a moving fine grid region can be locally defined to capture the

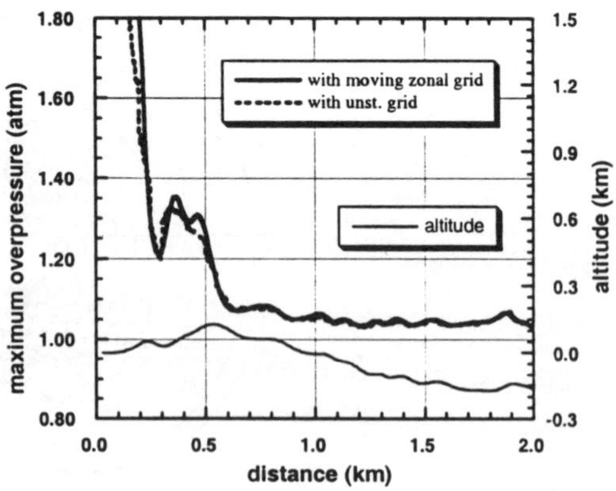

Fig. 8 Computed maximum overpressure distributions
- comparison of the zonal-grid and unstructured-grid solution -

blast wave without any difficulties. Second one is an unstructured grid with the solution-adaptive grid strategy. This method can automatically concentrate the grid cells to the region of blast wave. Both methods were tried [11-14]. Figure 8 shows the computed maximum overpressure distributions for these two methods in the two-dimensional problem. The moving grid solution seems slightly better (because it shows higher overpressure) than the unstructured grid solution, but they are almost identical in the region away from the explosion. So let's assume that these two results are as accurate each other. Although fair comparison is very difficult, we concluded that the former approach is more efficient for the present problem and conducted three-dimensional real flow simulations as was shown in the previous section. There are two main reasons. First, the unstructured solution-adaptive grid method requires large amount of memory. In this two-dimensional example, the moving overset grid computation required 1 hours on VPP500/1PE with 12,000 grid points in total. It required several MB of memory. On the other hand, the unstructured grid computation required less than 1 hour on the same computer with 100,000 grid points. It requires 20-25 MB of memory. The reason why the number of grid points are so different is that the shock wave is almost orthogonal to the ground surface, and therefore the unstructured grid (that requires the grid increase in both directions) needs many more grid points. Shock wave interaction makes the configuration of the frontal shock wave complicated only near the ground surface. The unstructured grid with triangles spends large number of grid points for the region out of our interest. This leads to the fact that the unstructured grid approach requires large amount of memory. In three-dimensional problems using tetrahedra, this drawback would be remarkably pronounced. Our estimation showed 5 to 10GB memory would be necessary for the unstructured-grid computations. On the other hand, overset zonal grid computations require at most 1 or 2 GB of memory in three dimensions, and the simulation was actually conducted with less than 256MB. This is the first reason for the choice of the overset zonal method. Second reason is rather important reason for having adopted the overset zonal moving grid approach and that is the main theme of the present paper. During the course of the study, we have found that the local grid

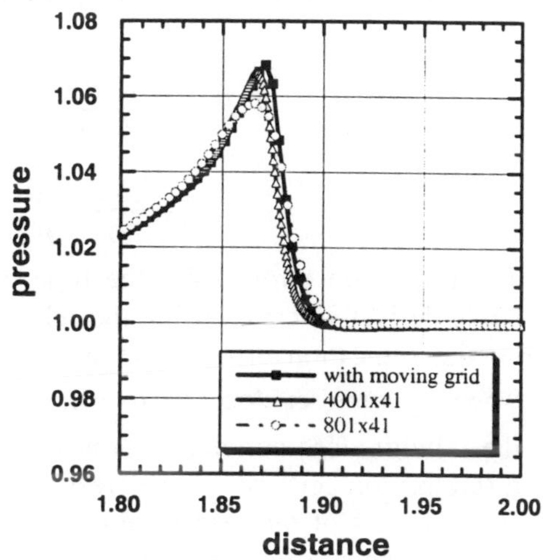

Fig. 9 Computed surface pressure distributions

resolution is enhanced when grids move with the shock wave. When the grids move with the frontal shock wave, the resultant shock wave was more crisply captured than that would be obtained with the stationary grid having the same spatial grid resolution. Let's take two dimensional example. Figure 9 shows the ground surface pressure distributions computed by the moving grid system. Only the blast wave region is closed up in the figure. The single-zone solutions with the very fine (4001 points along the ground) and fine grids (801 points) are plotted for comparison. The moving-grid solution shows the highest pressure peak at the shock wave. The interesting observation here is that the spatial grid resolution of the locally moving grid in the zonal solution is the same as the grid resolution of the 801 grid points but showed more accurate solution than that of 4001 grid points. The fact that the grid moves at the blast wave speed obviously improved the solution accuracy.

4 Effect of Moving Grid

To understand the reason for the accuracy improvement, a linear scalar wave equation is considered. When, the grid moves at the velocity u_g, the basic equation for a linear scalar equation becomes

$$\frac{\partial u}{\partial \tau} + (c - u_g)\frac{\partial u}{\partial \xi} = 0 \qquad (1)$$

Here c is constant. It is clearly noticed that the convective speed is changed from c in the stationary grid to $c - u_g$ in the moving grid system. All the truncation error terms for the space derivatives have the coefficient of this convective speed $c - u_g$ as shown in Eq. (2).

$$\begin{aligned} T.E. = &-\frac{\Delta t}{2}\frac{\partial^2 u}{\partial t^2} + \frac{(c-u_g)\Delta x}{2}\frac{\partial^2 u}{\partial x^2} \\ &-\frac{(\Delta t)^2}{6}\frac{\partial^3 u}{\partial t^3} - (c-u_g)\frac{(\Delta x)^2}{6}\frac{\partial^3 u}{\partial x^3} + h.o.t. \end{aligned} \qquad (2)$$

The truncation errors from the time derivative terms can be translated to the space derivative errors using the relations in the original equations. Therefore all the truncation errors have $c - u_g$ as their coefficient and would decrease when u_g were set closer to c. In fact, when the grid moves at the velocity c, the solution does not change at any grid point and correspondingly there are no discretization errors. The result indicates that the key reason for the improvement of the grid resolution in the moving grid system comes from the fact that the coefficients of the convective speed (eigenvalues for the system of equations) are reduced. The truncation error analysis shown here demonstrates that it can be explained as the change of the coefficients of the truncation errors. The grid resolution can be improved even when the grid moves not at the exactly same speed as the phenomenon interested in.

One-dimensional shock tube problem was tried to see the effect in the system of equations. Many cases were tried and some of them are shown here. Figure 10 shows the density plots when the computational grids move at 95 % of the shock speed. Note that this is not the zonal computation. All the grid points move at the same speed of 95% of the shock wave. Therefore, the modification of the computer program is only a few statements at most. The improvement is clearly seen when compared to the stationary grid solution (dotted line). On the other hand, the solution accuracy is degraded in the expansion region since the rarefaction wave proceed to the opposite direction to the grid motion. Figure 11 shows the result when the grids move at 100% of the rarefaction wave. In this case, expansion wave is better captured in the moving grid although the contact discontinuity and shock wave slightly smear out. For this particular problem, three overset grids, each of which moves at the speed of shock wave, contact discontinuity and rarefaction wave, respectively should be prepared to enhance the accuracy in the whole region.

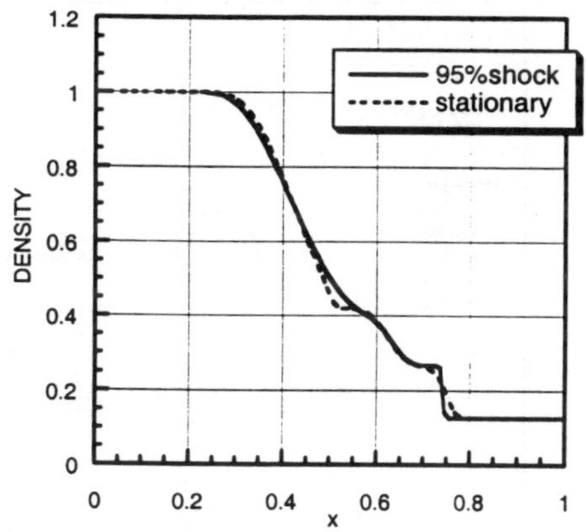

Fig. 10 Density distribution of shock tube problem moving grid(95% shock speed) .vs. stationary grid

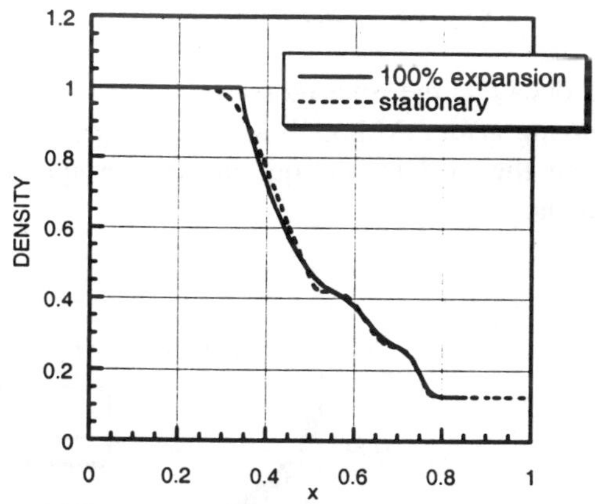

Fig. 11 Density distribution of shock tube problem moving grid(100% expansion) .vs. stationary grid

Next, the transport phenomenon of the isolated vortex is simulated. An isolated vortex is placed in the freestream and transported downstream as is shown in Fig. 12. The grid moves at the freestream speed. Since this is an Euler computation, the vortex strength would not essentially change. One of the results is shown in Fig. 13 where density of the center of the vortex core and the vorticity there are plotted versus time. Both the moving grid and stationary grid solutions are plotted. Due to the discretization errors, both the results show

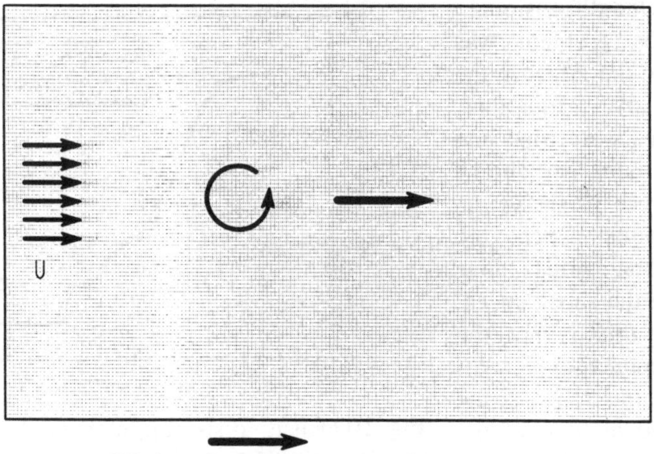

Fig. 12 Schematic picture of the vortex transport problem

vortex decay as the time increases but it is clearly weakened for the solution by the moving grid. Although we need to investigate the dependency of the vortex decay to the grid resolution, freestream speed and other factors, the result is promising.

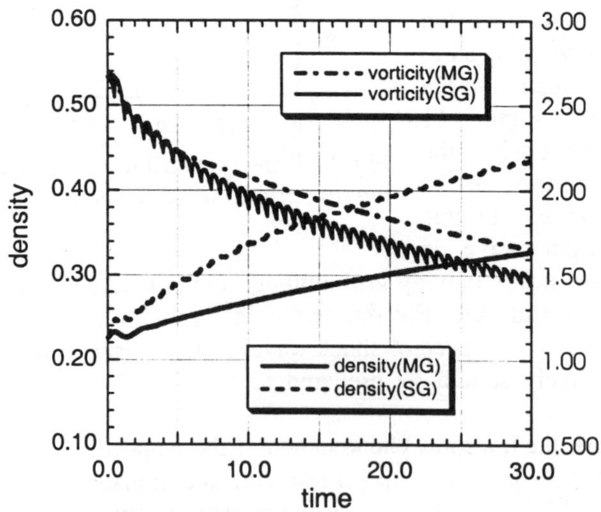

Fig. 13 Decay of the vortex:
- minimum density and vorticity of the center of the vortex core -

5 Conclusions

Blast wave propagation over a three-dimensional practical ground surface configuration was simulated using Euler equations. The computation was carried out with the overset moving zonal method and the decay of the blast wave is well captured by the simulation. The comparison of the result with that of the solution-adaptive unstructured grid approach showed some of the feature of the overset moving grid system. During the process of the simulation, it was found that the grid motion improves the accuracy of the solution. The feature of the moving grids was confirmed by another applications such as a shock tube problem and a vortex transport problem. The results indicate that moving grid system introduces the "Lagrangean" effect into the Eulerian computations. The truncation error analysis briefly presented here shows that it can be explained as the change of the coefficients of the truncation errors. The examples above clearly showed that the improvement of the solution occurs due to the movement of the computational grids. There are many applications where the solution can be improved by using the moving grid system. The computational overhead is simply the computation of the time metrics term (which is constant) and existing computer codes can be easily modified.

References

[1] Baker, W.E., "Explosion in Air," Univ. Texas Press, Austin and London, 1972.

[2] Kinnery, G.F. and Graham, K.J., "Explosirve Shocks in Air," Springer-Verlag, 1985.

[3] Kingery, C.N. and Pannill, B.F., "Peak Overpressure .vs. Scaled Distance for TNT Surface Bursts (Hemispherical Charges)," Ballistic Research Laboratories, MR.-1518, 1964.

[4] Brode, H.L., "Blast Wave from a Spherical Charge," The Physics of Fluids, Vol. 2, No. 2, pp. 217-229, 1959.

[5] Plooster, M. N., "Shock Wave from Line Sources. Numerical Solutions and Experimental Measurements," The Physics of Fluids, Vol. 13, No. 11, pp. 2665-2675, 1970.

[6] Fujii, K., "Unified Zonal Method Based on the Fortified Solution Algorithm," Journal of Computational Physics, Vol. 118, pp. 92-108, 1995.

[7] Roe, P.L., "Characteristic-Based Schemes for the Euler Equations," Annual Review of Fluid Mechanics, pp. 337-365, 1986.

[8] Thomas, J.L., van Leer, B. and Walters, B.W., "Implicit Flux-Split Schemes for the Euler Equations," AIAA Paper 85-1680, 1985.

[9] Obayashi, S., "Free-Stream Capturing in Fluid Conservation Law for Moving Coordinates in Three Dimensions," NASA CR177572, 1991.

[10] Fujii, K. and Shimizu, F.: Computations of Blast Wave Propagation Using an Overset Moving Zonal Method, to appear in CFD Review97,

Springer-Verlag, 1998.

[11] Shimizu, F., Fujii, K. and Higashino, F., "Ground Surface Effect on the Blast Wave Propagation in Two Dimensions," Transaction of the Japan Society for Aeronautical and Space Sciences, Vol. 36, No. 111, pp. 36-46, 1993.

[12] Shimizu, F., Fujii, K. and Higashino, F., "Three-Dimensional Blast Wave Propagating on the Realistic Ground Geometry," 6th International Symposium on Computational Fluid Dynamics, Collection of Technical Papers, pp.1148-1153, Sep., 1995.

[13] Sharov, D. and Fujii, K., "Unstructured Adaptive Mesh Method and Its Application to a Blast Wave Propagation Problem," The preprint of the 3rd World Congress on Computational Mechanics (WCCM III), May, 1994.

[14] Sharov, D. and Fujii, K., "Three-Dimensional Adaptive Bisection of Unstructured Grids for Transient Compressible Flow Computations," AIAA paper 95-1708, AIAA 12th Computational Fluid Dynamics Conference, June, 1995.